```
TD                                    47346
345
W25        Walker, Rodger

Water Supply, Treatment and Distrib
                                   ution
```

DATE DUE			
NOV 3 0 1988			
AUG 1 2 1990			
FEB 2 3 1991			
MAY 0 4 1999			
AUG 0 3 2004			
MAR 2 8 2012			

Waubonsee Community College

WATER SUPPLY, TREATMENT AND DISTRIBUTION

WATER SUPPLY, TREATMENT AND DISTRIBUTION

Rodger Walker

Consulting Engineer
Vancouver, B. C., Canada

PRENTICE-HALL, INC. *Englewood Cliffs, New Jersey* 07632

Library of Congress Cataloging in Publication Data
Walker, Rodger.
　　Water supply, treatment, and distribution.
　　Includes index.
　　1. Water-supply engineering.　I. Title.
TD345.W25　　　363.6′1　　　77-8206
ISBN 0-13-946004-7

© 1978 by Prentice-Hall, Inc., Englewood Cliffs, N.J. 07632

All rights reserved. No part of this book
may be reproduced in any form or
by any means without permission in writing
from the publisher.

Printed in the United States of America

10　9　8　7　6　5　4　3　2　1

PRENTICE-HALL INTERNATIONAL, INC., *London*
PRENTICE-HALL OF AUSTRALIA PTY. LIMITED, *Sydney*
PRENTICE-HALL OF CANADA, LTD., *Toronto*
PRENTICE-HALL OF INDIA PRIVATE LIMITED, *New Delhi*
PRENTICE-HALL OF JAPAN, INC., *Tokyo*
PRENTICE-HALL OF SOUTHEAST ASIA PTE. LTD., *Singapore*
WHITEHALL BOOKS LIMITED, *Wellington, New Zealand*

CONTENTS

1 WATER DEMAND 1

Planning Municipal Services *1*
Population *2*
Domestic Water Demand *7*
Fire Protection *11*
Industrial Water Requirements *11*
Irrigation *13*
Water Reuse *14*

2 WATER QUALITIES 16

Properties *16*
Synthesis of Water *16*
Medicinal Waters *17*
Microorganism Populations of Natural Waters *17*
Bacteriology *18*

3 WATERBORNE DISEASES 20

Waterborne Diseases 20
Source of Water 24
Removal of Turbidity 24
Disinfection 24
Clean Distribution Systems 25
Cross-Connection Control 25
Monitoring the Distribution System 26
Common Waterborne Diseases 26
Bacterial Diseases 27
    Cholera 27
    Bacillary dysentery 27
    Leptospirosis 28
    Paratyphoid 28
    Tularemia 29
    Typhoid 29
Helminthic Diseases (Parasitic Worms) 33
    Dracontiasis 33
    Echinococcosis 33
    Schistosomiasis 34
Protozoal Diseases 35
    Amoebiasis 35
    Giardiasis 36
Viral Diseases 36
    ECHO virus disease 37
    Infectious hepatitis 38
    Poliomyelitis 38

4 NONPATHOGENIC ORGANISMS 41

Iron Bacteria 41
    Lepthothrix 42
    Crenothrix 42
    Gallionella 42
Removal of Bacterial Slimes 43
Nematodes 44

5 ALGAE 46

Algae *46*

Blue Green Algae *60*

Green Algae *61*

Diatoms *62*

Flagellates *62*

Tastes and Odors from Algae *62*

Filter-Clogging Algae *64*

6 CHEMICAL AND PHYSICAL PARAMETERS 67

Introduction *67*

Proposed Drinking Water Quality Regulations and Proposals *68*
    Public water supply system *68*
    Water consumption *69*
    Canadian standards *69*
    Color *71*
    Odor *72*
    Tastes *73*
    Turbidity *73*
    Temperature *75*
    pH and pOH values *76*
    Alkalinity *78*
    Hardness *80*
    Temporary hardness *82*
    Permanent hardness *83*
    Carbonate hardness *83*

Toxic Chemicals *83*
    Arsenic (As) *83*
    Barium (Ba) *83*
    Boron (B) *84*
    Cadmium (Cd) *84*
    Chromium (Cr) *85*
    Cyanide (CN^-) *85*
    Lead (Pb) *86*
    Nitrates and nitrites (NO_3 and NO_2) *86*
    Selenium (Se) *86*
    Silver (Ag) *87*

Pesticides and Herbicides *87*

Non-Toxic Chemicals *89*
    Alumium (Al) 89
    Ammonia 89
    Calcium (Ca) 90
    Chlorides (Cl) 91
    Copper (Cu) 91
    Iron (as Fe) 92
    Magnesium (Mg) 93
    Manganese (as Mn) 93
    Methylene blue active substance (MBAS) 94
    Phenolic substances (as phenol C_6H_5OH) 94
    Phosphates (as PO_4) 94
    Total dissolved solids (TDS) 94
    Organic chemicals 95
    Sulfates (SO_4) 96
    Sulfides (SO_3) 96
    Uranyl ion (UO_2^{--}) 97
    Zinc (Zn) 97

7 RADIONUCLIDES *99*

Sources of Radioactivity *99*

Units of Radioactivity *100*

Fallout from Atomic Explosions *101*

Water Treatment Processes *101*

Monitoring Municipal Water Supply Systems *102*

8 WATER RESOURCES *104*

Hydrology *104*

Water Sheds *104*

Choice of Water Supplies *106*

Ground Water *107*

Ground Water Recharge *108*

Water Divining *108*

Location of Water Wells *109*

Advantages and Disadvantages
of Ground Water Supplies *109*
    Advantages 109
    Disadvantages 110

Lakes and River Water Supplies *111*

Water Sampling *111*

Runoff *113*

Probability Method *113*

9 GROUND WATER AND WELL PUMPS   118

Wells *118*

Artesian Wells *121*
    Nonflowing artesian wells 122
    Flowing artesian wells 122
    Control of flow 122
    Construction 122
    Corrosion problems 123
    Screens 123
    Pumps for artesian wells 123

Well Pump Selection *124*

Net Positive Suction Head (NPSH) *126*

Vertical Turbine Pumps *127*
    Submergence 127
    Well alignment 127

Submersible Pumps *128*
    Submergence 128
    Annulus space 128
    Well alignment 128

Suction Location *129*

Well Pump Operation *129*
    Carbonate scale 130
    Iron deposits 130
    Iron bacteria 130
    Sand 130

Horizontal Collectors *131*

Developing the Well *132*

Disinfection *132*

10 STRAINING AND SCREENING 134

Trash Racks *134*

Fish Screens *134*

Microstrainers *135*

Microstraining Surface Water Supplies *136*

Pretreatment by Microstrainers *137*

Pressure Strainers *137*

11 AERATION AND DEAERATION 139

Aeration *139*
   Theoretical concepts 140
   Design of aerators 140
   Natural and forced draft aerators 141
   Spray aerators 142
   Weirs and waterfalls 142
   Forced aeration 142
   Aerating reservoir 142
   Disadvantages of aeration 143

Deaeration *143*
   Steam deaeration 143
   Vacuum tower deaerators 143

12 MIXING, FLOCCULATION AND CLARIFICATION 148

Mixing *148*

Velocity Gradients *149*

Flocculation *151*

Theory of Coagulation *152*

Coagulation Control *154*

Clarification *155*

Tube Settlers *156*

Theory of Tube Settlers *160*

Reynold's Number (Re) *161*

Future Developments *164*

13 COLOR REMOVAL — 167

Definition of Color *167*

Organic Acids *167*

Color Removal Process *168*

14 LIME SODA SOFTENING — 172

Lime Soda Softening *172*

Should Water be Softened? *172*

Limitations of Lime Soda Softening *174*

Health Aspects *174*

Disinfection and Virus Inactivation *175*

Quality of Lime Softened Water *175*

Temporary and Permanent Hardness *175*

Equipment Used for Softening *175*

Effect of Inhibitors *177*

Filtration of Lime Softened Waters *178*

15 RECARBONATION — 180

Recarbonation *180*

Liquid Carbon Dioxide *181*

Submerged Combustion Burners *182*

Surface Combustion *184*

Production of Carbon Dioxide *185*

Adjustment of pH *185*

Threshold Treatment *185*

16 FILTRATION 186

Filtration *186*
Filtration Theory *186*
Upflow and Biflow Filters *188*
Dual and Mixed or Multimedia Filters *188*
Multimedia Filters *190*
Backwashing *193*
Air Scour *194*
Constant Rate and Declining Rate Filtration *195*
High Rate Filtration *196*
Slow Sand Filtration *196*
Filter Contamination *197*
Diatomaceous Earth Filters *197*

17 IRON AND MANGANESE REMOVAL 200

Iron and Managenese Removal *200*
   Removal of inorganic iron and manganese 201
   Iron in ground waters 201
   Aeration 202
   Chlorine and chlorine dioxide 202
   Potassium permanganate ($KMnO_4$) 203
   Lime softening 203
   "Organic" or chelated iron 203

Managenese Removal *204*
   Potassium permanganate 204
   Iron and manganese removal by ion exchange 205
   Sequestering the chelating process 206

18 CHEMICALS AND CHEMICAL FEEDING 208

Chemicals and Chemical Feeding *208*
Purity of Commercial Chemicals *209*
Equipment for Feeding Chemicals *209*
Chemical Feed Tables *210*
Rectangular Method of Making Dilutions *213*

Formula Method of Making Dilutions *215*
Solution Strength *215*
Handling of Dangerous Chemicals *216*

19 DISINFECTION AND FLUORIDATION   218

Disinfection *218*
    Chlorination 219
    Concentration 219
    pH 221
    Retention time 221
    Turbidity 221
    Free and combined residuals 221
    Flow measurement 222
    Breakpoint chlorination 222
    Self-contained hypochlorinators 224
    Chlorine dioxide 224
    Chlorination equipment 224
    Portable chlorination equipment 225
    Ozone (O_3) 225
    Ultraviolet light (UV) 227
    Criteria for the acceptability of
      An ultraviolet disinfection unit 230
    Potassium permanganate ($KMnO_4$) 232
    Silver 232

Fluoridation *232*
    Optimum fluoride levels 233
    Chemicals used in fluoridation processes 234
    Physical and chemical properties 235
    Small systems 237
    Medium sized systems 237
    Large systems 238

20 WASTE DISPOSAL   241

Waste Disposal *241*
Filter Backwash Water *242*
Coagulated Clarifier Sludges *243*
Lagoons *243*
Drying Beds—Freezing *243*
Vacuum Filtration *244*

Centrifuges *244*
Filter Presses *244*
Quantity of Sludge *245*
Alum Recovery *246*
Softening Sludges *246*
Drying Beds *246*
Thickening *247*
Size of Thickener *247*
Diameter of Thickener *252*
Depth of Thickener *253*

21 STABILIZATION *256*

Stabilization *256*
Well Waters *256*
Marble Chip Test *257*
Saturation Indices *258*
    Langelier index 258
    Ryznar stability index 259
Specific Conductivity *259*
Organic Deposits *260*

22 DEMINERALIZATION *262*

Demineralization *262*
Low Total Dissolved Solids (TDS) *262*
    Reverse osmosis (RO) 262
    Electrodialysis (ED) 263
    Ion exchange 265
    Base exchange softening 266
    Resin capacity 268
    Cation—hydrogen ion exchanger 270
    Strong basic anion exchanger 271
High Total Dissolved Solids (TDS) *272*
    Evaporation 272
    Vapor recompression 273
    Freeze—desalting processes 273
    Solar evaporation 274

Summary of the Present State of the Art *275*

Cost of Producing Demineralized Water *275*

23 DISTRIBUTION SYSTEMS *279*

Distribution Systems *279*

Legal Aspects *279*

Network Studies *282*

Hardy Cross Method *282*

Electric Analog *283*

Computer Programs *283*

Thawing Frozen Pipes *284*

Metering *285*
    Meter repairs 286

Leakage Surveys *286*

Heating of Water in Distribution Systems *286*

24 CROSS-CONNECTION CONTROL *289*

Cross-Connection Control *289*
    Cross-connection means 289
    Backflow means 289
    Backsiphonage means 290

Summary of Problems *290*

Air Vent Valves *291*

Hydrants *292*

Cross-Connections *292*

Backflow Prevention Devices *292*
    Approved air gap 292
    Vacuum breakers 293
    Reduced pressure backflow preventers 293
    Double check valves 293

25 PIPELINES *297*

Pipeline Friction *297*
    William and Hazen equation 297
    Manning equation 298
    Colebrook-White equation 298

Large Diameter Pipes *300*
Medium Sized Pipes *300*
Smaller Sized Pipes *300*
Cathodic Protection *301*

26 WATER HAMMER — 303

Water Hammer *303*
Allievi's Equation *305*
Gravity Systems *308*
Pumping Stations *309*
  Flywheel at the pump coupling 311
  Vacuum and pressure relief valve 311
  Air vent valves along the pipeline 311
  Surge buffer tank with an air inlet and relief valve 311
  Hydropheumatic tanks 312

27 RESERVOIRS — 313

Reservoirs *313*
  Equilizing storage 313
  Fire protection reserve 314
  Emergency reserve 314
Type of Reservoir *315*
Valving *316*
Controls *316*
Design of Reservoir *316*

28 HYDROPNEUMATIC TANKS — 319

Hydropneumatic Tank Systems *319*
Hydropneumatic Tank Capacity *320*
Pump Characteristics *322*
Size of Hydropneumatic Tanks *323*
Hydropneumatic Tank as Pressure Vessels *325*
Water Level Control Systems *326*

29 PUMPS — 328

Cost of Pumping *328*
  Installed costs 328

      Power costs *329*
      Supervision and maintenance *330*
      Down time and stand-by equipment *330*

Shaft Speed *330*

Pump Types *331*
      Low lift or intake pumps *331*
      High lift or service pumps *331*
      Booster pumps *332*
      Fire pumps *332*
      Transfer and plant services *334*
      Chemical pumps *334*

Net Positive Suction Head (NPSH) *335*

Vortexing and Submergence *335*

Specific Speed (Ns) *336*

Pump Selection *337*
      Invitation to tender *337*
      Testing procedures *338*

Maintenance *339*
      Mechanical seals *339*
      Stuffing boxes *339*
      Pipe stresses *339*
      Bedding *339*

30 DRIVES—MECHANICAL AND ELECTRICAL   *341*

Choice of Drives *341*

Electric Motors *341*
      Open motors *343*
      Totally enclosed motors *343*
      Water cooled motors *344*
      Reduced voltage starting *344*
      Wound rotor motors *345*

Variable Speed Drives *345*
      Wound rotor motors *345*
      Electro-magnetic couplings *345*
      Hydraulic couplings *345*
      Variable frequency drives (VFD) *345*
      Combustion engines *346*

Engine Selection *346*
      Dual fuel engines *347*

Thrust Bearings *348*

APPENDICES *349*

Table A-1 Gallons (UK)—Gallons (US)—Cubic Meters *350*

Table A-2 Cubic Meters—Gallons (UK)—Gallons (US) *364*

Table A-3 Gallons (US)—Cubic Meters—Gallons (UK) *378*

Table A-4 Atomic Weights *392*

Table A-5 Periodic Table of the Elements *393*

Table A-6 Temperature Conversions *394*

Table A-7 Chemical Feed Requirements for Flows Measured in US gpm *395*

Table A-8 Chemical Feed Requirements for Flows Measured in UK gpm *396*

Table A-9 Chemical Feed Requirements for Flows Measured in Liters per Second *397*

Table A-10 Chemical Feed Requirements for Flows Measured in Cubic Meters per Day *398*

Table A-11 Chemical Feed Requirements for Flows Measured in Cubic Meters per Day *399*

Table A-12 Conversion of Feet and Meters *400*

Table A-13 Pressure Conversions: Pounds per Square Inch (PSI) and Kilo Pascals (kPa) *401*

INDEX *411*

ABBREVIATIONS

mgd (U.S.)	million gallon day (U.S.)
mgd (U.K.)	million gallon day (U.K.)
10^6 m³ pd	million meter cubed per day
gpcd (U.S.)	gallon per capita day (U.S.)
gpcd (U.K.)	gallon per capita day (U.K.)
m³ pcd	meter cubed per capita day
lpcd	liters per capita day
J.A.W.W.A.	Journal of American Water Works association, Inc.
mg/ℓ	milligrams per liter
ppm	parts per million

WATER SUPPLY, TREATMENT AND DISTRIBUTION

1

WATER DEMAND

PLANNING MUNICIPAL SERVICES

The two most difficult questions to answer in the water supply industry are "how large will the service area be in 5 to 10 years time?" and "how much water will it require?"

There are also governmental and financial approvals necessary before a project can proceed. Even relatively small projects (under $1 million) require 3 years or more, before the utility is in operation. If land acquisitions and right-of-way agreements are to be negotiated before engineering designs can be completed, then another 1 or more years must be added to the time required for project completion. As choice land for utility development becomes scarce, negotiations become more complicated and time consuming.

Capital cost estimating is always a nightmare and, even when all the preliminaries of land acquisitions and the various approvals have been obtained, the cost estimate must be kept constantly under review. Changes in the economy, labor disputes, and political elections all tend to slow down municipal projects; these delays must be contemplated in the preliminary planning stages if the requirements of a growing metropolis are to be met even close to schedule. The only time variable left to the planner is the date on which he must begin in order that a particular deadline may be met. Many large metropolises continuously plan extensions to their present-day requirements. Zoning bylaws may be adopted to control population densities and limit water distribution system overloads before increased capacity can be provided. Design studies of this magnitude are detailed and complex, but are paramount to economic growth.

Every available source of information should be used to predict population growth. Urban development occurs at such a pace that communities are no longer isolated within their own boundaries. They are small sections of larger developments, each dependent on the other. This includes commerce and transportation services, as well as water supply and sewage disposal. There is scarcely a river on the continents of Europe or North America that does not intimately link the water supply and sewage disposal of several cities together. Cities often draw their water supplies from a river, treat and discharge their effluents back into the same river for use by the next city further downstream. Sewage treatment of city A is of vital importance to the water supply of city B. Water treatment costs for city B may be lower if city A, upstream, discharges a well-treated effluent.

The enormous changes in growth and life patterns of the last few decades promise to continue. Alvin Toffler[1] describes the situation in his book *Future Shock*,

> ... changes in the process by which man forms cities, for example. We are now undergoing the most extensive and rapid urbanization the world has ever seen. In 1850, only four cities on the face of the earth had a population of 1,000,000 or more. By 1900, the number had increased to 19. By 1960, there were 141, and today, world urban population is rocketing upwards at a rate of 6.5% per year, according to Edgar de Vries and J. R. Thysse of the Institute of Social Science in The Hague. This single stark statistic means doubling the earth's urban population within 11 years.

POPULATION Countries are fortunate if they have demographic data concerning births, deaths, and diseases. The United Nations Organization has compiled data concerning developing countries. World population estimates projected to the end of the century, prepared by the United Nations, are quoted in Table 1-1. If these predictions are correct, there will be between 54% and 91% more people on the earth in the year 2000 compared to 1970 depending upon which prediction proves the more accurate.

The increase in population of many developing nations is influenced by longer life expectancy, as well as an increase in birth rates. The infant mortality, that is to say, deaths under 1 year of age per 1000 live births, in North America is approximately 23. Infantile mortality ranges from 13.6 in Sweden to 86.8 in Albania. A number of African countries are well in excess of 100, several are over 200, and one of them is given as 259 per 1000 live births, and much of this problem is related to water quality. Improved water supply, sanitation, nutrition, medical facilities, and education will increase the population of persons between the ages of 15 and 65.

TABLE 1-1 United Nations Estimate of World Population[2] (in Thousands)

	"Low" Variant	"Medium" Variant	"High" Variant
1960	2,998,180	2,998,180	2,998,180
1965	3,265,555	3,280,522	3,305,862
1970	3,544,781	3,591,773	3,656,157
1975	3,840,439	3,944,137	4,070,083
1980	4,147,337	4,330,037	4,550,733
1985	4,462,720	4,746,409	5,096,198
1990	4,782,859	5,187,929	5,689,910
1995	5,109,362	5,647,923	6,325,593
2000	5,448,533	6,129,734	6,993,986

Age structure diagrams help to determine population trends. Examples of these diagrams from *Population Resources Environment* by Paul and Ann Ehrlich[2] are shown in Figs. 1-1 and 1-2 and present, in graphic format, the age structure of the population. Provided they are drawn to the same scales, they are identical in area.

The percentage of males in each age group is shown to the left of the center line, and the females to the right. The cross-hatched areas for ages zero to 15 at

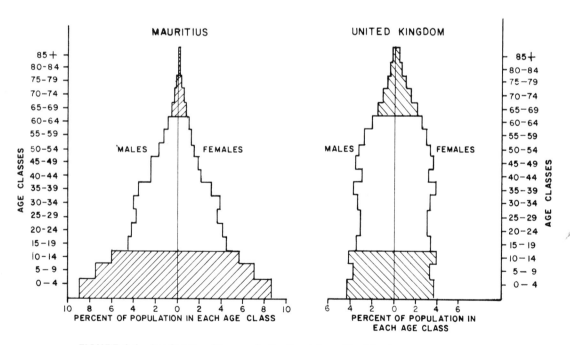

FIGURE 1-1 Age Structure Diagrams for the Population of the Island of Mauritius and of the United Kingdom in 1959 (From *Population, Resources, Environment: Issues in Human Ecology*, 2nd Ed., by Paul R. Ehrlich and Anne H. Ehrlich, W. H. Freeman and Company, 1972)

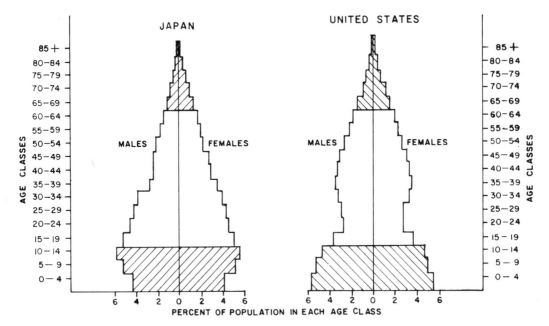

FIGURE 1-2 Age Structure Diagrams for the Population of Japan and the United States in 1960 (From *Population, Resources, Environment: Issues in Human Ecology,* 2nd. Ed., by Paul R. Ehrlich and Anne H. Ehrlich, W. H. Freeman and Company, 1972)

the bottom of the diagrams and from 65 upward at the top of the diagrams represent those who are dependent on the productive groups between the ages of 15 and 65 years.

These profiles contrast the age distribution in a rapidly growing, predominantly young population on the island of Mauritius with the slowly growing population in a well established country such as the United Kingdom. In Mauritius, in 1959, 44% of the population was under the age of 15 years as opposed to the United Kingdom where only 23% was under the age of 15 years. It would be reasonable to assume that the rate of growth in Mauritius is likely to increase at a greater rate within the next 10 to 15 years than the population of the United Kingdom.

Some nations have shown a reduction in the lower age groups as indicated by the population profiles for Japan and the United States (Figure 1-2). In 1960, some measure of birth control, economic recession, or other phenomena has slowed the rate of population increase. Comparing the 1960 population profiles for Japan and the United States, the United States should reasonably anticipate a higher rate of increase in the 1970's than Japan.

Birth and death rate relationships are not the only factors which influence population growth. Immigration from other countries and interstate migration of people substantially affect the equation of water supply and demand in the United States.[3] For example, between March 1964 and March 1965, 6,000,000 people changed states, with California and Florida being the largest recipients. Immigration statistics are almost as reliable as those on births and deaths, but interstate migration is more difficult to assess and can only be forecast by changes in the commerce of the area. The people who migrate interstate are nearly always people in the production ages of 15 to 65, and they are the ones who have a larger demand on the water supply.

The art of forecasting population trends is more than an exercise in simple extensions of statistics. Plotting populations from each census period on semi-logarithmic paper helps to provide a starting point for future population predictions. An example of this technique is shown in Figure 1-3.

If semi-logarithmic paper is not available, plain paper can be used with the log scale drawn from the B or C scale of an ordinary slide rule. It will be noted that population data plotted in this manner on semi-log paper usually produces a straight line. One of the advantages of this method is the fact that a few thousand people migrating into the area where the population is already 20,000 to 30,000 does not seriously upset the semilog curve in the upper regions of the graph. In other words, a new industry attracting 2000 to 3000 people, plus the secondary industries which accompany such an influx, will not affect a community of 20,000 as much as a community of 3000. Such an influx is clearly shown in Figure 1.3 for the city of Prince George, British Columbia. In the 10-year period between 1951 and 1962 the population rocketed from 4500 to approximately 14,000, due in large part to the rapid expansion of the pulp industry. From 1957 to 1965, the growth followed a more natural trend until another influx occurred in 1966 which continued to the census year of 1971. Three authorities have shown their projected future growth estimates for 1980 to range from about 48,000 to 62,000 people.

Semi-log graphs maintain their usefulness in helping to assess past performance. From this platform, future predictions can be evaluated in the light of local conditions, climate, industry, geographical location, world markets, and government policies. However, even with the best available data, predictions can easily be out 20 to 30%, particularly with the smaller towns and cities which may have passed their prime usefulness and are starting to decline. There are many examples of this in Western Canada where small villages appeared as the Canadian Pacific Railway (CPR) and the Canadian National Railway (CNR) moved westward to the coast. The railroad stores and accommodation shacks were located at 6 to 8 mile intervals and many of these little communities continued as people homesteaded the area. Grain elevators, houses, schools, and stores were built to service the community. However, as roads and automobiles improved, and a journey from one town to another could be accomplished in 8 to 10 minutes instead of 2 to 3 hours, the whole aspect of village life was altered. Once people were able to drive into the larger cities for their groceries and evening entertainment, the smaller villages could no longer economically maintain their previous status. Community lifestyle changed, growth came to a standstill, and in many cases started to decline. The changeover from steam to diesel locomotives brought many changes to the communities along the rail-

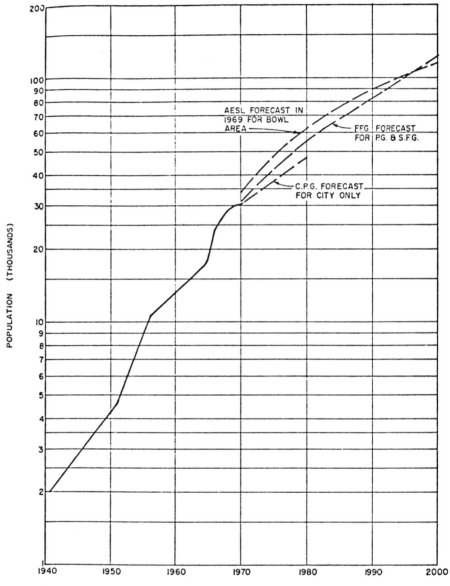

FIGURE 1-3 Population Forecast for the City of Prince George, B. C., Canada

roads. The old steam locomotives required servicing at frequent intervals but a diesel engine can haul a train from Montreal to Vancouver with only an occasional stop for oil and crew changes. As a result the old roundhouses were no longer needed and their employees moved elsewhere.

Industries have a considerable influence on populations and water demands. A new industry in one part of the world will change the lifestyles of people in other lands. In the 1950's naturally-occurring sodium sulfate was mined extensively in the province of Saskatchewan and found a ready market in the manufacture of glass, pulp, and paper. The development of the rayon industry for the manufacture of artificial textiles produced high grade sodium sulfate, a by-product which flooded the market. The demand for the lower grade natural sulfate dropped to a low ebb. Eventually, when the nylon industry replaced the rayon industry, natural sodium sulfate was again in demand. The working population follows the industrial climate, resulting in an ever-changing pattern from year to year.

DOMESTIC WATER DEMAND

Per capita per day water consumption figures can only be comparative from one area to another when it is determined if the flows are measured at the consumer's house meter, or as it enters the distribution system. There is a considerable difference between the water consumption per capita for metered supplies as compared to unmetered supplies. Distribution system leakage is also a major factor, and a system with only a 10% unaccountable loss is considered to be "bottle tight." Many distribution systems have much more than 10% loss. The Brussels Water Company in Belgium quoted 30% of average annual demand was leakage. Five Canadian towns also quoted an average of 30% leakage.[4] Losses from water mains and service connections in the U.S. are estimated to vary from 5% to 45% of the average flow. Unmetered services without exception result in water wastage. In many instances, the average demand of the city has been reduced by as much as 50% by installing meters on each service and charging customers for the water they use. Once meters have been installed and the metered readings totalled and compared to the water leaving the plant, then the unaccounted distribution system losses can be estimated, and maintenance programs can be instituted to locate and repair the broken mains and reduce the losses.

The cost advantages in metering each customer are accompanied by a reduction in water consumption, allowing the utility to serve more people than it could supply if water were sold on a flat rate. An example of the benefits of metering appeared in the August 1972 issue of *Water and Pollution Control*.[5]

> Consumption dropped noticeably year by year after water meters were installed at Repentigny, Quebec, despite the fact that the population grew steadily from 15,500 in 1964 to 26,000 in 1972, including 4200 in the neighboring municipality of Charlemagne, which gets its water from Repentigny. It wasn't until 1970 that the water use returned to the 1964 level, 8 years after the original crisis.

This is by no means an isolated instance. Chilliwack, in British Columbia, installed 7500 meters in 1963 and reduced their consumption by 40%; Penticton reduced their consumption by 40 to 45%; Nanaimo reported that the consumption in a completely metered part of their system was 60 (gpcapd) whereas in another unmetered area it was 160 (gpcapd).

The completely metered community enables the water department to periodically compare the demand against the summation of the individual service meters. If the difference is sufficiently large, a leak detection program should be instituted and the problem areas corrected. Small leaks over a period of time can amount to considerable losses. Assuming an average distribution system pressure of 50 psi, Table 1-2 shows the leakage rates that can occur from relatively small holes.

TABLE 1-2 Water Leakage at 50 psi

Hole of $\frac{1}{16}$ in. dia.	(1.587 mm) =	600 U.S. gp day. (2.27 m³/day)
Hole of $\frac{1}{8}$ in. dia.	(3.175 mm) =	2,400 U.S. gp day. (9.08 m³/day)
Hole of $\frac{1}{4}$ in. dia.	(6.350 mm) =	9,500 U.S. gp day. (35.96 m³/day)
Hole of $\frac{3}{8}$ in. dia.	(9.525 mm) =	21,300 U.S. gp day. (80.6 m³/day)
Hole of $\frac{1}{2}$ in. dia.	(12.700 mm) =	38,000 U.S. gp day. (143.8 m³/day)
Hole of $\frac{3}{4}$ in. dia.	(19.050 mm) =	86,000 U.S. gp day. (325.5 m³/day)
Hole of 1 in. dia.	(25.40 mm) =	150,000 U.S. gp day. (567.8 m³/day)

A 1-inch diameter leak at 50 psi loses 150,000 gpd (U.S.), enough to supply a population of 1200 at an average per capita consumption of 125 gpcd (U.S.). Table 1-3 shows a typical demand rate for a medium sized city, and the 24 hour demand curve is shown in Figure 1-4. It varies with the weather and seasons of

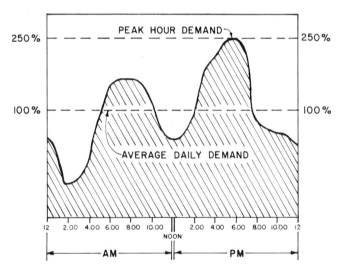

FIGURE 1-4 Typical Hourly Fluctuation in Percentage of Average Day for Small Communities

the year, most water being required in July and August. Records of peak month, peak week, maximum day, and peak hour are correlated with the annual averages. The following figures relate to the City of Edmonton, Alberta, in 1965–1966.

TABLE 1-3 City of Edmonton—Water Demand

	All Areas of the City		Londonderry Subdivision Only	
Average Annual	80 gpcd (U.K.)	100%	60 gpcd (U.K.)	100%
Peak Month	110 gpcd (U.K.)	137%	90 gpcd (U.K.)	150%
Peak Week	125 gpcd (U.K.)	156%	105 gpcd (U.K.)	175%
Peak Day	135 gpcd (U.K.)	169%	115 gpcd (U.K.)	192%
Peak Hour	250 gpcd (U.K.)	312%	210 gpcd (U.K.)	350%

The Londonderry subdivision was relatively new, and many lots required landscaping resulting in fairly high demands during the spring and summer when lawns were planted. Peak demands on hot, dry, summer days are much higher than during a rainy period. The city of Edmonton supports a fairly large industrial area which accounts for the higher per capita for the city as a whole as compared to the subdivisions.

A water main under reduced pressure conditions, that is to say, less than atmospheric pressure, is a potential health hazard of the first magnitude and every effort must be made to prevent this from occurring. Polluted ground water can readily enter the water mains through leaking joints under these conditions with disastrous consequences.

Where new townsites are developed, arbitrary water demand parameters must be evaluated during the early stages of design in order to size the water treatment, storage, and pumping facilities. Designers of sewage collection systems have similar problems, and have developed an interesting graph,[6] shown in Figure 1-5, which is of value in predicting the peak flows.

The water requirements of flush toilets and automatic home washers vary considerably.[7] There currently exists quite a range of water use between different types of toilets, from about 3.2 to 8 U.S. gallons or more per flush. Home laundry washing machines use between 32 to 59 U.S. gallons per 8-lb load. The design of shower heads will also materially influence water consumption since it is estimated that showers account for approximately 18% of the total in-house water consumption.

There is merit in using every available means to induce economy in the use of water without restricting its legitimate use. Economy in water use curtails capital investment in the waterworks, sewers, and sewage treatment. The best way to keep the water demand under reasonable control is to meter all services and to apply an equitable rate per unit volume. If people must waste water, they must pay for the privilege.

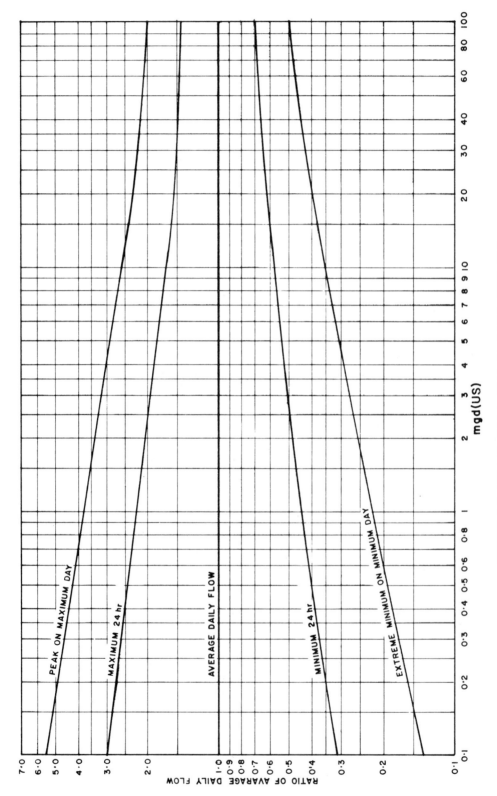

FIGURE 1-5 Ratio of Extreme Flows to Average Daily Flow (Courtesy, Water Pollution Control Federation)

FIRE PROTECTION

The National Fire Protection Association (NFPA) of the United States has compiled a series of standards containing minimum water flow and storage requirements for fire protection. The Canadian Underwriters' Association has similar regulations.

Fire hazards and control facilities vary with each community. The designer of water supply systems must check with the local authorities and insure that he has full knowledge of the area requirements. It may not be financially feasible to fully comply with all of the many facets that constitute a perfectly satisfactory system from the fire insurance underwriters' point of view, but at least the major aspects should be covered.

Very few municipal fire protection systems meet the underwriters' requirements in every aspect, since it would be necessary to have duplicate power supplies, transmission piping, storage, and pumping facilities. Failure to meet the conditions of the underwriters will result in higher fire insurance premiums, and for these reasons it is necessary for the engineers and the underwriters to consult with each other at the design stage. The consumer has to pay in either case, whether it be increased fire insurance or higher water rates, and if by good design he can have an adequate fire protection system at minimum cost, so much the better.

In industries where there are high fire risks such as distilleries, petroleum refineries, bulk loading stations, and lumber yards, it pays to completely comply with the requirements of the underwriters, not only from the viewpoint of fire insurance premiums, but also with regard to legal and moral responsibilities to neighbors and surrounding communities.

INDUSTRIAL WATER REQUIREMENTS

Almost every town, and city, has some industries within its boundaries. Since the economy is often dependent upon the industries, it is in the interest of the community to encourage industry. Industry usually has four water requirements:

Process Water
  Water which appears in the finished product, for example, in breweries, distilleries, soft drink, and chemical plants.

Cooling Water
  Used for makeup water to cooling water systems or in smaller installations; once through systems discharging directly to the sewer, creek, or river.

Boiler Water
  Used for makeup feed water to the boilers.

Potable Water
  For cafeterias, washrooms, and drinking fountains.

Some changes in the water may take place within the plant, either by softening or various chemical additions. In many instances industrial plants have their own wells to supplement their supplies, and possibly river intakes with pumps for fire protection. The important issue, as far as the municipal water purveyor is concerned, is to insure that it is absolutely impossible for water that is inside the plant to get back into the town's distribution system. (See Cross-Connection Control.) There are cases on record where industrial plants, having a supplementary well water supply of inferior quality, use city water during the day for process purposes and pump their industrial water through the meter, in the reverse direction, into the city's distribution system at night to avoid paying for the water they use during the day. But this is the least of the crimes as far as the municipality is concerned. The city's distribution system now contains water over which it has no quality control. It is probably unchlorinated, may be raw water contaminated with sewage, or may even contain industrial chemicals used in the process plant!

United States law comes from three basic roots, English, Spanish, and French. Under English law, case histories indicate that a purveyor who sells potable drinking water must accept responsibility for the wholesome quality of the water he delivers to his customers. The fact that one customer may inadvertently allow contaminated water to flow back into the distribution system does not relieve the purveyor of his responsibilities to his other customers. It is therefore in the interest of the purveyor, as well as his customers, to insure that an industrial plant has suitable backflow prevention devices installed at the point where the water enters the consumers' premises. For continual uninterrupted flow of water into the municipal system, it is customary for the water purveyor to insist on the installation of two suitable backflow preventors to be installed in parallel, so that one can be isolated from the system for periodic testing and inspection without having to deprive the customer of the use of the utility during these testing periods. This ruling should not be limited to industrial plants, but should be applied wherever there is a danger of contamination due to backflow or cross-connections. This principle applies to hospitals, mortuaries, nursing homes, laboratories, and other establishments not normally categorized as industry, but just as serious a risk whenever the water mains are under reduced pressure.

Some municipalities have agreed to allow the return of cooling water to the distribution system or to the town's reservoir. It has been argued that there has been no chemical or physical change during the cooling process, other than an increase in temperature, and the water is therefore unchanged. However, this practice should not be allowed since the municipality has no control over the water from the time it leaves its distribution system until it is returned. Furthermore, changes in water temperature are one of the major causes of breakages and leakages due to expansion and contraction of the piping system.

Many industries require "tailor made" water for their processes and will often use ion exchange plants to completely demineralize, and then add chemicals to provide the quality required for their needs. These processes are, of course, the responsibility of the customer and not the municipality. The water purveyor is responsible for providing a consistent water quality to his customers. (However it should be pointed out that nearly all city water mains contain some silt deposits, particularly in the larger diameter pipes where velocities are frequently less than the scouring velocities necessary to keep the deposits moving.) The water may be in the distribution system for several days from the time it leaves the treatment plant or pump station until it flows through the customer's tap. During this period there will be times during the night and other off-peak periods when the suspended matter will settle out only to be stirred up again whenever the mains are flushed. This can occur during a fire when maximum flows are likely to exceed the scouring velocities.

Sediment in the water is unlikely to be pathogenic or hazardous to normally healthy people, but it can be disastrous to certain people and industries unless guard filters are used to protect against such an eventuality. There was, in the early 1960's, a famous court case in Vancouver, British Columbia, Canada, where a photographic firm lost an entire batch of a customer's negatives in the developing tank due to silt from a flushed water main. Warnings should be published well in advance of a main-flushing program to advise people that disturbance of deposits will occur; however, in the event of a fire no advance warnings are possible. It is therefore in the interest of good public relations to advise the customers of these possibilities in order that they can take the necessary steps to protect their own interests.

The municipality should be acquainted with the industrial water usage within its jurisdiction. If the customer must treat the city water before it can be used, the treatment process should be known to the city in order that an assessment can be made of the impact of the waste disposal on the sewage system. The industrial customer should also pay his fair share of any extra costs involved in providing him with an adequate service, either directly or in the water rate. These problems should be resolved in the early stages of development so that the industry can take full advantage of the facilities offered by the city, and at the same time the city can enjoy the added prosperity brought into the area by the new industry without burdening the existing consumers with an unfair share of the costs involved.

IRRIGATION

Irrigation systems are normally independent of the municipal water supply. The technical, legal, and administrative problems of irrigation differ from those of a municipal utility, and it is far better to divorce the two even if the water comes from the same source. Growers in a particular area usually form regional irrigation districts, and appoint a manager and an engineer. The water quality criteria are somewhat different from other water supply schemes.[8] It is reported that sufficient sodium has been absorbed by citrus leaves from a single sprinkling with water containing 69 mg/ℓ to 190 mg/ℓ of sodium to cause serious leaf burn and defoliation. Sodium forms swelling colloids, reducing soil permeability to both water and air, and increases the pH of the soil-water solution to dangerously high levels. The deterioration of soil quality because of sodium in the irrigation water is a steady, cumulative process, with increasingly impaired drainage.

It is therefore essential that the water used for irrigation be compatible to the soil as well as to the crops. Fortunately, in North America there are numerous universities and institutions which specialize in these subjects and are prepared to advise according to each specific situation. Their help should be solicited to insure that the chemical constituents of the proposed water supply are compatible to the soil and crops.

WATER REUSE

The reuse of water by the process industries is a question of economics rather than technology. Apart from the recirculation of cooling water, there are few process industries that could economically justify the additional capital and operating costs that a reuse plant would incur. However, this situation is continually changing and the concept of reuse may eventually become financially attractive. Industries in highly competitive markets have considerable overheads for research, advertising, and marketing and since there is always a substantial risk of obsolescence for one or more of its products, the return on capital expenditure to cover the overhead costs must necessarily be high in order for the industry to remain in business.

With the relatively low cost of water and the large volumes usually involved, it is obvious that there is very little justification for capital expenditure for water economy. There is no point in spending capital to acquire 10 to 15% return on water reuse when the same capital could be used for a manufacturing project which would yield 40 to 50% annual return. However, this situation can change wherever there are problems of waste disposal or shortage of supply. If a trade effluent has to be treated before it can be released to the environment, then it is possible that very little additional treatment may make it suitable for reuse; but this may be more a question of ecology than of economics.

REFERENCES

1. TOFFLER, A. *Future Shock*. New York: Random House, Inc., 1970.
2. EHRLICH, P. R. and A. H. EHRLICH. *Population Resources Environment*, 2nd Ed. San Francisco: W. H. Freeman and Company, 1972.
3. BAXTER, S. S. "Tomorrow is Today." Joint Discussion. *J.A.W.W.A.*, Vol. 58, p. 929, 1966.

4. TWORT, A. C. *A Textbook of Water Supply*. London: Edward Arnold Ltd., 1963.
5. "Water and Pollution Control." Vol. 110 (8) August, 1972.
6. "Design and Construction of Sanitary and Storm Sewers." *W.P.C.F. Manual of Practice*, No. 9 (1970) published by the Water Pollution Control Federation, 3900 Wisconsin Avenue, Washington, D.C.
7. HOWE, C. W. and W. J. VAUGHAN. "In-House Water Savings." *J.A.W.W.A.* Vol. 64 (2) p. 118, 1972.
8. MCKEE and WOLFE. *Water Quality Criteria*. The Resources Agency of California—State Water Quality Control Board, Sacramento, Ca. Publication No. 3-A, 2nd ed., 1963.

2

WATER QUALITIES

PROPERTIES   Pure water is an odorless, colorless, tasteless, transparent liquid, but due to light scattering by small particles it exhibits a bluish tinge in lakes and reservoirs. It is essential to life, and without it in one form or another life ceases to exist. Water is almost indestructible, and the same water you used to wash your breakfast dishes will in all probability find its way back to the oceans from whence it came and in due course, those same molecules may be reevaporated into clouds and return as rain on some other part of the planet.

    The hydrological cycle where ocean water is evaporated and falls on earth as precipitation is one example of the enormous capability of the solar machine. Imagination in this vein can run riot. Thoughtfully contemplate the ice cube in your next drink! Could Christopher Columbus have bailed this same water out of the bilges of the Santa Maria when he sailed across the Atlantic to the West Indies in 1492? Or could Cleopatra, Queen of Egypt, have used it in any way?

SYNTHESIS OF WATER   The amount of water on the earth is probably constant but there are continual changes in its usable form and distribution—ice caps, salinity, water of crystallization, etc. The combustion of hydrocarbons results in the synthesis of water when the hydrogen atoms in the fuel combine with oxygen. For example, 1000 cubic feet of methane, the principal component of natural gas, when measured at atmospheric pressure will produce 95.5 pounds of water equivalent to 11.4 U.S. gallons (43.2 liters) on combustion. It is not suggested that the burning of hydrocarbons is significantly changing the hydrological cycle or the world water reserves as a whole, but it does indicate where the water comes from when boiler room stack gases are cooled below their dew point.

MEDICINAL WATERS

Many natural mineral waters are considered to be particularly wholesome and are believed by some to be beneficial in the cure of certain illnesses and diseases. These waters are featured at health spas, natural hot springs, and other convalescent and recreational resorts. They are normally outside the professional interests of the majority of water supply engineers.

MICRO-ORGANISM POPULATION OF NATURAL WATERS

From the viewpoint of the municipal water supply industry, the most important aspect of water quality control is the continual assurance that the water delivered to the consumer is free from pathogenic microorganisms. Geldreich has noted that "The waters in some of the community water supply systems in the U.S. often contain a myriad of microorganisms that carry past the disinfection barrier. Although the majority of those that survive and flourish are not pathogenic, the situation presents a potential danger."[1]

Pathogenic bacteria and viruses are difficult to isolate and identify routinely and, unless there is a specific problem, where illness has resulted, these analyses are not normally performed. However, pathogenic organisms causing such diseases as typhoid fever, paratyphoid fever, cholera, bacillary dysentery, amebic dysentery, poliomyelitis, and infectious hepatitis, are carried and excreted from the intestines of infected people together with many other microorganisms known as *fecal coliforms*. The confirmative fecal coliform test-results of a water sample is an indication that pathogenic bacteria or viruses may also be present. Fortunately, the tests for fecal coliform bacteria are necessarily of fecal origin, although the name *coliform* has been derived from the bacteria which are present in the colon. Coliform bacteria are by no means confined to the human intestine; the organisms were reported by Smith (1895) in the intestines of dogs, cats, pigs, and cattle.[2] Furthermore, it has since been discovered that coliform organisms are by no means confined to the animal body (including humans) but are widely distributed elsewhere in nature. The finding of a few coliforms in a water supply has no special significance. However, the detection of coliforms in a number of 1-mℓ samples is an indication that recent sewage pollution of the water supply may have occurred. Further testing should then be done to

Chap. 2   *Water Qualities*

determine whether or not the coliforms are of fecal origin. For detailed instructions on procedural methods the reader is referred to the following standard texts: *Standard Methods for the Examination of Water and Wastewaters*[3] and *The Bacteriological Examination of Water Supplies.*[4]

BACTERIOLOGY

It is only in the last century, with the pioneer work of early bacteriologists Downes and Blunt (1877),[2] Louis Pasteur, Robert Koch (the German bacteriologist who in 1884 isolated the organism responsible for Asiatic cholera), and many others, that a scientific understanding of waterborne diseases has been developed. There is probably more research in progress today in this field than ever before. The question then arises, "How did anybody manage to live to a ripe old age before the advent of pasteurization, chlorination, and all the many medical and sanitary engineering advances that we have today?" The answer is that a lot of people didn't and still don't in many countries where modern standards of hygiene do not exist. Mr. M. N. Baker[5] in his book *The Quest for Pure Water* cites many interesting references to the attempts of previous generations to overcome the problems of polluted water supplies.

It is easy to appreciate that the habit of drinking tea and coffee has undoubtedly saved countless millions of lives which would have succumbed to one of the many waterborne diseases if it had not been necessary to boil the water before the tea or coffee could be made. King Cyrus (535 B.C.) used silver flagons for transporting water supplies during his many battles and the use of silver may have further significance than as a simple, relatively non-corrodible and unbreakable container, when it is realized that silver is an effective water bactericide in the Katadyn process.[6]

One of the saving graces of the last century was the advent of the slow sand filter. The history of this development has been well researched and documented by Baker[5] and others. Perhaps one of the best documented accounts of the results of slow sand filtration is reported for the 1892 cholera epidemic in Hamburg.[7] During the Hamburg epidemic the deaths in the several cities were as shown in Table 2-1.

TABLE 2-1 Death and Death Rates from Cholera[7]: Three Cities in 1892

City	Population	Deaths	Deaths per 10,000 Inhabitants
Hamburg	640,000	8,605	134.4
Altona	143,000	328	23.0
Wandsbeck	20,000	43	22.0

Although the Hamburg cholera epidemic is unique in that sufficient statistics were available for Professor Koch to correlate the evidence and present his findings, it is not historically unique. Many cases of waterborne diseases had occurred in populated communities but were not so well documented.

Baker[5] has compiled a very full account of the development of filtration from the 17th century onward. James Simpson, the Engineer of the Chelsea Water Works Company, in 1829 completed the first one-acre slow sand filter in metropolitan London. The results of Simpson's experiments and those of other engineers of their day proved that if water is treated in this manner, consumers would be less likely to be stricken with disease than those who drank unfiltered water. Much of the water used in metropolitan London was taken from the River Thames. It was reported that the river stank so badly that scented drapes were hung in front of the open windows of Parliament.

This and other events eventually led to the Metropolis Water Act of 1852, whereby the filtration of all Thames river water for metropolitan London became a legal requirement, and the use of slow sand filters as a method of water treatment soon became generally accepted in the countries of Europe.

REFERENCES

1. GELDREICH, E. E. et al. "The Necessity of Controlling Bacterial Population in Potable Waters. Community Water Supply," *J.A.W.W.A.*, Vol. 64, page 596. 1972.

2. PRESCOTT, S. C. et al. *Water Bacteriology*, 6th Ed. New York: John Wiley & Sons Inc., 1950.

3. *Standard Methods for the Examination of Water and Wastewater*, 13th Ed. Published jointly by A.W.W.A., American Public Health Association, and Water Pollution Control Federation, 1971. Available from the Publications Sales Department. American Water Works Association, 2 Park Avenue, New York, N. Y.

4. "The Bacteriological Examination of Water Supplies, Report No. 71." Department of Health and Social Security. Welsh Office. Ministry of Housing and Local Government. Available from Her Majesty's Stationery Office (H.M.S.O.) London, 1969.

5. BAKER, M. N. *The Quest for Pure Water*, published by the American Water-Works Association, 1949.

6. HOLDEN, W. S. *Water Treatment and Examination*. London: J. & A. Churchill, 1970.

7. MAXCY, K. F. *Preventative Medicine and Public Health*, 8th Ed. New York: Appleton-Century-Crofts Inc., 1956.

3

WATERBORNE DISEASES

WATERBORNE DISEASES In developed countries such as the United States, Canada, and Western Europe, waterborne diseases on the epidemic scale have been almost eradicated. When they do occur, they are usually attributed to relatively small communal or private water supply systems using untreated surface waters or shallow wells without disinfection.

There are always isolated instances where chlorinators or water treatment facilities have failed; Kinshasa in Zaire had a serious epidemic during a civil disturbance when the normal degrees of vigilance were not observed. A similar outbreak also took place in Switzerland. In under-developed countries waterborne diseases are common and frequently spread to epidemic proportions; and the loss of life can be horrendous.

When an epidemic occurs there are two principal lines of attack; one is to determine the cause of the outbreak and the other is to curtail its effects, treat the sick, and stop the spread. Unfortunately, these two approaches are almost diametrically opposed to each other. If the second group descends on the scene with tons of hypochlorite and completely disinfects everything in sight, they will probably destroy some of the essential evidence vital to the first group that is endeavouring to find the root cause of the outbreak. But when lives are in danger it is difficult to imagine any authority wishing to prolong suffering. The water supply is nearly always the number one suspect and usually the first area to be highly disinfected. In many cases the true source of infection is not always defined since it could result from an intermittent cross-connection between a polluted source and the water supply, or from food, inadequate personal hygiene, or bodily contact with an infected person.

Epidemiologists are frequently frustrated by the masking of essential data during the investigation of an epidemic. They are often loath to state categorically whether or not certain diseases are waterborne since in many cases the suspected organisms are unable to survive for any appreciable period of time in direct contact with water. However, since many pathogenic organisms are from the intestines of humans and animals, it follows that any suspended matter present in water in the form of turbidity, silt, and even in a coagulated color could be host to pathogenic organisms, and the chance of chlorine being able to reach all the organisms and kill them is remote. Furthermore, chlorine is not very effective against viruses and particularly hepatitis, even with unhindered access to the organism. The effectiveness of the chlorine may also be nullified by carbonaceous material and the organism is then relatively safe from disinfection. In order to insure reasonable security from waterborne diseases in a municipal supply, the following extract[1] is worth quoting.

> Although the detection of coliform bacteria is the primary concern in potable-water-quality measurements, attention must also be directed to controlling the general bacterial population. The bacterial flora of finished water reflects the microbial flora characteristics of the raw water and the filter bed. Better monitoring of the filter barrier for both viral and bacterial pathogen removal appears to be dependent upon the continuous monitoring system for turbidity which should never exceed the recommended limit of 0.2 units. Once microorganisms enter the distribution system, they may be harbored in protective slime and sediments that develop in portions of the system. Some of these organisms may be a factor in creating health problems among the very young, the debilitated, and the senior citizens of a community. In addition, high noncoliform populations in finished water have been implicated in supressing coliform growth tests media. The critical level for such suppression occurs when the general bacterial population exceeds $1000/m\ell$. This bacterial population can be effectively controlled to a level below $500/m\ell$ by maintaining a residual chlorine level in the distribution system.

The water purveyor has both the moral and the legal responsibility for the water supplied to a community. He must make every possible effort to insure that his "disinfection barriers" are always in place since pathogenic organisms are almost always present in raw water and if given suitable opportunities can easily become of epidemic proportions. Fortunately, in North America, the disinfection barriers are reasonably well controlled and waterborne outbreaks of this magnitude do not occur very often. Unfortunately, both management and public may be lulled into a false sense of security. The fact that waterborne outbreaks do occur and that pathogenic organisms do in fact exist is exemplified by the recent review by Craun and McCabe.[2]

Currently, about 14 waterborne disease outbreaks occur each year in the U.S. and cause on an average 1600 illnesses and one death per year. This is not a leading cause of the American public's illnesses, but it represents a residual that should have been eliminated in this age of sanitation.

Only outbreaks associated with water used for drinking or domestic purposes are included in this analysis. To be considered an outbreak at least two cases of infectious disease must be reported.

With reference to Table 3-1 and 3-2 it is interesting to note that there are more outbreaks with private water supply systems than there are with public systems. But there are probably more private systems and they are obviously much smaller than the public utilities. It is also obvious that some private systems will not meet the same standards of construction and operational management of the larger public water supply systems. The instances of outbreaks attributed to specific organisms are shown in Table 3-3.

TABLE 3-1[2] Average Annual Number of Waterborne Outbreaks (1938–1970)

	Outbreaks	
Years	Public Systems	Private Systems
1938–45	12	26
1946–50	6	17
1951–55	3	7
1956–60	5	7
1961–65	3	8
1966–70	4	10

TABLE 3-2[2] Cases of Waterborne Disease per Outbreak (1938–1970)

	Illness per Outbreak	
Years	Public Systems	Private Systems
1938–45	1,000	50
1946–50	292	43
1951–55	333	33
1956–60	207	23
1961–65	2,603	39
1966–70	166	93

TABLE 3-3 Waterborne Disease Outbreaks[2] (1961–1970) by Type of Illness and System

	Private Systems		Public Systems		Total	
Illness	Outbreaks	Cases	Outbreaks	Cases	Outbreaks	Cases
Gastroenteritis	25	4,498	14	22,048	39	26,546
Infectious hepatitis	22	664	8*	239	30*	903
Shigellosis	16	939	3	727	19	1,666
Typhoid	14	104			14	104
Salmonellosis	4†	96	5	16,610	9†	16,706
Chemical poisoning	7	42	2	4	9	46
Enteropathogenic E. coli	4	188			4	188
Giardiasis	1	19	2	157	3	176
Ambeiasis	2	14	1	25	3	39
Total	95	6,564	35	39,810	130	46,374

* One gastroenteritis outbreak also included seven cases of infectious hepatitis.
† One gastroenteritis outbreak was preceded by outbreak of 38 cases of salmonellosis.

Once the disinfection barriers have broken down, the door is open to all pathogenic organisms that may be present and a severe outbreak may manifest several diseases at one time. Diseases appear to move across the world in waves and then appear to be relatively dormant for a period of time only to blossom forth again. It has been said that cholera is again on the rampage. Cholera follows a fairly definite timetable, about four to six major outbreaks per century. One occurred in 1920. A smaller one about 1950. One in China and surrounding countries early in the 1970's, and statistically, we can expect another major outbreak between now and the turn of the century.

It is interesting to note that reported cases of typhoid have been on the decline in recent years and infectious hepatitis has been on the increase (Figure 3-1).[2] Infectious hepatitis outbreaks are cyclic, peaking at 7 to 9 year intervals; presently showing a relatively low profile, but likely to become pre-eminent again in 1978–1980.

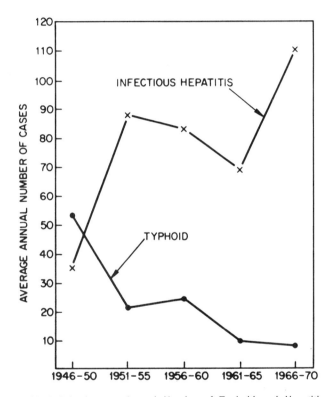

FIGURE 3-1 Average Annual Number of Typhoid and Hepatitis Cases Occurring in Waterborne Outbreaks, 1946-70[2]

The decline in the number of typhoid cases is partly due to better water treatment facilities, but the present treatment technology is less able to substantially reduce the numbers of viruses present in water compared to the carnage of bacteria.

Outbreaks of viral hepatitis have been demonstrated to be waterborne but the usual mode of transmission is person-to-person contact, presumably by the fecal-oral route. It is also reported[2] that the "majority of infectious hepatitis outbreaks in public systems occurred as the result of contamination of the distribution system, primarily through cross-connections and back siphonage; for private systems, contamination of untreated ground-water supplies was the important factor."

At this stage of development the state-of-the-art appears to indicate that the most effective disinfection barriers are as follows:

SOURCE OF WATER

The choice of source water is of primary importance; its monitoring and upstream control if at all possible are paramount considerations in the operation of a safe and wholesome water supply. Most water treatment processes only reduce the number of organisms present in the finished water, particularly if the source of supply comes from a river and there is residual turbidity in the finished water. But the finished water quality can still be maintained at its previous high standard if adequate monitoring of raw water exists and control systems are used to detect the changes that occur and take the necessary action in the treatment processes. Settling chambers and clarifier volumes are large and present a time lag in process control, particularly with the older water treatment plants, and early warning devices to detect higher turbidities of the source waters become important.

REMOVAL OF TURBIDITY

If the water contains suspended solids, it is highly probable that the solids will contain colonies of bacteria and viruses. The problem is to arrange suitable meetings and introductions so that the chlorine may inactivate the organisms. But unless the turbidity is reduced to acceptable standards, there can be no guarantee that the bacteriological quality is within the limits of tolerance. It has been reported[3] that flocculation and precipitation of 300 mg/ℓ of magnesium hardness using the lime-soda-ash softening process inactivates 99.88 % of the viruses present in raw water. The high pH and the hydroxide alkalinity necessary to precipitate the magnesium would have a significant influence on the inactivation of the viruses. Nevertheless, recent research seems to indicate that clarification followed by filtration is, at least, one important component of the disinfection barriers.

DISINFECTION

Having monitored the source water and treated it to reduce turbidity, the next barrier is usually a dosage of free residual chlorine. The size of the dose and the period of contact time necessary for it to be effective depends on a number of circumstances, and hard and fast rules cannot be applied. Under ideal conditions a free chlorine residual of 0.3 mg/ℓ at a pH of 7 or less after a 20-minute contact period may be considered reasonable for the inactivation of most free-swimming pathogenic bacteria, and the cholera *Vibrio* is said to be the fastest swimmer in the world considering the organism's length/time ratio.

However, a free chlorine residual capable of inactivating most bacteria would probably not even slow down a virus since they are more resistant to disinfection. On the other hand, if the water turbidity is low, the possibility of large numbers of viruses being present is minimal.

CLEAN DISTRIBUTION SYSTEMS

How stable is the water leaving the clear-water well of the treatment plant and entering the distribution system? There are very few waters which do not undergo some measure of change in the distribution system. Indeed in many water supply systems where surface waters are impounded in reservoirs and the only treatment they get is chlorination, and possibly fluoridation, before they enter the distribution system, the water mains act as a series of elongated cylindrical settling chambers. The evidence of this is obvious when a water main is flushed by opening the hydrants, since during the quiet hours of the night when flows are reduced, settling of suspended matter in the mains will occur.

In most cases the deposits are not particularly hazardous to health and, apart from the nuisance they cause in plugging washing machine strainers and ruining the odd batch of laundry, they do not normally cause too much distress from the public health point of view. Nevertheless, these silts are possible locations where microorganisms, worms, and other unwanted creatures can habitate and be reasonably safe from the hazards of chlorine. Since chlorine is readily absorbed by organic matter, the presence of silt will almost certainly remove any traces of chlorine that may otherwise have been present.

CROSS-CONNECTION CONTROL

Having ensured that the water entering the distribution system is wholesome and free from pathogenic organisms and toxic substances, the water purveyor must insure that it is not contaminated as a result of cross-connections. This is far easier said than done, even in cities where strict surveillance is practiced. If a water supply system is always under positive pressure there is little danger from the usual type of cross-connections. But whenever the mains are under reduced pressure, as in the case of a fire when pumpers are attached to the hydrants or a break in the main necessitates its being isolated and drained, then the problem due to cross-connections becomes important. A main break in the lower part of a distribution system resulted in a number of complaints from consumers on an upper level who reported that air was being drawn into their faucets. Two hotels lost the water from their heating boilers, and a laundry complained that the water in the washtubs disappeared, which undoubtedly entered the distribution system. One of the most significant cross-connections of recent times was reported from Holy Cross College located in Worcester, Massachusetts.[13]

MONITORING THE DISTRIBUTION SYSTEM

Although monitoring the distribution system by taking spot samples from various locations and testing them for coliform and certain chemical contaminants may not prevent outbreaks, nevertheless they are indicators of possible sources of contamination.

Monitoring a water supply is not a disinfection barrier, and there are severe limitations on the number of samples that can be taken and tested. Most health authorities have minimum standards concerning the numbers of samples that must be taken per month depending on the population of the serviced area. Many cases of contamination have been detected by systematic monitoring programs and corrected by the waterworks staff before they have become public knowledge.

COMMON WATERBORNE DISEASES

A closer look at the various types of waterborne diseases would be advantageous. They are generally grouped into the following categories; arranged in the order of organism size starting with the smallest (References two to 15 inclusive):

Viruses
    (a) Echo virus diseases
    (b) Infectious hepatitis
    (c) Poliomyelitis

Bacteria
    (a) Cholera
    (b) Bacillary dysentery
    (c) Leptospirosis
    (d) Paratyphoid fever
    (e) Tularemia
    (f) Typhoid

Protozoa
    (a) Amebiasis
    (b) Giardiasis

Helminths (parasitic worms)
    (a) Dracontiasis
    (b) Echinococcosis
    (c) Schistosomiasis

The list above presents a very simplified synopsis and is in no way a complete catalogue of waterborne diseases.

Cholera

BACTERIAL DISEASES

Alternative Names Asiatic cholera, Indian cholera, El Tor.

The *Vibrio comma* is a short, curved, comma-shaped rod, 1 to 4 microns in length by 0.2 to 0.4 microns in diameter, actively motile with a single terminal flagellum, non-sporing and gram-negative. It has little resistance to disinfecting agents or to drying. It is rapidly overgrown by other organisms and is not active for very long in water heavily polluted with sewage. In ordinary clean river water, however, it will survive for 1 or 2 weeks and for as long as a month in spring water. The organisms remain viable in exposed stools for 1 or 2 days in summer and up to a week in cold weather.

Asiatic cholera is an acute, infectious, disease acquired by ingestion of the *Vibrio comma* with food or drink. The incubation period is a few hours to 5 days, usually 3 days.

Distribution The disease is widely endemic in Asia and the Far East, especially in India and the delta of the Ganges. It is presently pandemic in parts of Indonesia having originated in the Celebes. With the possible exception of Australia and New Zealand, there is scarcely a country of the world that has not seriously suffered.

Since cholera is a carrier disease there are no countries immune from the organism. It is only good sanitation, sewage collection, treatment, and disinfection of the water supplies, that prevents an outbreak from becoming widespread with devastating results.

Bacillary Dysentery

Alternative Names Shigellosis. Causative organisms are *Shigella flexneri, Shigella dysenteriae, Shigella sonnei* and *Shigella boydii*. The first two organisms are common in the tropics and *Shigella sonnei* is found in temperate zones. The resistance of the dysentery bacteria is similar to that of other intestinal bacilli. They are readily killed by heat and disinfectants. They are reported to be capable of surviving in sterile water for periods up to 24 days, especially when stored at low temperatures, but death occurs more speedily in natural and impure waters, presumably by bacteriophages as well as by anti-bacterial substances. Efficient chlorination is a reliable method of preventing waterborne dysentery.

Bacillary dysentery is an acute or chronic inflammatory disease of the colon. The incubation period is 1 to 7 days but usually less than 4 days.

Distribution Bacillary dysentery is one of the chief causes of death in areas where people are living under primitive conditions, particularly among the very young and the enfeebled. It is a "filth" disease and it has been demonstrated that when good sanitary conditions exist, epidemics can be curtailed although sporadic outbreaks of limited extent still continue.

Leptospirosis

Alternative Names   Weil's disease, hemorrhagic jaundice, canicola fever, mud fever and swineherd's disease.   *Leptospirae* are comparatively long, spiral, corkscrew-shaped, motile, organisms. There is considerable variation in length (4 to 40 microns), but they are usually 6 to 12 microns long and approximately 0.1 microns in diameter.

Taylor[7] warns that since there are frequently large populations of harmless *leptospira* in water, great care must be taken in the investigation of an outbreak of Weil's disease before blame is leveled at the water supply. *Leptospirae* are readily destroyed by heat. In natural river waters the *leptospirae* survive from 3 to 9 days depending on the temperature. Chlorination dosage of 0.3 mg/ℓ at pH of 5 and 25°–26°C (77.0 to 78.8°F) will inactivate the organism in 3 minutes; but at pH 8 a residual of 3 mg/ℓ is required to achieve the same result.

Definition   The *leptospirae* causing disease to man are excreted from infected mammals such as rodents, small carnivores, and certain domestic animals. The organisms enter the body through small breaks in the skin or the mucous membranes, or by ingestion of contaminated water. The disease is accompanied by damage to the kidneys and enlargement of the liver, fever (102 to 104°F), vomiting and possibly jaundice. The incubation period is 4 to 19 days (average 9 to 10 days).

Distribution   The *leptospirae* are considered to be native to animal hosts, reaching man through contaminated food or water. In the case of water it is possible to contract the disease by contact with contaminated water as well as by ingestion.

Paratyphoid

Alternative Names   Enteric fever.

There are over 1500[15] different *salmonella* organisms but only about 200 different types are detected in any given year and only the typhoid and the paratyphoid, A and C, are restricted to man. All *salmonella* organisms could be waterborne, but they are usually in food.

Typhoid and paratyphoid fevers are similar to each other in their symptoms but paratyphoid is much milder and less dangerous. Causative organisms include:

(a) *Salmonella paratyphi* A, a natural pathogen of man not known to be a natural pathogen of other animals;

(b) *Salmonella paratyphi* B, a natural pathogen of man but also occasionally found in cattle, swine, sheep, and chickens; and,

(c) *Salmonella paratyphi* C, a natural pathogen of man.

Paratyphoid, similar to typhoid, results in a high fever and rose-colored spots on the abdomen caused by the infection entering the bloodstream. The incubation period is usually 1 to 10 days, which is less than the usual period for typhoid. The other numerous *salmonella* diseases may also be waterborne but this is not usually the route they use to infect man; the food chain is more prevalent.

Distribution The disease occurs throughout the world and the *salmonella* organisms are found in the intestinal flora of many domestic and wild animals but typhoid and paratyphoid are only common to humans.

Tularemia

Alternative Names Deer fly fever.

This infectious disease is caused by the organism *Francisella tularensis*, also called *Pasteurella tularensis*. It is a small nonmotile, nonspore-bearing, aerobic bacillus occuring regularly in colloidal and bacillary forms, and very rarely bipolar. The incubation period is from 3 to 5 days.

Distribution The infection is basically one of wild mammals, particularly rodents and rabbits; man is only an incidental host. In many cases, outbreaks have occurred when untreated river and pond waters have gained access to the distribution system either during an emergency, cross-connection, or failure of the disinfection barriers. Inadequately constructed wells have also been contaminated by rodents causing outbreaks of the disease. The water supply is not the only cause of infection. It can be transmitted by fly bites, ticks and other insects, as well as by direct contact with an infected animal or its carcass. Occasional cases have occurred from skinning and dressing tree squirrels and other rodents.

Typhoid

Alternative Names Enteric fever.

The causative organism, *Salmonella typhi*, is hosted by human beings and spread by the ingestion of foods and liquids contaminated by human feces. No animals are known to suffer from typhoid fever or to naturally harbor the bacteria. The organism is able to survive an entire winter in frozen soil and for as long as 7 days in well water. Studies have shown that they can survive in raw river water from 5 to 9 weeks; however 99.9% of them are inactivated in the first week. When sanitation principles are not followed and infected excreta are left exposed on the ground, the first washing rains will carry the pathogens into the nearest watercourse. In acid soils 50% of the typhoid bacilli will perish in the first 48 hours, but the remainder may live for a period of several months. The organism will leave the body in both the feces and urine and occasionally the sputum. Fortunately, typhoid bacteria have about the same resistance to free chlorine residuals as *Escherichia coli*, the organism used for monitoring the bacteria content of water supplies. If *E. coli* is absent there is little likelihood of infection from typhoid; however, the same typhoid organisms have been known to survive for 211 days in unchlorinated tap water exposed to sunlight at room temperature Normally, since the motile, flagellated, gram-negative organism has no spore, it is relatively easy to kill with free chlorine. The only difficulty is for the chlorine to reach the bacillus when it is imbedded in fecal masses. The organisms are known to live for 12 days in raw sewage and for 14 days in a septic tank.

Definition Typhoid fever is a very serious illness. The case fatality rate, that is to say, the number of people who contracted the disease and died as a result of it, was between 10 and 20%. Recently introduced antibiotics have reduced the case fatality rate to 1%. However, antibiotic resistant strains of the organism have evolved.

Unfortunately there is a long incubation period that varies in different outbreaks of the disease and makes it difficult to determine where and when the infection actually occurred. In mild cases of the illness a patient may not have consulted a physician and, after recovery, is in complete ignorance of the fact that he or she is a carrier, honestly disclaiming any history of typhoid or enteric fever. Any person who is a carrier of the organism should never be employed in a waterworks or food manufacturing plant.

Incubation Period The long incubation period for typhoid is one of the problems facing the epidemiologists who try to pinpoint the source and time of the infection. A typical waterborne outbreak of typhoid has three phases:

1. A period of invasion lasting about a month. There is frequently a prevalence of diarrhea and gastroenteritis for 2 or 3 weeks followed by a few cases of typhoid fever.

2. A period of decline and secondary infection which may be prolonged. There will be a certain number of primary cases directly contracted from the infected water, but most cases will be due to infection acquired from existing cases before diagnosis is established and before protective measures are organized.

3. A period of intensive reaction from both primary and secondary infections unless drastic measures are taken to eliminate the sources of infection and institute adequate levels of personal hygiene.

Distribution Typhoid is widespread throughout the world, but it has been almost eradicated on the North American continent and in Western Europe with the introduction of good water supplies and improved standards of sanitation and personal hygiene.

Current practice is to recommend inoculation against typhoid for all people working in water and sewage treatment plants, but does not guarantee immunity.

Croydon Typhoid Outbreak The Croydon outbreak in 1937, near London, England, is a typical example of how a typhoid outbreak can occur, and the lessons learned should never be forgotten. The events are briefly reported as follows:[7]

The Addington Well had been in use for 49 years (built about 1888) without epidemiological incidence. The well was approximately 205 feet deep, and 10 feet in diameter and in all probability partially lined with masonry. Horizontal tunnels, called *adits* are cut into the chalk aquifer in a radial fashion from the central casing, resembling the spokes of a wheel. In the Addington Well there were two rows of adits known as the upper and lower, which were approximately 150 feet and 162 feet below the engine room floor respectively. The tunnels were large enough for a man to enter with a small wheeled skip to bring out the mined chalk. The steam-driven reciprocating plunger pump had a long pump rod extending down to the bottom of the well casing. Water would flow through the adits into the central well and be pumped up to the surface.

The Addington Well produced an average of 1.5 mgd (U.K.) (1250 U.S. gpm) and served a section of the Borough of Croydon consisting of approximately 40,000 people (i.e., 45 gpcd (U.S.)). The water, as it left the pump, passed through pressure filters without precoagulation or chemical addition. The water was hard, normally clear and bright, but of satisfactory organic and bacterial purity. However, occasionally the water would become turbid due to heavy rains, no doubt carrying surface waters directly into the aquifer, and on these occasions the filtered water was intermittently chlorinated with about 0.1 mg/ℓ of chlorine to correct the bacteriological condition of the water when the turbidity was high and the filters were in use.

On September 24, 1937, work was started in one of the adits and continued for just over a month. Until October 15, the well water was pumped to waste during the day while the men were at work and to the service reservoir at night. From October 16, however, the water was pumped to the reservoir as required, whether work was in progress underground or not. Open buckets, periodically removed in a skip, were used in the adit for urination but defecation underground was forbidden. On October 27 a case of typhoid fever was discovered in the area served by the Addington Well. Allowing an incubation period of 14 days, the date of the infection would be early October. There followed 341 cases of typhoid (including 19 secondary cases) indicating the highest infectivity at the end of October. In subsequent experiments it was found that cultivated typhoid bacteria remained viable in the unchlorinated Croydon water for about 4 weeks at 10 to 15°C (50 to 59°F). Of the 14 workmen employed in the adit, one had suffered from typhoid fever during the 1914–1918 war and was, unknown to himself, a chronic carrier. Profuse typhoid cultures were consistently obtained from the feces, but repeated examination of the urine gave negative results. The outbreak which resulted in 341 cases of typhoid fever culminated in 43 deaths (12.6%). The Ministry of Health Report (1938) on the inquiry into this outbreak arrived at the following conclusions: "The epidemic arose from an avoidable and external cause and not from anything in the well itself. In fact the infection was due to an unfortunate and rare incidence of three factors:

(a) Constructive changes taking place in the well.

(b) One of the workmen being a typhoid carrier, and

(c) The process of chlorination being in abeyance."

The subsequent legal proceeding into the cause of the outbreak makes interesting reading. The carrier was excreting the same phage type (D) as the typhoid bacteria excreted by several of the patients. All the workmen denied that defecation ever took place in the adits, but it was a long walk back from the place of work to the toilet facilities at the surface. It was possible that if a man was so inclined he could defecate in the long dark galleries where water was flowing without the risk of being detected and thus save himself a long and tedious walk to the surface.

It was also possible that a bucket of urine could accidentally have been spilled into the well. Typhoid organisms are often present in the urine of carriers on an intermittent basis. The fact that they did not isolate the organism from the urine of this particular carrier was no proof that it did not exist. As in most typhoid outbreaks, the evidence pointing to the water supply was circumstantial. In spite of much effort, typhoid bacteria were not isolated from the numerous samples of the water, the deposit in the reservoir, or the chalk removed from the adit of the well in the course of the work carried out therein. The investigation did not, however, commence until November 3rd when, as later evidence indicated, the water ceased to be infected.

The distribution of cases was very irregular. In a residential boys' school situated near the reservoir, 17 cases occurred among 150 boys (11.3%), no doubt due to drinking large quantities of water. However, a mental hospital nearby had only one case of fever among over 1000 occupants. The avoidable errors in waterworks practice revealed by this outbreak were listed as follows:

(a) Lack of recognition of the potential dangers to the source.

(b) Defective maintenance of works.

(c) Inefficient application and supervision of treatment of water. Intermittent chlorination according to the results of monthly samples was ill-advised and hazardous, even had the chlorine dose been sufficient, since much harm might have arisen in the intervening periods. The removal of all avoidable pollution from the gathering grounds, continuous and efficient filtration and chlorination, with regular, competent laboratory control, followed by careful protection of the treated water in the service reservoir and mains, were needed to ensure a safe supply.

(d) Failure to exclude the well from service while the work was in progress or, alternatively, to ensure safety of the water by efficient treatment confirmed by close laboratory observations.

(e) The employment of workmen without medical examination, careful selection, and thorough instruction as to cleanliness, changing of footwear, etc.

(f) The use of open buckets for urination.

It was finally concluded that "in spite of all other circumstances, the Croydon typhoid outbreak would have been avoided had the water been efficiently chlorinated, namely, by such a dose of chlorine that an excess of not less than 0.1 mg/ℓ remained in the water in the service reservoir (or otherwise after a minimum contact of one hour)."

Dracontiasis

HELMINTHIC DISEASES (PARASITIC WORMS)

Alternative Names Dracunculiasis, dracunculosis, medina, serpent, dragon, or guinea worm infection.

The disease is caused by *Dracunculus medinensis*, an elongated cylindrical, threadlike worm. Males of the species are rare, ranging in size from 12 to 40 millimeters. The female has a smooth cuticula and is much larger than the male, averaging about 1 meter in length.

Human infection results from drinking water containing infected *Cyclops* which serve as intermediate hosts. In endemic areas, open wells and surface water supplies provide a source of infected copepods.

The symptoms of the infection gradually manifest themselves over the entire 8 to 12 month incubation period and become pronounced a few hours before the appearance of the worm beneath the skin. These symptoms consist of abnormal redness of the skin, intense itching, giddiness, difficulty in breathing similar to asthma, and sometimes vomiting and diarrhea.

Distribution The disease produced by the guinea worm has been recognized for many centuries. It is highly endemic in a number of regions in tropical Africa and over large areas of India. It also occurs in Arabia (especially along the Red Sea), Iran, Afghanistan, and Russian Turkestan.

Echinococcosis

Alternative Names Hydatidosis, granulosus, dog tapeworm.

Echinococcus is caused by a small worm only 4 to 5 mm in length consisting of a head and two or three segments.

Infection results in the formation of cysts in the tissue of the liver, and lungs. The disease may give few symptoms but can result in death. The incubation period is variable, sometimes several years.

Distribution Echinococcosis had not been recognized as a waterborne disease until the September-October (1961) issue of *World Health*, the magazine of the World Health Organization (WHO). That issue included contaminated water as a mode of human infection of this disease.

Schistosomiasis

Alternative Names   Manson's intestinal schistosomiasis, bilharziasis *mansoni*, intestinal bilharziasis, bilharzial dysentery, schistosomal dysentery, and commonly referred to by the British Eighth Army troops, stationed in North Africa during World War II, as "Bill Harris" or blood fluke disease.

There are three species of blood fluke trematodes which infect humans and are the chief producers of this disease:

Schistosoma mansoni (*S. mansoni*), *Schistosoma japonicum* (*S. japonicum*), and *Schistosoma haematobium* (*S. haematobium*), The first two organisms give rise primarily to intestinal infection and the third to urinary infection.

Humans are the principal hosts of *S. mansoni* and *S. haematobium* and, for the most part, the eggs of the former are passed out of the human body in the feces and of the latter organism in the urine. When the eggs from any of the three parasites reach fresh water, they release free-swimming larvae or miracidia which must penetrate into an appropriate snail, the intermediate host, within 48 hours or they will perish.

In the construction of dams to impound water in artificial reservoirs, consideration must be given to the ecology in order to avoid the creation of habitats favorable to snail production if the spread of this disease is to be curtailed. The organisms are visible to the naked eye and in clear waters they have the appearance of minute white hairs.

Definition   The cercariae emitted by the snails gain access to a human or animal host either by direct penetration through the skin or by ingestion in the mouth. They then enter the bloodstream and are carried to the liver and portal veins where they develop into adult male and female worms in about 2 months. When this stage of their life cycle is completed, their eggs are excreted in the feces or urine of the host. The symptoms are those of dysenteric, pulmonary, and abdominal pains. Chronic itching of the skin and dermatitis are also important symptoms of the disease.

Incubation Period   1 to 3 months or longer.

Distribution   Schistosomiasis is a very ancient disease and Egyptian mummies of the period of 1250–1000 B.C. show indications of the disease. Although the direct mortality rate is low, it may cause many years of severe debility and through weakness the patient is subject to many other infections which accumulatively prove to be fatal. Unfortunately, up until 1962, schistosomiasis was not a notifiable disease in many countries of the world, and specific data showing its current distribution is not readily available. It is known to occur in Africa, Arabia, the Middle East, northeastern and eastern South America, China, Japan, the Philippines, and the Celebes.

The larvae of certain other schistosomes of birds and rodents may penetrate the human skin causing dermatitis known as swimmers' itch. This is prevalent amongst bathers in North American lakes and certain coastal seawater beaches. These schistosomes do not mature in man.

Amebiasis

PROTOZOAL DISEASE

Alternative Names Amebic dysentery, amebic enteritis, amebic colitis.

The organism responsible for this disease is the protozoa *Entamoeba histolytica*, and the infection is derived from contaminated food or water. There are believed to be at least two species of this organism, the large (11 to 12 microns) and the small (7 to 9 microns). An infection may result from either, or from a combination of both. The protozoa are single-cell microscopic animals. They form very resistent cysts which are able to survive for the protection of the organism under adverse environmental conditions. Humans are the reservoirs of the infectious agent.

The organism will survive in water at 10°C for as long as a month. It has been demonstrated that coagulation and filtration will generally provide an effective barrier, but not always. It has been stated that slow sand filtration followed by chlorination is completely effective. However, chlorine on its own, or even ammonia chloramines at 2.0 mg/ℓ and a pH of 4.0 will usually inactivate the organism in 30 minutes; however at higher pH levels, 4.0 mg/ℓ were needed at pH 7.6 and 5.5 mg/ℓ at pH 11.0. Free chlorine in this case is about twice as effective as chloramines at pH values below 8.0 but less effective than chloramine above pH 11.0. Thus it is suggested that superchlorination followed by dechlorination and rechlorination should be used to safeguard the water supply if this organism is present in endemic proportions. Other halogens such as iodine and bromine have been used, but chlorine is the cheapest to manufacture and use as a water disinfectant.

Definition The infection takes place primarily in the colon of the larger intestine. The patient may not show any symptoms or may have either diarrhea or constipation, loss of appetite, abdominal discomfort, and blood and mucous in the stools. The disease can cause much suffering but is usually not fatal.

Incubation Period Usually 2 to 3 days but can be longer (up to 4 weeks).

Distribution Amebic infection is worldwide, often to the extent of 50% of the population in localized areas with primitive sanitation. Clinical amebiasis is prevalent in hot countries and is not so frequent in temperate zones probably due to better conditions of sanitation. Cyst carriers may continue for years excreting the organism without showing any visible evidence that they have the disease.

Giardiasis[15]

Alternative Names Giardia enteritis, lambliasis.
The organism responsible for this disease is a flagellated protozoan known as *Giardia lamblia*.

Definition An infection of the small bowel which may be associated with a variety of intestinal symptoms, chronic diarrhea, steatorrhea, abdominal cramps, floating, greasy, or malodorous stools, fatigue and loss of weight.

Incubation Period In a waterborne epidemic in the United States, clinical illnesses occurred 1 to 4 weeks after exposure.

Distribution Clinical cases have been reported in many parts of the world; children are more susceptible to the disease than adults and it is reported that the carrier rate may range between 1.5 and 20%. It is also reported that of a total of 1419 tourists visiting Leningrad between 1970 and 1974, 23% of them suffered with giardiasis epidemiologically related to drinking tap water.
The organism is hosted in man and possibly domestic animals and the mode of transmission is through fecal contamination of water and hand to mouth transfer of cysts from the feces of an infected individual.

VIRAL DISEASES

Viruses are the smallest and simplest living organisms in existence and measure from 10 to 450 millimicrons in size (a millimicron is one millionth of a millimeter). The small viruses are visible only by electron microscope and will pass through bacteriological tight filters. Strictly parasitic, viruses grow only inside appropriate host cells in a living organism.

Where it is possible to culture bacteria in various nutrient broths and on agar, viruses are only cultured on animal tissue such as monkey kidney, chick embryos, and other animal cells. Virus growth in tissue cultures results in degeneration of host cells. Not all viruses infect laboratory animals, chick embryos, or tissue cultures; for example, the infectious hepatitis virus infects only humans and therefore does not lend itself to laboratory investigations. Although water is far from being an ideal medium for sustaining viruses, they can exist in it for substantial periods of time. The movement of water in creeks and streams causes rapid dilution and thereby effectively disperses viruses, but virus concentrations in stagnant lake waters remain relatively unchanged for considerable periods of time. The virus responsible for infectious hepatitis is said to be capable of passing through groundwater aquifers for considerable distances from the source of contamination into a water well. Poliovirus, coxsackie, ECHO, and adenoviruses have been isolated from the feces of infected humans, and many of these viruses can be found in sewage treatment plant effluents. Therefore, their presence must always be considered as a possibility in surface water supplies immediately downstream of sewage outfalls. It is reported by Chang that chlorination of water, with a free chlorine residual of 1.0 mg/ℓ and maintained for 30 minutes, will destroy the virus providing it is not inside particulate material. However, since the viruses are submicroscopic, it follows that the turbidity of the water must be held to extremely low limits to insure that they are not protected from the disinfecting action of the chlorine. Thayer and Sproul, in 1966, reported the results of their experiments concerning the sensitivity of pH in respect to virus inactivation. Other workers have found that considerable reductions in the number of viable organisms can be achieved by flocculation and clarification with alum.

The use of ultraviolet irradiation of the treated water prior to leaving the water treatment plant is a possible alternative. Under conditions of test, ultraviolet (UV) treatment would appear to be capable of producing a virus-free effluent from a treated water low in suspended solids and turbidity, but which may still contain a high ammonia or organic content thereby reducing the effectiveness of the chlorine dosage. Under these circumstances, UV treatment could substantially reinforce the disinfection barriers.

It has also been pointed out that slow sand filters are reasonably effective in removing viruses, but their success is due to the biological actions that take place on the top layer of sand rather than any mechanical straining capabilities the slow sand filter may possess.

ECHO Virus Diseases

Alternative Names   The term ECHO was coined from the initial letters of "Enteric cytopathogenic human orphan." The ECHO virus diseases are aseptic meningitis, epidemic exanthem, and infantile diarrhea.

The ECHO virus closely resembles the polio viruses and coxsackie viruses, and can be propagated in the laboratory only on tissue cultures prepared from human or monkey cells. The various strains of virus are identified by numbers.

Definition   Epidemic exanthem which manifests itself as a skin rash may also be accompanied by aseptic meningitis, a more serious form of meningitis of the nervous system.

Infantile diarrhea has been caused by ECHO virus type 18 resulting in an outbreak of watery diarrhea among newborn infants in a nursery.

In the respiratory-enteric syndrome, the symptoms consist of a fever lasting for 2 days, nasal discharges and vomiting in infants less than 2 years of age and is due to ECHO virus types 8 and 20.

Distribution   ECHO virus type 9 was responsible for widespread epidemics of aseptic meningitis with or without rash in Europe and the United States in 1956 and 1957. In nearly all cases the virus can be recovered from the feces of infected patients.

Infectious Hepatitis

Alternative Names Infective hepatitis, epidemic jaundice, catarrhal jaundice.

This disease is presumed to be caused by a virus, more resistant to chlorination than the virus of poliomyelitis; it will survive 1 mg/ℓ of chlorine for 40 minutes and a dosage as large as 15 mg/ℓ is not always sufficient to destroy it completely. Although the disease is normally spread by person-to-person contact, there have been numerous cases where the explosiveness of the epidemic has indicated that it is waterborne, but usually only where there is gross fecal pollution. Shellfish outbreaks tend to confirm waterborne virus.

Definition The patient suffers from fever, nausea, loss of appetite, vomiting, fatigue, headache, restlessness, mental confusion, and sometimes coma. The stools frequently become clay-colored. The liver becomes enlarged and tender and the skin and whites of the eyes develop a yellowish color.

Incubation Period 9 to 38 days, usually averaging 23 to 24 days. The prolonged period of incubation makes it particularly difficult in many cases to determine the source of infection since much of the vital evidence is not in existence by the time the outbreak is known.

Distribution The disease presently referred to as hepatitis has been known for thousands of years. It is reported that Hippocrates, in the 5th century B.C., mentioned it in his writings, but the modern concept of the virus etiology of the disease is largely due to the results of medical research during World War II when there were 170,000 cases reported in the U.S. Army alone.

The disease is present thoughout the world and is endemic in some areas. Outbreaks frequently occur in camps and places where large numbers of people congregate under poor sanitary conditions.

As previously mentioned, outbreaks of specific diseases often appear to take place after a period of relatively quiet coexistence with humanity as a whole, rising to a crescendo of epidemics and then tapering off again while another organism appears to take off on a worldwide rampage. Figure 3-1 shows the relationship between typhoid and hepatitis.

Poliomyelitis

Alternative Names Acute anterior poliomyelitis, infantile paralysis.

The are three distinct types of poliomyelitis virus, but they are all approximately 28 millimicrons in diameter and their principal differences are in their surrounding protein coatings. These differences are important when endeavoring to locate the source of infection. The virus is fairly resistant, surviving for 180 days in artificially infected river water stored in the dark at 4°C (39.2°F). The effect of chlorine depends upon the presence of organic matter, but according to Taylor[7] in the absence of extraneous organic matter 0.05 mg/ℓ of free chlorine is sufficient to inactivate the virus in 10 minutes, and with chloramine a dosage of 0.75 mg/ℓ will inactivate it after a 2 hour contact period.

Definition Symptoms include fever, headache, gastrointestinal disturbance, stiffness of neck and back, with or without paralysis. The virus invades the central nervous system and paralysis usually occurs in lower portions of the body, confining the patient to the use of a wheelchair. The severity of the disease appears to be in proportion to the degree of infection. If the paralysis has not been acute, some mobility can be achieved, but the patient is usually incapacitated.

Incubation Period The mouth is probably the portal of entry and the primary virus multiplication takes place in the throat or in the intestines or both. It is thought possible that the virus multiplies in the tonsils and then enters the bloodstream. The incubation period is usually between 1 and 2 weeks, but may range between 4 days and 5 weeks. The patient may be a carrier of the virus for several months.

Distribution The disease is worldwide in its distribution and it occurs both sporadically and in epidemics. Incidence seems to be highest in the summer and early fall but varies from year to year and region to region. Children from 1 to 16 years of age are more frequently attacked than adults, but in several areas this pattern in changing to include older children and young adults. In countries where artificial immunization has been widely practiced, epidemics have been confined to the least vaccinated population groups. The reservoirs of infection are humans, most frequently children suffering from clinically unrecognized or nonapparent infections. There has been some doubt over the years as to whether or not poliomyelitis is a waterborne disease. However, the outbreak in the city of Edmonton, Alberta, Canada, in 1952 and 1953 strongly suggested that the disease was waterborne.[7] Edmonton had a population of 183,500 in 1952 when there were 95 cases resulting in seven deaths; in 1953, there were 322 cases and 16 deaths from poliomyelitis. The city takes its raw water supply from the North Saskatchewan River, 20 miles downstream from the small town of Devon, which had a population of 1600. The river water is hard and contains a fair amount of suspended solids. The water at Edmonton is softened by the cold-lime process, clarified, recarbonated, filtered, and chlorinated. This treatment was probably sufficient in preventing the spread of typhoid fever some 2 years before when a number of cases of this disease occurred along the river bank above Edmonton and below Devon, but was not sufficient to prevent the poliomyelitis in Devon just prior to the outbreak in Edmonton. It also appears that the chlorination equipment at the sewage treatment plant in Devon had failed on both occasions when there were poliomyelitis cases in the town. It is believed that the chlorine residual at the Edmonton water treatment plant was not sufficient to kill the virus which had survived its journey down the river, through the softening process into the Edmonton distribution system.

REFERENCES

1. GELDREICH, E. E. et al. "The Necessity of Controlling Bacterial Population in Potable Waters." "Community Water Supply." *J.A.W.W.A.*, Vol. 64, page 596, 1972.
2. CRAUN, G. F. and L. J. MCCABE. "Review of the Causes of Waterborne Disease Outbreaks." *J.A.W.W.A.*, Vol. 65 (1), p. 74, 1973.
3. SPROUL, O. J. "Virus Inactivation by Water Treatment." *J.A.W.W.A.*, Vol. 64 (1), p. 31, 1972.
4. MAXEY-ROSENAU *Preventative Medicine and Public Health*, 10th Ed. New York: Appleton-Century-Crofts Inc., 1973.
5. MILLER, A. P. "Water and Man's Health," published by the Office of Human Resources and Social Development Agency for International Development. Washington, D.C. 1962. Technical Series No. 5.
6. HUNTER, G. W. et al., *Tropical Medicine*, 5th Ed. Philadelphia and London: W. B. Saunders Co., 1976.
7. TAYLOR, E. W. *The Examination of Waters and Water Supplies*, 7th Ed. London: J. & A. Churchill Ltd., 1958.
8. SALVATO, J. A. *Environmental Sanitation*. New York: John Wiley & Sons Inc., 1958.
9. CHANG, S. L. "Viruses, Amebas and Nematodes and Water Supplies." *J.A.W.W.A.*, Vol. 53 (3), p. 288, 1961.
10. THAYER, S. E. and O. J. SPROUL. "Virus Inactivation in Water Softening Precipitation Processes." *J.A.W.W.A.*, Vol. 58 (8), p. 1063, 1966.
11. Committee Report. "Viruses in Water," *J.A.W.W.A.*, Vol. 61 (10), p. 491, 1969.
12. BERG, G. *Transmission of Viruses by the Water Route*. New York: Inter-Science Publishers (a division of John Wiley & Sons), 1965.
13. TAYLOR, F. B. "The Holy Cross Episode." *J.A.W.W.A.*, Vol. 64 (4), p. 230, 1972.
14. VAJDIC, A. H. "The Inactivation of Viruses in Water Supplies by Ultra-Violet Irradiation." Published by Ontario Water Resources Commission, Division of Research, Paper No. 2015 1969.
15. BENENSON, A. S. *Control of Communicable Diseases in Man*, 12th Ed. Washington, D.C.: The American Public Health Association, 1975.

4

NONPATHOGENIC ORGANISMS

Nonpathogenic bacteria are also found in water distribution systems. They form slimes by secreting gluey materials either as capsular structures or as extracellular excretion products.

Iron bacteria are one of the most common types of nuisance organisms in the water industry. They transform the soluble iron content of the water into insoluble compounds which precipitate on well screens, on the pipes of the distribution system, and on the walls and floors of reservoirs. Iron bacteria will discolor the water, and produce undesirable tastes and odors. Sulphate-reducing bacteria can also produce extremely low localized pH levels immediately adjacent to the pipe with resulting corrosion accompanied by the liberation of gaseous hydrogen sulphide.

IRON BACTERIA

So called "iron bacteria" in water supplies are nonpathogenic but cause nuisance situations in the water industry. They are able to abstract iron either from the water itself or from pipes or surfaces with which they come in contact. The abstraction, oxidation, and storage of iron by these organisms is continuous and in the course of time produces large accumulations of gelatinous, slimy brown-color deposits of ferric hydroxide. Some of these bacteria are able to abstract manganese as well as iron.

There are usually three organisms grouped under the general description of iron bacteria:

(a) *Leptothrix*
(b) *Crenothrix*
(c) *Gallionella*

Crenothrix and *Leptothrix* are difficult to differentiate under ordinary microscopic examination.

Leptothrix

Leptothrix is a rod-shaped bacterium (1 micron diameter by 2-6 microns in length) and may exist naked in chains or enclosed in a sheath, the cells being gram-negative. Reproduction occurs by fragmentation of naked filaments or by the production of swarmer cells which are motile by means of flagellae.

Crenothrix

These organisms are similar to threads which attach themselves at one end to some foreign object, such as a particle of dirt or the rough surface of a pipe wall. A rosette of threads may be attached to other objects and they grow into quite large tufts of organisms. The threads may measure 5 microns in thickness and 250 microns in length. The sheath consists of an organic matrix upon which iron oxide deposits as the threads age and become stained more deeply, eventually turning a dark brown. The sheaths do not completely disappear when treated with hydrochloric acid. Reported infestations of water supply systems usually refer to this organism as the chief offender. The success of this species in water systems may be due in part to the production and release of vast numbers of spores. It is claimed that this organism grows readily in water containing organic matter, irrespective of the iron content. The indication that the organisms only exist in waters containing up to 4 mg/ℓ of oxygen would possibly explain the association of this organism to wells producing water deficient in oxygen.

Gallionella

This is a bean-shaped organism and can be clearly differentiated from the other two organisms, *crenothrix* and *leptothrix*. For further details on the identification of these organisms reference should be made to Holden.[1] *Gallionella* is often the predominant organism in waters where light is excluded, in wells, water mains, and particularly in dead ends. The bean-shaped cells continually secrete ferric hydroxide as numerous fine threads of variable length become twisted together to form a stalk.

REMOVAL OF BACTERIAL SLIMES

The presence of the bacteria is first manifested by the appearance of organic matter when fire hydrants or dead ends are flushed, or from well water, surge tanks, and reservoirs. If laboratory facilities are available the slimes should be examined both microscopically and chemically to determine the composition of the deposit and to identify the responsible organisms. However, the presence of the brown slimes is usually sufficient indication that iron bacteria are present. Black deposits of iron sulfide and the objectionable odor of hydrogen sulfide would indicate the presence of sulfate-reducing bacteria. The recommended treatment is to use copper sulfate and/or chlorine. A contaminated well can be the most difficult problem and the best approach is to remove the pump and dose the well with 50 to 100 gallons of water solution containing not less than 1.0 mg/ℓ of copper sulfate. Force this solution through the screen into the formation with a well driller's surge plunger, usually a solid plunger frequently made of wood with rubber rings cut from rubber jointing sheets to form a tight seal around the casing. If the well is inside a building or with limited headroom, this method may not be possible. Wherever possible, wells should be designed with removable enclosures to enable a well driller's "work over rig" to be brought to site to enable this surging to be done. Care must be exercised when operating the rig that the upstroke of the surge block does not cause the inward collapse of the well screen.

Many cases of iron bacterial infection have been temporarily resolved by various disinfection techniques using copper sulfate, sodium hypochlorite, potassium permanganate, inhibited hydrochloric acid, hydrogen peroxide (H_2O_2), hexameta-phosphate (Calgon) and no doubt many other chemicals, only to prove disappointing a few months later by a return of the "rusty monsters."[2] To avoid this occurrence several authorities[3] recommend that the treatment, either copper sulfate or chlorine, be continued for about 1 week to prevent the reseeding of the system.

The question is often asked where the iron bacteria come from in the first place. It is believed that they are associated with groundwater rather than surface water supplies. *Crenothrix* does not like water with more than 4 mg/ℓ of dissolved oxygen, and *Gallionella* prefers dark places, and most of these organisms do not reproduce prolifically in waters below 10°C (50°F). Where the organisms have been isolated from spring waters,[2] evidence would suggest that they are already present in the aquifer and manifest themselves in the way that we encounter them once their environmental conditions are suitable for reproduction. There is also a possibility that they are introduced into the aquifer from the surface by the drilling rigs and the bailing and surging operations that take place in the development of the well.

Another difficulty which frequently occurs, together with the problem of iron bacteria, is the presence of iron and manganese bicarbonates in the well water. These chemicals can readily be removed by suitable treatment when the water reaches the surface; however, the introduction of chlorine as a disinfecting agent into the well on a continuous basis may keep the bacteria in check, but may also oxidize the iron and possibly the manganese in the well casing to the detriment of the well screen, which would then require acidizing. If the situation continues to deteriorate with the necessity for frequent workovers, it is probably advisable to consider abandoning the well and drilling another. Over a period of time the well may recover and become usable again, but this is not very likely once it has been idle for some period of time. If the screen can be removed and replaced with a new one without too much expense, the well may be restored to service, but expert advice should be obtained before this work is undertaken.

NEMATODES[4,5,6] This nuisance organism is a small worm capable of being collected on a 5 micron millipore filter paper (i.e., not smaller than five thousandths of a millimeter or larger than 0.002 inch). The worms are present in the outfall of most biological sewage treatment plants and also from surface runoff from grasslands.

Much research has been done during the last 10 years and the reader should refer to the voluminous literature on this subject in the journals of the American Water Works Association (A.W.W.A.) and other scientific and industrial publications on water. It is believed that the worms are not pathogenic, but on the other hand they have proven to harbor viruses and other pathogenic organisms, carrying them in their intestines through the disinfection barriers of flocculation, clarification, filtration, and chlorination processes into the distribution system. They also excrete a gummy substance which imparts a distinct odor to water when diluted to 1 part in 20,000 parts of water. It is possible that their eggs are able to pass through sand filters and hatch into motile organisms in the filter underdrains and clear-water wells. Unfortunately, a chlorine dosage of 200 mg/ℓ failed to kill nematodes even after a 2.5 hour contact period. Chang[5] found that nematodes could ingest and carry human enteric bacteria and viruses. He concluded that, although a majority of the pathogens may disappear 1 to 2 days after ingestion, 5–16% may remain viable if the worms are carried to the finished water supply within 24 hours after departure from this habitat. Fortunately, the probability of nematodes ingesting pathogens is small, as pathogens present only a small fraction of the organisms found in domestic drinking waters and waste treatment plants. However, regardless of the effects of nematodes in treated water, municipal water supply systems which contain these organisms in their distribution systems cannot be considered to be of high quality.

Maffitt[6] describes an investigation into the causes of a nematode infestation of a public water supply which showed that water wells infested with the iron bacteria *crenothrix* were the source of food supply for the nematodes. When the wells were cleared of *crenothrix*, the worm population also disappeared.

REFERENCES

1. HOLDEN, W. S. *Water Treatment and Examination.* London: J. & A. Churchill, 1970.
2. GRAINGE, J. W. "My Encounters with the Rusty Monsters" Paper prepared for 9th Annual Convention—Alberta Water Well Drilling Association, Red Deer, Alberta, Canada. April 1966.
3. COX, C. R. "Operation and Control of Water Treatment Processes." Published by World Health Organization, Geneva, 1969.
4. CHAUDHURI, N. et al. "Source and Persistence of Nematodes in Surface Waters." *J.A.W.W.A.*, Vol. 56 (1), p. 73 1964.
5. CHANG, S. L. "Survival and Protection Against Chlorination of Human Enteric Pathogens in Free Living Nematodes Isolated from Water Supplies." Published in Journal of Tropical Medical Hygiene, Vol. 9 (136), 1960.
6. MAFFITT, H. C. "Elimination of Nematodes, *Crenothrix* and other Organisms." *J.A.W.W.A.*, Vol. 58 (1), p. 119 1966.

5

ALGAE

ALGAE Algae are small aquatic plants. They require light for their growth, and many species are able to produce oxygen. This accounts for the highly dissolved oxygen content of algae-infested waters during the daylight hours, which often falls below the saturation level during the night as decaying and dying algae blooms consume the oxygen in the process of decomposition.

The algae population of lakes and reservoirs is increasing annually due to the increase in nutrients entering the streams from sewage and the nutrient-rich surface runoff waters from cultivated agricultural lands. This process of "aging" is known as *eutrophication*. Algae are not known to be pathogenic to humans when ingested in small quantities but some are harmful to many animals and others cause taste, odor, and nuisance problems when in full bloom. During the early stages of their decomposition many of these organisms impart a very noticeable and disagreeable taste to the water. These tastes are objectionable enough in the natural state but they are sometimes increased in magnitude when the water is chlorinated. Consumer complaints are telephoned into the waterwork's office, and the plant operator is frequently tempted to reduce or stop the chlorine dosage in an attempt to improve a bad situation. Decreasing the chlorine dosage may improve the taste and odor problem, but the most important disinfection barrier to pathogenic organisms is diminished and in many cases it is better to increase the chlorine dosage up to "breakpoint" proportions and so oxidize the offensive component.

Reviews of the fascinating subject of algae are contained in Palmer's book *Algae in Water Supplies*,[1] Holden,[2] and *Standard Methods*.[3]

Algae of most importance to the water industry according to Palmer[1] are listed in Table 5-1 and illustrated in Figure 5-1 to 5-6.

TABLE 5-1 Taste and Odor Algae[1]
(Representative Species)

Group and Algae
Blue-Green Algae (*Myxophyceae*):
    Anabaena circinalis
    Anabaena planctonica
    Anacystis cyanea
    Aphanizomenon flos-aquae
    Cylindrospermum musicola
    Gomphosphaeria lacustris, kuetzingianum type
    Oscillatoria curviceps
    Rivularia haematites
Green Algae (nonmotile *Chlorophyceae*, etc.):
    Chara vulgaris
    Cladophora insignis
    Cosmarium portianum
    Dictyosphaerium ehrenbergianum
    Gloeocystis planctonica
    Hydrodictyon reticulatum
    Nitella gracilis
    Pediastrum tetras
    Scenedesmus abundans
    Spirogyra majuscula
    Staurastrum paradoxum
Diatoms (*Bacillariophyceae*):
    Asterionella gracillima
    Cyclotella compta
    Diatoma vulgare
    Fragilaria construens
    Stephanodiscus niagarae
    Synedra ulna
    Tabellaria fenestrata
Flagellates (*Chrysophyceae, Euglenophyceae*, etc.):
    Ceratium hirundinella
    Chlamydomonas globosa
    Chrysosphaerella longispina
    Cryptomonas erosa
    Dinobryon divergens
    Euglena sanguinea
    Glenodinium palustre
    Mallomonas caudata
    Pandorina morum
    Peridinium cinctum
    Synura uvella
    Uroglenopsis americana
    Volvox aureus

TASTE AND ODOR ALGAE

Species Names	Linear Magnifications
Anabaena planctonica	250
Anacystis cyanea	250
Aphanizomenon flos-aquae	250
Asterionella gracillima	250
Ceratium hirundinella	250
Dinobryon divergens	250
Gomphosphaeria lacustris, kuetzingianum type	500
Hydrodictyon reticulatum	10
Mallomonas caudata	500
Nitella gracilis	1
Pandorina morum	500
Peridinium cinctum	500
Staurastrum paradoxum	500
Synedra ulna	250
Synura uvella	500
Tabellaria fenestrata	250
Uroglenopsis americana	125
Volvox aureus	125

FIGURE 5-1 Taste and Odor Algae

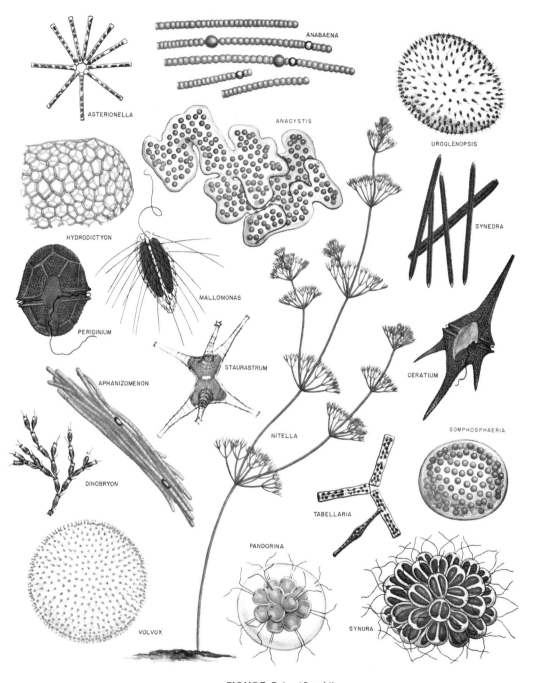

FIGURE 5-1 (Cont'd)

FILTER CLOGGING ALGAE

Species Names	Linear Magnifications
Anabaena flos-aquae	500
Anacystis dimidiata	1000
Asterionella formosa	1000
Chlorella pyrenoidosa	5000
Closterium moniliferum	250
Cyclotella meneghiniana	1500
Cymbella ventricosa	1500
Diatoma vulgare	1500
Dinobryon sertularia	1500
Fragilaria crotonensis	1000
Melosira granulata	1000
Navicula graciloides	1500
Oscillatoria princeps (top)	250
Oscillatoria chalybea (middle)	250
Oscillatoria splendida (bottom)	500
Palmella mucosa	1000
Rivularia dura	250
Spirogyra porticalis	125
Synedra acus	500
Tabellaria flocculosa	1500
Trachelomonas crebea	1500
Tribonema bombycinum	500

FIGURE 5-2 Filter Clogging Algae

FILTER CLOGGING ALGAE

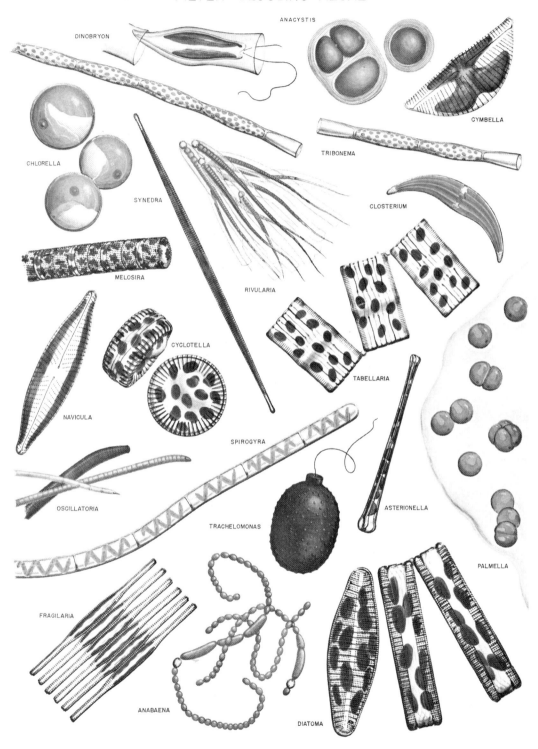

POLLUTED WATER ALGAE

Species Names	Linear Magnifications
Agmenellum quadriduplicatum, tenuissima type	1000
Anabaena constricta	500
Anacystis montana	1000
Arthrospira jenneri	1000
Carteria multifilis	2000
Chlamydomonas reinhardi	1500
Chlorella vulgaris	2000
Chlorococcum humicola	1000
Chlorogonium euchlorum	1500
Euglena viridis	1000
Gomphonema parvulum	3000
Lepocinclis texta	500
Lyngbya digueti	1000
Nitzschia palea	2000
Oscillatoria chlorina (top)	1000
Oscillatoria putrida (middle)	1000
Oscillatoria lauterbornii (bottom)	1000
Phacus pyrum	1500
Phormidium autumnale	500
Pyrobotrys stellata	1500
Spirogyra communis	250
Stigeoclonium tenue	500
Tetraedron muticum	1500

FIGURE 5-3 Polluted Water Algae

POLLUTED WATER ALGAE

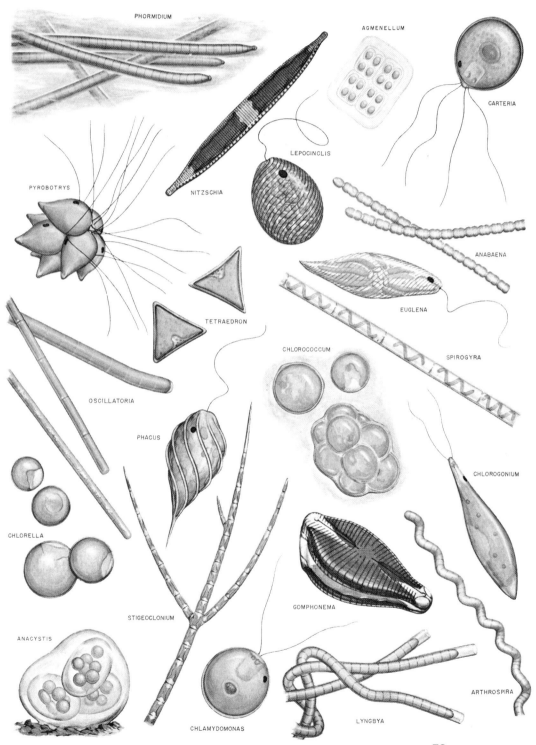

CLEAN WATER ALGAE

Species Names	Linear Magnifications
Agmenellum quadriduplicatum, glauca type	250
Ankistrodesmus falcatus var. acicularis	1000
Calothrix parietina	500
Chromulina rosanoffi	4000
Chrysococcus rufescens	4000
Cladophora glomerata	100
Coccochloris stagnina	1000
Cocconeis placentula	1000
Cyclotella bodanica	500
Entophysalis lemaniae	1500
Hildenbrandia rivularis	500
Lemanea annulata	1
Meridion circulare	1000
Micrasterias truncata	250
Microcoleus subtorulosus	500
Navicula gracilis	1000
Phacotus lenticularis	2000
Pinnularia nobilis	250
Rhizoclonium hieroglyphicum	250
Rhodomonas lacustris	3000
Staurastrum punctulatum	1000
Surirella splendida	500
Ulothrix aequalis	250

FIGURE 5-4 Clean Water Algae

CLEAN WATER ALGAE

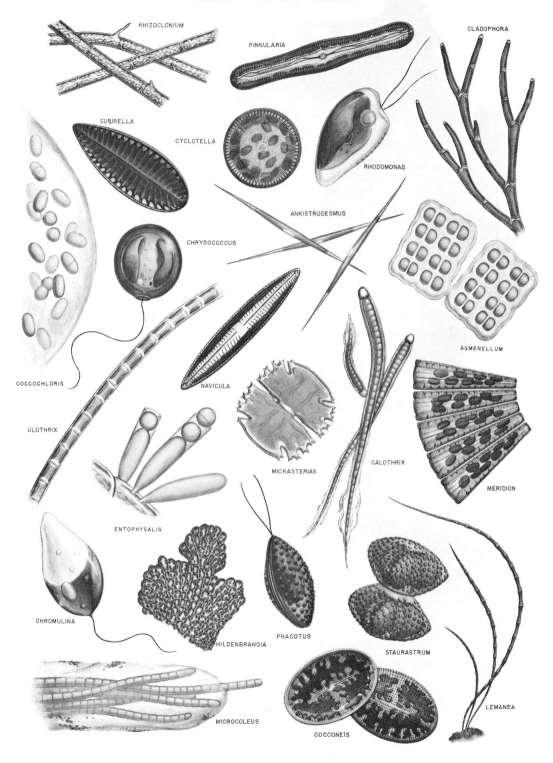

PLANKTON AND OTHER SURFACE WATER ALGAE

Species Names	Linear Magnifications
Actinastrum gracillimum	1000
Botryococcus braunii	1000
Coelastrum microporum	500
Cylindrospermum stagnale	250
Desmidium grevillei	250
Euastrum oblongum	500
Eudorina elegans	250
Euglena gracilis	1000
Fragilaria capucina	1000
Gomphosphaeria aponina	1500
Gonium pectorale	500
Micractinium pusillum	1000
Mougeotia scalaris	250
Nodularia spumigena	500
Oocystis borgei	1000
Pediastrum boryanum	125
Phacus pleuronectes	500
Scenedesmus quadricauda	1000
Sphaerocystis schroeteri	500
Stauroneis phoenicenteron	500
Stephanodiscus hantzschii	1000
Zygnema sterile	250

FIGURE 5-5 Plankton and Other Surface Water Algae

PLANKTON AND OTHER SURFACE WATER ALGAE

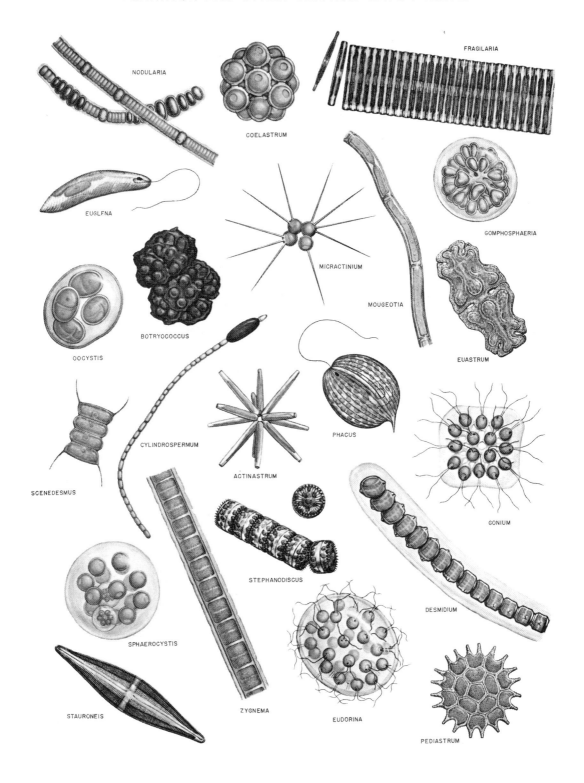

ALGAE GROWING ON RESERVOIR WALLS

Species names	Linear magnifications
Achnanthes microcephala	1500
Audouinella violacea	250
Batrachospermum moniliforme	3
Bulbochaete insignis	125
Chaetophora elegans	250
Chara globularis	4
Cladophora crispata	125
Compsopogon coeruleus	125
Cymbella prostrata	250
Draparnaldia glomerata	125
Gomphonema geminatum	250
Lyngbya lagerheimii	1000
Microspora amoena	250
Oedogonium suecicum	500
Phormidium uncinatum	250
Phytoconis botryoides	1000
Stigeoclonium lubricum	250
Tetraspora gelatinosa	125
Tolypothrix tenuis	500
Ulothrix zonata	250
Vaucheria sessilis	125

FIGURE 5-6 Algae Growing on Reservoir Walls

ALGAE GROWING ON RESERVOIR WALLS

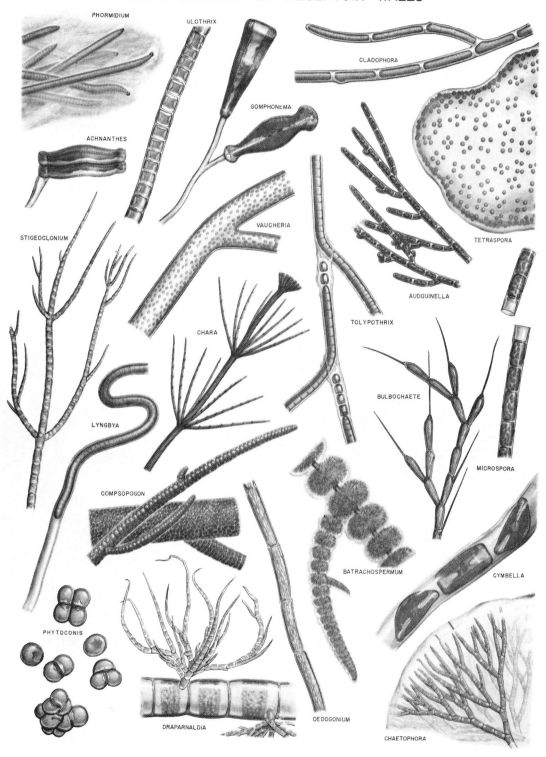

The highly dissolved oxygen in the upper levels of a lake is due to the consumption of carbon dioxide by the action of sunlight on the chlorophyll of the algae. The absence of oxygen in the lower layers is due to the absorption of oxygen by the decomposition of the decaying organic matter, as well as algae, as it settles to the bottom of the lake. Usually, as one species has started to wane, another will be in the preliminary stages of decay and it is at this stage that the taste and odor problems associated with actinomycetes are most prevalent. It is better to remove live algae from the water prior to treatment by microstraining and filtration, rather than destroy it with chemicals and then have to handle dead algae. Unfortunately, certain species such as *Oscillatoria* will pass through a microstrainer since they are very small rods and can easily pass through a strainer if they approach it head-on rather than sideways.

BLUE-GREEN[2] ALGAE

Alternative Names   Cyanophyta, Myxoplyceae.

This species of algae is responsible for the major problems of water quality and plant capacity of surface water treatment. The larger organisms quickly clog rapid and slow sand filters, and many species impart unacceptable tastes which are accentuated by chlorination. Blue-green algae are unique in that there are no distinctive nuclei, nor are there any flagellated reproductive cells involved in the life history. The food reserve is described as cyanophycean starch, a substance of uncertain nature but akin to starches in higher plants. Blue-green algae include unicellular, colonial, and filamentous forms, any of which are to be found in tropical, temperate, or even cold regions. They may be planktonic, bottom-living, and some species encrust rocks and the higher plants. Characteristically, they occur in greatest abundance in eutropic waters, and in an impoundment may color the water green, bluish-green, or even red. In anticyclonic weather, some species form surface scums which drift rapidly under the influence of wind; the genera which may be involved include, *Microcystis (Anacystis)*, *Oscillatoria*, *Anabaena*, and *Aphanizomenon*.

Fortunately, most of the species of blue-green algae are susceptible to small doses of copper sulfate, but treatment should be phased to coincide with the early stages of a bloom. Treatment of a well-established bloom may provide major problems of taste in an impoundment with the possibility that the whole water mass may become deoxygenated. An example of this was Buffalo Pound Lake which provides the water supply for the cities of Regina and Moose Jaw in Saskatchewan and normally has a fairly high algae content, but also supports a large fish population. In the late 1960's unseasonably warm late fall weather prevailed long enough to reactivate the algae into the active growth stage of a fresh series of blooms. About the time they were in full bloom the cold weather froze the surface of the lake, thus depriving the algae of the oxygen needed to assist in its decomposition. The decaying algae deoxygenated the water resulting in a nearly complete fish kill. The lake became anaerobic and breakpoint chlorination with chlorine dosages well in excess of 25 mg/ℓ was necessary over a prolonged period of time to produce reasonably palatable water.

In a situation such as this there is very little that a water treatment operator can do to ward off the problems other than to ensure that sufficient chlorination capacity can be put into operation to reach breakpoint. This type of problem is likely to become more and more frequent as the process of eutrophication proceeds. The only real answer is to reduce and, if possible, eliminate the nutrients (organics, phosphates, and nitrogenous compounds) which enter rivers and lakes and encourage the prolific growths of algae.

GREEN ALGAE[2]

Alternative Names *Chlorophyceae* and *Chlorophyta*. Green algae is very diverse in form, ranging from spherical, motile, unicellular organisms which are only a few microns in diameter to filaments which may be many millimeters in length. Photosynthetic pigments are localized in chromatophores which are characteristically a grass-green color, due to the predominance of chlorophyll a and b. The reserve products of carbon assimilation are usually stored as starch, and the group is very common in fresh water.

The genus *Chlamydomonas* with some 600 species may be found in all kinds of habitat. This small unicellular form can be troublesome at a waterworks because it may not be retained by filters and, in these circumstances, is a common source of complaint from consumers, particularly if water is left standing in carafes. Some species of *Chlamydomonas* form gelatinous masses on the sand surfaces of filters as "palmelloid growths" and are difficult to remove during filter cleaning operations. With rapid sand filters, surface wash sweeps (also known as filter agitators) help to break up the crusts that form on the sand surface, but if the filter is drained and dried, the combination of mucous, sand, and calcium carbonate crystals may result in cementation of the filter surface.

The genus *Ankistrodesmus* is extremely resistant to the normal dosage of copper sulfate which may be applied at an impoundment in order to control growths of diatoms. When the competition for the survival of the diatoms has been eliminated, species of *Ankistrodesmus* flourish as "after growths" and some are small enough to pass through sand filters.

Other small green algae such as *Scenedesmus* are also "filter passers," whereas the larger colonial forms such as *Volvox* and *Pediastrum* are effectively removed even by rapid sand filtration or by microstraining. Filamentous forms such as *Cladophora*, *Enteromorpha*, and *Hydrodictyon*, will, in midsummer, reproduce very rapidly in the pond-like environment of slow sand filters and also in the shallow margins of storage reservoirs. Although they do not necessarily impede filtration, mass mortality may produce undesirable tastes and odors.

DIATOMS[1] Diatoms are organisms belonging to the *Bacillariophyceae* group and are characterized by the presence of silica in the cell walls. The cell walls are sculptured with a narrow streak, band, or other markings, and by the presence of a brown pigment associated with the chlorophyll. They can be a very serious treatment problem.

FLAGELLATES[1] Flagellae are microscopic whip-like extensions which provide an organism with motility. Many species of algae and protozoa have this type of appendage enabling them to be free-swimming organisms.

TASTES AND ODORS FROM ALGAE Tastes and odors are always difficult to describe and classify. The nose can play tricks depending upon the extent of exposure to a particular smell. For example, people have died as a result of exposure to hydrogen sulfide simply because prolonged exposure to the gas has resulted in temporary immunity from smelling its presence. Odors are most noticeable at about 60°C (140°F). One ingenious device for water treatment plant operators who are continually plagued by algae taste and odor problems consists of a two-compartment spray aerator. In one compartment a fine spray of filtered, treated water passes through a small activated carbon filter and is heated to 60°C. The other compartment has alternatively sources of water from the intake as well as various parts of the plant, both before and after post-chlorination also heated to 60°C. The operator acclimatizes his nose to the "odorless" water and then sniffs the other waters to obtain some measure of the magnitude of the problem. An arrangement to mix the two waters together (the odorless water and any one of the other waters) in a direct ratio by observing the flow through small rotameters, enables the operator to measure the "Threshold Odor Number" of the water in question. The threshold odor number represents the number of dilutions with odorless water necessary to just sense the odor. There are lists describing odors and tastes of various species of algae; however, when the water is chlorinated they can be quite different, resembling a medicinal and sometimes a "celluloid" taste. Table 5-2 lists odors and tastes according to C. Mervin Palmer.[1]

One of the worst offenders is the group of flagellates known as *Synura*. This is probably the most troublesome genus in waterworks practice. Relatively small numbers will produce tastes described as *cucumber* and the problem is accentuated by chlorination. Fortunately, these flagellates are fairly easily removed by conventional filtration but problems do occur if they are fragmented and pass through the filter media. Unfortunately filtration does not remove the odor. Synura odor must be removed at source.

TABLE 5-2 Odors, Tastes, and Tongue Sensations Associated with Algae in Water

		Odor When Algae Are—			Tongue
Algal Genus	Algal Group	Moderate	Abundant	Taste	sensation
Actinastrum	Green		Grassy, musty		
Anabaena	Blue-green	Grassy, nasturtium, musty.	Septic		
Anabaenopsis	Blue-green		Grassy		
Anacystis	Blue-green	Grassy	Septic	Sweet	
Aphanizomenon	Blue-green	Grassy, nasturtium, musty.	Septic	Sweet	Dry.
Asterionella	Diatom	Geranium, spicy	Fishy		
Ceratium	Flagellate	Fishy	Septic	Bitter	
Chara	Green	Skunk, garlic	Spoiled, garlic		
Chlamydomonas	Flagellate	Musty, grassy	Fishy, septic	Sweet	Slick.
Chlorella	Green		Musty		
Chrysosphaerella	Flagellate		Fishy		
Cladophora	Green		Septic		
(Clathrocystis)	See *Anacystis*.				
Closterium	Green		Grassy		
(Coelosphaerium)	See *Gomphosphaeria*.				
Cosmarium	Green		Grassy		
Cryptomonas	Flagellate	Violet	Violet	Sweet	
Cyclotella	Diatom	Geranium	Fishy		
Cylindrospermum	Blue-green	Grassy	Septic		
Diatoma	Diatom		Aromatic		
Dictyosphaerium	Green	Grassy, nasturtium	Fishy		
Dinobryon	Flagellate	Violet	Fishy		Slick.
Eudorina	Flagellate		Fishy		
Euglena	Flagellate		Fishy	Sweet	
Fragilaria	Diatom	Geranium	Musty		
Glenodinium	Flagellate		Fishy		Slick.
(Gloeocapsa)	See *Anacystis*.				
Gloeocystis	Green		Septic		
Gloeotrichia	Blue-green		Grassy		
Gomphosphaeria	Blue-green	Grassy	Grassy	Sweet	
Gonium	Flagellate		Fishy		
Hydrodictyon	Green		Septic		
Mallomonas	Flagellate	Violet	Fishy		
Melosira	Diatom	Geranium	Musty		Slick.
Meridion	Diatom		Spicy		
(Microcystis)	See *Anacystis*.				
Nitella	Green	Grassy	Grassy, septic	Bitter	
Nostoc	Blue-green	Musty	Septic		
Oscillatoria	Blue-green	Grassy	Musty, spicy		
Pandorina	Flagellate		Fishy		
Pediastrum	Green		Grassy		
Peridinium	Flagellate	Cucumber	Fishy		
Pleurosigma	Diatom		Fishy		
Rivularia	Blue-green	Grassy	Musty		
Scenedesmus	Green		Grassy		

TABLE 5-2 Odors, Tastes, and Tongue Sensations Associated with Algae in Water (Cont'd)

Algal Genus	Algal Group	Odor When Algae Are— Moderate	Odor When Algae Are— Abundant	Taste	Tongue sensation
Spirogyra	Green		Grassy		
Staurastrum	Green		Grassy		
Stephanodiscus	Diatom	Geranium	Fishy		Slick.
Synedra	Diatom	Grassy	Musty		Slick.
Synura	Flagellate	Cucumber, muskmelon, spicy.	Fishy	Bitter	Dry, metallic, slick.
Tabellaria	Diatom	Geranium	Fishy		
Tribonema	Green		Fishy		
(*Uroglena*)	See *Uroglenopsis*.				
Uroglenopsis	Flagellate	Cucumber	Fishy		Slick.
Ulothrix	Green		Grassy		
Volvox	Flagellate	Fishy	Fishy		

Peridinium is frequently responsible for strong fishy tastes in water supplies. These flagellates thrive in northern temperate and arctic regions.

FILTER-CLOGGING ALGAE

Although most forms of algae tend to reduce filter runs, 43 of the more important filter-clogging algae are listed according to Palmer[1] in Table 5-3.

TABLE 5-3 Filter-Clogging Algae

Group and Algae

Blue-Green Algae (*Myxophyceae*):
    Anabaena flos-aquae
    Anacystis dimidiata (*Chroococcus turgidus*)
    Gloeotrichia echinulata
    Oscillatoria amphibia
    Oscillatoria chalybea
    Oscillatoria ornata
    Oscillatoria princeps
    Oscillatoria pseudogeminata
    Oscillatoria rubescens
    Oscillatoria splendida
    Rivularia dura

Green and Yellow-Green Algae (nonmotile *Chlorophyceae*, etc.)
    Chlorella pyrenoidosa
    Cladophora aegagropila
    Closterium moniliferum
    Dichotomosiphon tuberosus
    Dictyosphaerium pulchellum

TABLE 5-3 Filter-Clogging Algae (Cont'd)

Hydrodictyon reticulatum
Mougeotia sphaerocarpa
Palmella mucosa
Spirogyra porticalis
Tribonema bombycinum
Ulothrix variabilis
Zygnema insigne

Diatoms (*Bacillariophyceae*):
Asterionella formosa
Cyclotella meneghiniana
Cymbella ventricosa
Diatoma vulgare
Fragilaria crotonensis
Melosira granulata
Melosira varians
Navicula graciloides
Navicula lanceolata
Nitzschia palea
Stephanodiscus binderanus
Stephanodiscus hantzschii
Synedra acus
Synedra acus var. radians (*S. delicatissima*)
Synedra pulchella
Tabellaria fenestrata
Tabellaria flocculosa

Pigmented Flagellates (*Chrysophyceae*, etc.):
Dinobryon sertularia
Peridinium wisconsinense
Trachelomonas crebea

Most of the microscopic organisms present in waters filtered through a rapid sand filter will be caught in the top half inch of the sand. A few of the organisms will penetrate deeper into the filter bed while others disintegrate quickly as they come in contact with the sand. The longer the filter runs, the greater the percentage of organisms that will penetrate below the top half inch of filter bed into the supporting media. Mixed media or dual media filters have a larger capacity for suspended solids because they allow deeper penetration into the filter media and as a result, they will achieve longer filter runs between backwashes.

Some algae pass through the beds of both rapid and slow sand filters into the treated water. These include *Synedra, Oscillatoria, Chlamydomonas, Euglena, Navicula, Nitzschia, Phacus,* and *Trachelomonas.* The ease with which the algae penetrate depends upon several factors, the principal ones being chemical conditioning, the rate of flow, the grade of sand used, the conditions of the filter, and the type of organism. Very minute algae and flagellates penetrate with greater facility than other types. When the penetration is slow, it may be a few hours before the organisms reach the underdrains. Frequent backwashing, even when the filter is not clogged, will tend to remove the algae and reduce the number that reach the filtered water. A filter bed that cannot be thoroughly cleaned at each backwash can be damaged beyond repair by influxes of algae and other forms of pond life, such as water beetles, fresh-water shrimp, etc., that usually inhabit algae-containing waters.

In one water treatment plant the filter effluent rate control valve leaked rather badly with the result that when the plant clear-water well was full and the plant shut down, the leaky rate control valve slowly drained the filter beds. The organisms, including the insect life, were trapped in the filter media and died when they were completely exposed to air. Over a period of several years, sufficient oils had been excreted from the dead organisms to form a mass of dark-brown sticky deposit on the filter media that could only be removed by organic solvents, barring it from use in a municipal water treatment plant. The only solution to this problem was to completely replace the filter media and to repair the rate control valve to prevent it from leaking.

The action of chlorine on algae, apart from causing accentuated taste problems, also tends to break down the algae to a rather sticky mess which will be likely to gum a filter bed very much like an overdose of polyelectrolyte.

The mixed media filters, as marketed by Neptune MicroFLOC Incorporated, have a reverse gradation compared to the normal sand filter. With normal sand filters the top half inch usually consists of the finest grain sizes becoming gradually larger deeper into the filter. Therefore the top half inch or so quickly becomes plugged, particularly when pre-chlorination is practiced, causing some of the algae to break down. Mixed media filters with reverse particle size gradations have larger voids in the top 8 to 12 inches of the filter media than in the lower part of the bed. As a result they have a much higher capacity for algae than ordinary sand filters or even dual media filters. Prechlorination tends to break down certain species of algae into other substances which appear to act as filter aids. The turbidity of the filtered water may be appreciably improved by the use of chlorination, but the length of filter run can be significantly reduced, particularly with the use of high prechlorine dosages.

REFERENCES

1. PALMER, C. M. "Algae in Water Supplies." U. S. Department of Health, Education and Welfare; Public Health Service, Division of Water Supply and Pollution Control, Washington, 25 D.C. 1962. (Now U.S. Environmental Protection Agency, Cincinatti, Ohio).

2. HOLDEN, W. S. *Water Treatment and Examination*. London: J. & A. Churchill, 1970.

3. *Standard Methods for the Examination of Water and Wastewater*, 13th Ed. Published jointly by *A.W.W.A.*, American Public Health Association, and Water Pollution Control Federation, 1971. Available from the Publications Sales Department. American Water Works Association, 2 Park Avenue, New York, N. Y.

6

CHEMICAL AND
PHYSICAL PARAMETERS

INTRODUCTION

The chemistry of water is as fascinating as the study of its biology. Of the 103 elements presently known, there is little doubt that most of them are present in natural waters in one form or another. Modern analytical technology has not only opened several new doors, but has enabled routine chemical tests that previously took days to be performed in minutes. Many elements are now checked in a routine fashion that a few years ago would have tied up a whole laboratory staff for weeks. This enables many more parameters to be measured and compared. Some results are alarming when it is discovered that elements like selenium and mercury are higher in certain streams than previously realized; but, it is often found that the natural background levels of these elements are also higher than at first believed. One interesting development resulting from advances in analytical chemistry has been the capability to analyze, in minute detail, the flesh and internal organs of fish caught in the fjords of the West Coast of British Columbia. In this way, the Department of Geology of the University at British Columbia has been able to discover mineral outcroppings in these extremely rugged and difficult areas. Water, the universal solvent, picks up minute traces of the various elements and the fish and other aquatic organisms concentrate some of them in their bodies. The chemist is able to analyze the minerals for the geologist to evaluate.

From the public health viewpoint, considerable research (both practical and statistical) into the causes of illness and disease, related to the chemical composition of water supplies, has indicated the need for the numerous regulations from which our present standards have been compiled. The generally accepted standards for both chemical and bacteriological parameters in Canada and the U.S. have been as follows:

Canadian Drinking Water Standards and Objectives 1968[1]

U.S.A.—Public Health Service Drinking Water Standards (*Revised 1962*)[2]

A review of the older standards is given in *Water Quality and Treatment*[3] (3rd Ed.), which also includes some of the standards of the World Health Organization (WHO), both the European (1961) and the International (1963), as compared to the AWWA Recommended Potable Quality Goals—1969.

The Environmental Protection Agency (EPA) of the U.S. has recently published the interim primary standards of the Safe Drinking Water Act. After June 1977, with some possible minor adjustments, the "interim" label will be removed and the regulation standards will represent the minimum requirements to be satisfied by all public water systems in the U.S., unless they are specifically authorized for exclusion.[4]

Where water supplies are scarce, it is not always possible to adhere to the recommended standards, particularly with the very soluble elements like sodium and potassium, which can only be reduced by demineralization processes. Where doubts exist as to the suitability of the water for a particular purpose, reference should be made to McKee and Wolf—*Water Quality Criteria*.[5]

In addition to the above standards, most states and provinces have their own specific standards, but wherever "common carriers" such as railways, passenger ships, and aircraft travel from state to state, the national standards apply. When the new Safe Drinking Water Act becomes mandatory in the U.S. in June 1977 it will supersede all other standards, unless for specific reasons the local standards are more stringent than those recommended by the EPA. No doubt many other countries will revise their own regulations to comply with the EPA requirements.

PROPOSED DRINKING WATER QUALITY REGULATIONS AND PROPOSALS

The essential elements of the new U.S. Safe Drinking Water Act have been carefully thought out and well defined. They have been reported in the Journal of the AWWA[4] and the main issues are noted below. Once the Act becomes law it is possible that more definitive details will be published.

Public Water Supply System

The Safe Drinking Water Act will apply to each public water supply system and may be defined as "a system for the provision to the public of piped water for human consumption, if such a system has at least 15 service connections or regularly serves at least 25 individuals."[4] It is expected that where a water supply is used only on a seasonal rather than a year-round basis, for example camping sites and other occasional resorts, where people are not exposed to the regular use of the water, that some modifications to the requirements of the Act may be authorized providing safety is not jeopardized.

Small communities where the cost of complete treatment would be exorbitant may also be permitted some modifications of the Act providing there is no relaxation of safety.

Water Consumption

The maximum contamination levels (MCL) for the interim standards are based on an ingested consumption of 2 liters of water per person per day (equivalent to 0.528 U.S. gallons or approximately 2 U.S. quarts). In general, with the possible exception of the MCL for fluorides, the ambient temperature is not taken into account.

Canadian Standards

Present Canadian standards are given in the *Canadian Drinking Water Standards and Objectives (1968)*[1] and the following Tables 6-1, 6-2, 6-3 and 6-4 relate to the more important parameters to be considered when assessing a new water supply in Canada. However new *Canadian Standards and Objectives* are presently being prepared and will probably be more stringent than the 1968 version.

TABLE 6-1   Physical[1]

Parameter	Objective	Acceptable
Color—T.C.U.[1]	< 5	15
Odor—T.O.N.[2]	0	4
Taste	Inoffensive	Inoffensive
Turbidity—J.T.U.[3]	< 1	5
Temperature °C	<10	15
pH Units	—	6.5–8.3

[1] T.C.U. — True Color Unit, Platinum Cobalt Scale
[2] T.O.N. — Threshold Odor Number
[3] J.T.U. — Jackson Turbidity Unit

TABLE 6-2 Drinking Water Standards for Toxic Chemicals[1]

Toxicant	Objective mg/ℓ	Acceptable Limit—mg/ℓ	Maximum Permissible Limit—mg/ℓ
Arsenic as As	Not Detectable	0.01	0.05
Barium as Ba	Not Detectable	<1.0	1.0
Boron as B	—	<5.0	5.0
Cadmium as Cd	Not Detectable	<0.01	0.01
Chromium as Cr^{+6}	Not Detectable	<0.05	0.05
Cyanide as CN	Not Detectable	0.01	0.20
Lead as Pb	Not Detectable	<0.05	0.05
Nitrate + Nitrite as N	<10.0	<10.0	10.0
Selenium as Se	Not Detectable	<0.01	0.01
Silver as Ag	—	—	0.05

TABLE 6-3 Recommended Limits for Other Chemicals in Drinking Water[1]

Chemical	Limit—mg/ℓ	
	Objective	Acceptable
Ammonia as N	0.01	0.5
Calcium as Ca	<75	200
Chloride as Cl	<250	250
Copper as Cu	<0.01	1.0
Iron (dissolved) as Fe	<0.05	0.3
Magnesium as Mg	<50	150
Manganese as Mn	<0.01	0.05
Methylene Blue Active Substances	<0.2	0.5
Phenolic Substances as Phenol	Not Detectable[1]	0.002[3]
Phosphates as PO_4 (inorganic)	<0.2	0.2
Total Dissolved Solids	<500	1,000
Total Hardness as $CaCO_3$	<120	
Organics as CCE + CAE[2]	<0.05	0.2
Sulphate as SO_4^-	<250	500
Sulphide as H_2S	Not Detectable	0.3
Uranyl Ion as UO_2^-	<1.0	5.0
Zinc as Zn	<1.0	5.0

[1] "Not detectable" by the method described in the latest edition of "Standard Methods" (AWWA, APHA, WPCF) or by any other acceptable method approved by the control agency.

[2] Total of carbon chloroform and carbon alcohol extractibles.

[3] Based on taste and odor considerations. Concentration greater than 0.05 mg/l may be objected to by the majority.

TABLE 6-4 Classification of Drinking Water Quality Based on Hardness[1]

Quality Classification	mg/ℓ CaCO$_3$	Grains (CaCO$_3$)/imp. gal.	Grains (CaCO$_3$)/U.S. gal.
Very Good	<80	<5.6	<4.7
Good	81–120	5.6–8.4	4.7–7.0
Fair	121–180	8.4–12.6	7.0–10.5
Poor	>180	>12.6	>10.5

Color

Pure water has no color. The presence of dissolved foreign substances in solution alters the colors to blue, green, yellow, or brown according to the amount and nature of the material present and is referred to as *true color*. When the water is turbid as well as colored, it has *apparent color*, which is usually higher than true color. Before the color of a turbid water can be accurately measured, the turbidity should be removed (preferably by centrifuging) since some of the color intensity may be disturbed by filtration. In theory, true color should be filterable without change, but in many cases filtration cannot be used to clarify a colored turbid water since the particles of color may coagulate in the pores of the filter and be removed.

The present standard unit of measurement is known as the True Color Unit (T.C.U.) where 1 unit is equivalent to the color produced by 1 mg/ℓ of platinum in the form of chloroplatinate ion. See *Standard Methods*[7] for details of variations in tinting. In the U.S. the unit was known as an *APHA unit* (1955) (American Public Health Association) and consists of dilutions of the chloroplatinate solution as viewed through Nessler tubes. In the United Kingdom, similar solutions were prepared and called *Hazen Color Standards*.

One of the most important conditions of color measurement is pH, since color is pH-sensitive. Generally, the higher the pH, the more intense is the color; but if the solution is sufficiently acidified, the color will be appreciably reduced. Many standard pieces of equipment (generally colored glass comparators) are available for measuring color and are convenient for field investigations. Since color is susceptible to change during transit, a color test in the field is more useful than a test in the laboratory.

A British Columbia muskeg swamp water had 500 APHA color units and 5 mg/ℓ of iron. When the sample was transported 700 miles to the laboratory it was reported as 15 APHA color units and 0.3 mg/ℓ of iron. The analyst failed to mention that there was a little dark-colored deposit on the bottom of the bottle into which the iron and organic matter had coagulated and settled.

Colored waters are objectionable to domestic consumers and industrial users alike, not only because of their appearance but for the problems caused with washing clothes, manufacturing food stuffs, paper manufacturing, etc. It is believed that the organic compounds causing true color are not harmful to health, but they will absorb chlorine and thereby seriously reduce the effectiveness of chlorine as a disinfectant. Whenever color and iron are present together in one sample, it is not unusual to find that the iron has been chelated by the organic matter which will prevent it from being removed by aeration. Chelation is described as maintaining metallic ions in a solution by a liquid with which the metallic ions are strongly bound in a relatively stable and inactive complex.

The acceptable limit for color is usually given as 15 T.C.U. with an objective of less than 5 T.C.U. Color is a good visual indication of pollution or post-treatment contamination. Highly colored waters usually have an oxygen demand, either biological or chemical, and they are likely to become devoid of dissolved oxygen when held in storage.[8,9]

Odor

The detection of odor is usually fairly simple, but to determine it quantitatively is very difficult and cannot be reduced to an exact science. Some people are blind to certain colors but react normally to other combinations of colors, and so it is with odors.

Similarly, the intensity of the odor plays various tricks; for example, hydrogen sulfide in very small concentration has, to some people, a sweet, not unpleasant odor. However in stronger concentrations, it has a most unpleasant smell and, unfortunately, the nasal organs responsible for detecting and classifying odors appear to be turned off when the exposure exceeds a certain limit. Many chemical and oil refinery operators have died from exposures to lethal dosages of hydrogen sulfide simple because they have become immune to its odor, particularly in large concentrations.

In water treatment, odors usually originate from biological sources such as algae, decaying organic matter, and various side reactions initiated by bacteria. *Standard Methods*[7] suggests a water temperature of 40°C which approximates body temperature, but when there is excessive heat, another standard threshold odor test at a temperature of 60°C (140°F) can be used.

The qualitative nature of odor is the most difficult to determine, and in order to root out the causative element a full knowledge of the water source and treatment system is essential. In many cases trace elements or complex organic compounds secreted by algae and other living organisms will change appreciably with storage, treatment, and particularly chlorination, so that the true nature and location of the problem is most difficult to determine. A list of odors from various algae has been presented in Table 5-2. The smell of hydrogen sulfide may result from sulfate-reducing bacteria. The electrolysis of impure well waters due to the use of dissimilar metals in a well casing and pump can give rise to odors which are sometimes difficult to account for in biologically sterile water.

The measurement of odor is described as the Threshold Odor Number (T.O.N.) which is the number of times a sample of water must be diluted with odor-free water to be just detectable by the odor test. Details of the apparatus to be used and the procedure to be adopted are described in *Standard Methods*.[7]

Where odors in water supplies exist, forced-draft aeration is sometimes but not always effective. Activated carbon is also used to absorb odor-producing compounds, but the pH is somewhat critical and compounds tend to desorb at higher pH values. Much practical technical advice is available from the manufacturers of activated carbon.

Cox[10] has some good comments and advice on this subject which is recommended reading. The acceptable limit on the Threshold Odor Number (T.O.N.) is usually given as 4 with the objective of 0.

Tastes

The general comments on the subject of odors will apply to taste, which is the third of the five physical senses used by humans to judge the quality of water. Strictly speaking, there are only four possible tastes, namely, acid or sour, sweet, salt, and bitter; but tastes are also associated with odors. Tastes and odors together are associated with appearance.

If synthetic orange juice was any other color than orange, the public would insist that it didn't taste like orange. The same is true with water. In one town, people were quite used to drinking water which had a very strong earthy taste due to decomposed algae. When a new plant was installed producing sparkling water, clear, bright, tasteless, and odor-free, there were complaints that it was not as good as it used to be! Chlorine plays as many tricks with the taste as it does with the odor problem, but they are more difficult to isolate and treat. For example, phenols in water supplies should not exceed 0.002 mg/ℓ which is equivalent to 2 parts per billion. It is unlikely that phenol in very small quantities would ever be a public health hazard since triclorphenol is the active agent in an objectionable-tasting disinfectant gargle for sore throats. However, in view of the ability of phenolic compounds to produce substitution products in the presence of chlorine, the limits of phenol compounds must be kept extremely low to avoid unpleasant medicinal tastes and odors.

Gas-chromatographic analysis is now used to identify the complex organic chemicals causing both taste and odor problems. The equipment requires calibration with known quantities of the various chemicals, but once this has been completed, it is capable of analyzing many samples in rapid succession.

Turbidity

Turbidity in water is caused by the presence of suspended matter such as clay, silt, finely-divided organic and inorganic matter, plankton, and other microscopic organisms. It can also be caused by finely-divided air bubbles. It should be noted that turbidity is an expression of the optical property of a water which causes light to be scattered and absorbed rather than transmitted in straight lines through the water sample. Attempts to correlate turbidity with the weight of the suspended matter are impractical, as the size, shape, and refractive index of the particulate particles are optically important but bear little direct relationship to the concentration and the density of the suspended matter. For example, a cubic centimeter of pure gold in one regular cubic lump, placed in a beaker of water, would not indicate any turbidity. However, if the same cubic centimeter were in the form of colloidal gold, which is reported to have more surface area than a couple of tennis courts, it would fill the beaker with minute suspended particles and would represent a very high degree of turbidity, but its weight is no different than the first cubic centimeter which registered no turbidity at all.

The standard method for the determination of turbidity[7] has been based on the Jackson candle turbidimeter. This instrument consists of a long vertical glass tube closed at the bottom end. It sits in a stand with a lighted candle underneath the glass. Turbid water is slowly poured into the tube until the image of the candle flame, when viewed through the water, becomes indistinguishable against the general background illumination. The longer the light path, i.e., the more water there is in the tube, the lower the turbidity. A light path of only 2.3 cm would be equivalent to a turbidity of 1000 Jackson Turbidity Units (J.T.U.). On the other end of the scale, a light path of 72.9 cm would be equivalent to a turbidity of 25 J.T.U.'s. The Jackson candle turbidity is limited to turbidities of 25 to 1000 J.T.U.'s, and for readings less than 25 J.T.U.'s, more sophisticated equipment is necessary. Several types of apparatus are available but probably the best type of equipment is known as a nephelometer. This consists of a light source and photoelectric cell complete with millivolt meter calibrated against standard turbid suspensions of formazin. It is further simplified by using a calibrated plastic rod which can be inserted into the light path to standardize the equipment. Nephelometers are usually capable of measuring from approximately 0.1 to 40 J.T.U.'s and sensing differences of $\pm 10\%$ with repeated accuracy within these limits.

Since, however, there is no direct relationship between the intensity of light scattered through 90° as compared to viewing the image of a candle flame through a column of water, there is no valid basis for the practice of calibrating nephelometers in terms of candle units and in many cases the name Jackson is not used in relationship to turbidity units.

The importance of being able to measure turbidities to less than 1 unit has increased in significance with modern filtering techniques and the discovery that viruses are frequently hosted by minute particulate matter in suspension. The recent literature on the subject of waterborne diseases has stressed very strongly the need to reduce the turbidity as much as possible. The presently accepted limit of turbidity for domestic water is 5 J.T.U.'s with the objective of 0.1 J.T.U. It is almost certain that very shortly turbidities of 5 units will be unacceptable to most authorities. It is impossible to produce a sparkling clear and bright water unless the turbidity is less than 1.0 J.T.U.

When the new U.S. Drinking Water Quality regulation comes into force, it will be mandatory to maintain a monthly average not exceeding 1 Turbidity Unit (T.U.) and not exceeding 5 T.U.'s for more than a 48-hour period. If a sample from a water supply exceeds 1 T.U. it will be resampled within the hour. If repeat samples confirm that the limit has been exceeded it must be reported to the State Authorities within 48 hours.[4] The act may also require that the public be informed if the average of 5 T.U.'s is exceeded in the 48-hour period.

Chlorination can never become a fully effective disinfection barrier when there is suspended particulate matter in the water providing safe and secure habitats for the various pathogenic organisms which may be present. It is therefore a prerequisite that a reasonably safe and wholesome drinking water must have a very low turbidity for it to qualify as a first-class water in any domestic water supply system.

Temperature

The temperature of drinking water has often been an important issue in the merits of a town's water supply system. The cooler the water the higher the merit, with temperatures below 15°C (59°F) being preferred. If the source of water is a deep lake, a deep water intake will often insure a more uniform lower temperature. If, on the other hand, the source of the supply is a shallow lake affected by the seasons, there is not very much that can be done. With the advent of ice machines, domestic water coolers, and refrigerators, the need for cold water at the tap is not really great.

At the lower end of the temperature scale (near freezing) problems occur, i.e., frazil ice at the intake, very slow chemical reaction rates in the flocculators, and poor settling in the clarifiers due to high viscosities, broken distribution mains due to pipe contraction or water expansion, frozen services, and ground movements from frost penetration. Wherever these conditions exist it is customary to artifically heat the water a few degrees to about 40°F (4.4°C), if possible, prior to treatment. Try to prevent too much fluctuation in temperature by ensuring that the mains are sufficiently below the frost level to avoid freezing. A few degrees of temperature difference in the water is likely to result in expansion and contraction stresses in the water mains which frequently result in broken pipes. Concern was felt in one city which has its water treatment plant alongside a river next to a thermal power station. The water treatment plant receives its raw water supply partly from the river and partly from the heated condenser cooling water with a resulting temperature of approximately 55°F (12.8°C). This is almost ideal for a cold-lime softening process and uniform temperature in the distribution system; however, a new plant extension is now needed and there is no available space sufficiently close to the power plant to install the new addition. The new location is several miles away from the original plant site, so there will be sections of the distribution system alternately supplied from both the old and the new plants. Appreciable differences in the temperature of the water in the distribution system could significantly increase the work of the maintenance crews.

pH and pOH Values

Many chemical compounds dissociate in aqueous solutions by a process known as ionization. The ions produced are electrically charged. Those with a positive charge are known as *cations*, i.e., sodium (Na^+); and those with a negative charge are called *anions*, i.e., chloride (Cl^-). A cation is an electropositive ion of an electrolyte, which moves to the *cathode*, or negative electrode of an electrolytic cell. An anion is an electronegative ion of an electrolyte which moves towards the *anode*, or positive electrode of an electrolytic cell.

A typical example would be an aqueous solution of sodium chloride:

$$Na^+Cl^-$$

The sodium has a positive charge and is known as a cation because it will seek the negative electrode, while the chlorine ion is a negatively charged anion and will seek the positive electrode.

Water also partially dissociates into two ions called the hydrogen ion (H^+) which will migrate towards the cathode and is therefore called a cation, and the hydroxyl ion (OH^-) which moves towards the anode and is referred to as an anion.

The dissociation of water into its respective cations and anions is really very slight, and one liter of neutral water (i.e., pH = 7) contains only 1/10,000,000 gm of hydrogen ions. In its neutral state each mole contains an equal quantity of hydroxyl ions. This fraction would be too awkward for every-day use, so the term pH is used to determine the hydrogen ion activity and pOH to determine the hydroxyl ion activity. Pure water at 25°C dissociates to yield a concentration of hydrogen ions equal to 10^{-7} moles per liter, and since it also produces a hydroxyl ion for each hydrogen ion, it is obvious that 10^{-7} moles of hydroxyl ion are produced simultaneously.

Water dissociates into ions in accordance with the equation:[11]

$$H_2O = H^+ + OH^-$$

By substitution into the mass action equation we obtain:

$$\frac{(H^+)(OH^-)}{H_2O} = K$$

This is known as the "ion product" or "ionization constant" for water. When an acid is added to water, it ionizes in the water and the hydrogen ion concentration increases. Consequently, the hydroxyl ion concentration decreases in conformity with the ionization constant of 10^{-14}. For example, if acid is added to increase the (H^+) to 10^{-1}, the (OH^-) must decrease to 10^{-13}.

$$10^{-1} \times 10^{-13} = 10^{-14}$$

Likewise, if a base is added to water, for example caustic soda (NaOH), to increase the (OH^-) to 10^{-3}, the (H^+) decreases to 10^{-11}.

Expression of hydrogen ion concentrations in terms of molar concentrations is cumbersome. In order to overcome this difficulty, the values of the hydrogen ion concentrations are expressed in terms of their negative logarithms and designated to symbol pH for hydrogen ion concentration and the symbol pOH for hydroxyl ion concentration. The sum of their negative logarithms at 25°C is always 14. At 5°C it is 14.7 and at 50°C it is 13.3.

Then, by definition, the pH is the negative logarithm of the reciprocal of the hydrogen ion concentration or, more precisely, of the hydrogen ion activity in moles per liter. The term may be represented by:

$$pH = -\log(H^+)$$

or

$$pH = \log \frac{1}{(H^+)}$$

The pH scale is usually represented as ranging from 0 to 14, with pH 7 representing neutrality.

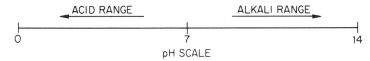

pH SCALE

Acid conditions increase as pH values decrease, and alkaline conditions increase as the pH increases. pH 7 has little significance as a reference point in water chemistry. About the only significance it has is the equality in the hydrogen ion and hydroxyl ion activity (see Table 6-5).

TABLE 6-5 Relationship Between pH Scale and Hydrogen Ion and Hydroxyl Ion Cencentration[10]

pH Value (Approximate)	Gram-Equivalent Weight of (H^+) Per Liter of Solution	Gram-Equivalent Weight of (OH^-) Per Liter of Solution
0.0	1.0	0.00000000000001
1.0	0.1	0.0000000000001
2.0	0.01	0.000000000001
3.0	0.001	0.00000000001
4.0	0.0001	0.0000000001
5.0	0.00001	0.000000001
6.0	0.000001	0.00000001
7.0 ----Neutral----	0.0000001 ----------Neutral----	0.0000001
8.0	0.00000001	0.000001
9.0	0.000000001	0.00001
10.0	0.0000000001	0.0001
11.0	0.00000000001	0.001
12.0	0.000000000001	0.01
13.0	0.0000000000001	0.1
14.0	0.00000000000001	1.0

Few people who work with pH values could accurately define its meaning and the definition is unnecessary in order to operate a water treatment plant.

The term pH is used universally to express the intensity of the acid or alkaline condition of a solution. It is a method of expressing the hydrogen ion concentration or activity which is important in almost every phase of water engineering practice. It should be noted that the hydrogen ion concentration at pH 4.0 is 10 times more than it is at pH 5.0 and similarly for all other values of hydrogen ion and hydroxyl ion activity. The term pH is frequently referred to in flocculation, coagulation, softening, disinfection, and corrosion control.

Approximate measurements of pH can be made by the use of color comparators or even litmus papers. In the case of a color comparator a few drops of an indicator solution are added to the water which causes it to change color. The color is then compared to a colored glass disc or tube which represents the pH value of the indicator solution used. This method is fairly simple and is useful for field analysis. However, the colors can be distorted by interference from trace elements, and the color standards may also fade.

Electric pH meters are the most popular method of obtaining pH measurements, and they can be readily built-in to a process control system which automatically controls acid and alkaline dosages to maintain preset controlled pH values. Temperature changes in the water will alter the pH meter calibration, but temperature compensation can be built into the system to overcome this variable. It is usual to standardize the meter and the electrode by immersing the electrode into a buffer solution with a high degree of alkalinity at a constant pH value.

Many foods and beverages have wide ranges of pH values; for example, carbonated soft drinks are often as low in pH as 3.0. The normal acceptable limit for municipal water supplies is between pH 6.5 and 9.5, but some water supplies have values as low as pH 5.0 and as high as pH 10.0. The solubility of ions in water is strongly influenced by the pH value of the water. In general, waters of low pH tend to be corrosive or aggressive to certain metals, asbestos cement, and cement mortar-lined pipe, while waters of high pH tend to be scale-forming. However, there are so many other factors that affect these reactions that a water should not be condemned on the basis of its pH value alone without reference to other parameters.

Alkalinity

The alkalinity of water is usually a measure of the hydroxide $((OH)^-)$, carbonate $((CO_3)^{--})$, and bicarbonate $((HCO_3)^-)$ radicals but other forms of alkalinity may be present in solution. They are measured by titrating with known strengths of acid to specific pH values, usually with the use of indicators such as phenolphthalein (sometimes known as "Phenol Alkalinity" and abbreviated to "Alk P" or "P" alkalinity) and methyl orange (abbreviated to "Alk M" or "M.O. Alkalinity"). Phenolphthalein gives a pink coloration to water at pH values of 8.3 and above. Titrating with acid until the "end point," where the pink color disappears, is a measure of the amount of acid needed to neutralize the hydroxide and carbonate alkalinity down to pH value of 8.3. To the same sample a few drops of methyl orange indicator are then added which changes the color of the sample to orange. Continuing to titrate with acid changes the color from orange to red at pH of approximately 4.2. The milliliters of acid added between the phenolphthalein end point at pH 8.3 and the methyl orange end point at pH 4.2 is a measure of the carbonate and bicarbonate alkalinities. The total number of milliliters of acid used in the titration from the start of the test through the phenolphthalein end point to the methyl orange end point is a measure of the total alkalinity of the water. In many laboratories these tests are done in conjunction with an electric pH meter, noting the milliliters of acid used to reach end point at pH 8.3 and pH 4.2. It is not the intention of this book to attempt to provide instructions on how to perform these tests since this is well covered in many other publications, particularly in *Standard Methods*.[7]

Distilled water has no alkalinity and it is therefore easy to change its pH value. For example, if the electrodes of a pH meter are placed in a beaker of distilled water, the pH might be almost anywhere other than pH 7. If you blow into the water by means of a pipette so that air bubbles to the surface, it will be noticed that the pH drops sharply to the acid end of the pH range due to the absorption of the carbon dioxide out of the breath. If, on the other hand, you tried the same experiment on a highly mineralized water with an alkalinity of several hundred milligrams per liter as calcium carbonate, the pH might be 7.3 at the time the experiment is first started and you may have to bubble your breath through the pipette for an hour before you are able to significantly change the pH. This sample would be described as a "buffered" water, in that it contains sufficient alkalinity, among other elements, to resist pH change. Providing the water did not precipitate any of its constituents, it would be regarded as "stable." Whereas the distilled water in the first experiment had no buffering action and while it could not precipitate out any of its contents (since it had none), it can easily swing across the pH scale from acid to alkali and back to acid again with very little effort. The water is therefore described as unbuffered.

It is generally accepted that only two of the three types of alkalinity can be present in any one sample of water, since a reaction would occur between the hydroxide and the bicarbonate alkalinities to form carbonate alkalinity:

$$OH^- + HCO_3^- = H_2O + CO_3^-$$

It is generally believed that the types of alkalinity are oriented in accordance with the pH scale as follows:

pH 11.0—9.4 Hydroxide Alkalinity and Carbonate Alkalinity.
No Bicarbonate Alkalinity.

pH 9.4—8.3 Carbonate Alkalinity and Bicarbonate Alkalinity.
No Hydroxide Alkalinity.

pH 8.3—4.6 Bicarbonate Alkalinity but no Carbonate Alkalinity or Hydroxide Alkalinity

pH 4.6—3.0   Mineral Acids.

The alkalinity relationships in Table 6-6 are useful in determining how much of each type of alkalinity is present in the water particularly when treating with lime for softening.

TABLE 6-6 Alkalinity Relationships

Case	Type of Alkalinity Present	Derivation
P = M	OH^- only	P measures OH^- only M measures OH^- only $\therefore$ P = M = OH^-
2P − M (Positive)	OH^- and CO_3^{2-}	2P measures 2OH + 2 (1/2CO_3) M measures OH + CO_3 $\therefore$ 2P − M = OH^- or 2 (M − P) = CO_3^{2-}
2P = M	CO_3^{2-} only	2P measures 2 (1/2CO_3) M measures CO_3 $\therefore$ 2P = M
2P − M (Negative)	CO_3^{2-} and HCO_3^-	2P measures 2 (1/2CO_3) M measures CO_3 + HCO_3 $\therefore$ 2P − M = −HCO_3
No P M only	HCO_3^- only	M = HCO_3

P is the titration to pH 8.3 and is equivalent to phenolphthalein alkalinity or "alk P." M is the titration to approximately pH 4.2 and is equivalent to methyl orange alkalinity or "alk M."

The acceptable alkalinity for municipal water is generally between 30 and 500 mg/ℓ as $CaCO_3$, but there are many water supplies well above and below these limitations.

Hardness

The terms "soft" waters and "hard" waters refer to the process of forming a lather when washing hands or clothes with soap. Soft water lathers easily, leaving the hands soft and no "ring" in the washbasin or bathtub. Hard water lathers with difficulty, leaving the hands rough and forming a ring or scum.

The cations calcium and magnesium are principally responsible for water hardness. The type of characteristics of the hardness caused by the calcium and magnesium are influenced by the alkalinity, the sulfate, chlorides, and other chemicals that might also be present.

For the purpose of uniformity, the hardness of water, similar to the alkalinity, is usually expressed in milligrams per liter (mg/ℓ), as calcium carbonate ($CaCO_3$), in spite of the fact that the hardness may be caused by magnesium sulfate having no calcium or carbonate whatsoever.

The terms "soft" and "hard" are used in the U.S. and Canada as follows:

Soft: Hardness less than 50 mg/ℓ as $CaCO_3$

Moderately hard: Hardness of 50—150 mg/ℓ as $CaCO_3$

Hard: Hardness of 150—300 mg/ℓ as $CaCO_3$

Very hard: Hardness over 300 mg/ℓ as $CaCO_3$

People accustomed to water of 100 mg/ℓ hardness would complain of a water of 200 mg/ℓ hardness but would judge that water of 50 mg/ℓ hardness was quite soft, whereas people who are accustomed to water of less than 10 mg/ℓ hardness would consider 50 mg/ℓ as hard.[12]

The hardness-producing metals, calcium and magnesium, must first be precipitated by the sodium oleate in the soap before a lather can be obtained. This type of water softening is both expensive and objectionable, but the advent of household synthetic detergents (syndets) has minimized the problems associated with washing clothes and dishes in hard water.

Hard water forms scales in boilers which appreciably reduces their efficiency. Waters may also be corrosive to certain metals and still deposit scales. The questions of the degree of hardness must be assessed for each industry; treating the water to a suitable degree of hardness, as opposed to using an alternative water supply, is a question of economics.

From the public health viewpoint, there are no reported instances where substances contributing to hardness are directly implicated as health problems, although some investigators have demonstrated that there is a statistical correlation between low water hardness and certain heart diseases. The inference is that there is a greater percentage of people suffering from cardiovascular diseases in soft water areas as compared to communities accustomed to hard waters. However, so far there has been no strongly documented evidence to suggest that soft water should be made harder. The only comments thus far are to the effect that a hardness of 80 mg/ℓ, as $CaCO_3$, is considered to be an acceptable goal for all-round domestic use.

The measurement of hardness in mg/ℓ as $CaCO_3$ is used in Europe as well as in North America, and the conversion from 1 unit to another can be facilitated as follows:

    1 Clark or English degree $= 14.3$ mg/ℓ as $CaCO_3$

    1 grain per U.K. gallon $\ \ = 14.3$ mg/ℓ as $CaCO_3$

    1 grain per U.S. gallon $\ \ = 17.1$ mg/ℓ as $CaCO_3$

    1 French degree $\ \ \ \ \ \ \ \ \ \ \ \ = 10.0$ mg/ℓ as $CaCO_3$

    1 German degree $\ \ \ \ \ \ \ \ \ \ = 17.8$ mg/ℓ as $CaCO_3$

There are four categories of hardness: temporary, permanent, carbonate, and non-carbonate.

Temporary Hardness

When the pH of the water is 8.3 or less, the alkalinity consists of bicarbonates. If the calcium and magnesium (in terms of equivalent mg/ℓ) are equivalant, or less than the total alkalinity, (also expressed in equivalent mg/ℓ), the hardness is said to be "temporary" on the basis that most of the bicarbonate will decompose to CO_2 on boiling.

$$Ca(HCO_3) \xrightarrow{\text{Boiling}} CaCO_3 + H_2O + CO_2$$

Since the solubility of calcium carbonate is only 14 mg/ℓ in cold water and 19 mg/ℓ in hot water, the calcium carbonate precipitates on boiling, the CO_2 is given off with the vapor, and the hardness is reduced from its original value to less than 20 mg/ℓ once all the calcium carbonate has settled out of the water.

In the case of magnesium:

$$Mg(HCO_3)_2 \xrightarrow{\text{Boiling}} Mg(OH)_2 + 2CO_2$$

and the solubility of magnesium hydroxide is approximately 9 mg/ℓ in cold water and 40 mg/ℓ in hot water. Solubility data taken from handbooks[13] will also be strongly influenced by the pH of the water.

The liberation of carbon dioxide on boiling is the cause of many corrosion problems in steam-heating installations using waters high in bicarbonates for the initial filling of the boilers and for subsequent make-up feed water.

The carbon dioxide liberated in the boiler passes over with the steam and is again absorbed by the condensate forming carbonic acid,

$$H_2O + CO_2 \longrightarrow H_2CO_3$$

which corrodes the condensate return lines. The carbonic acid is a very weak acid and is in very dilute solution, but since it breaks down again into carbon dioxide and water on boiling, the corrosion, although very slight, is continuous. Unless the CO_2 gas is purged out of the system through an air vent or neutralized with an alkali, it will continue to corrode.

Permanent Hardness

When the alkalinity is less than the calcium and magnesium expressed in terms of equivalents, it is usual to find that calcium and magnesium are associated with chlorides or sulfates or other radicals forming compounds which are not decomposed on boiling. This is known as "permanent hardness" or non-carbonate hardness.

Carbonate Hardness

This is similar to temporary hardness and refers to carbonate or bicarbonates associated with the calcium and magnesium.

Arsenic (As)

TOXIC CHEMICALS

Arsenic should be absent from water supplies as severe poisoning can result from the ingestion of as little as 100 mg. Chronic effects can also appear from its accumulation in the body at low intake levels. It is reported that arsenic is a possible cause of cancer. The arsenical concentration of drinking waters should not exceed 0.05 mg/ℓ. The element may be present in some natural waters or may gain access to shallow wells from surface washings or percolation where arsenical weed-killers or tree-sprays have been used in the vicinity of the well. Sheep-dipping may contaminate surface runoff waters and lakes and reservoirs may also be affected in this way. Arsenic can also come from natural sources. A crack developed[18] in a tank containing arsenical sheep-dipping solution. The escaping liquid contaminated a nearby well resulting in a concentration of 12 mg/ℓ and causing six deaths.

An unusual case of arsenical poisoning occurred to a trapper who lived in the Northwest Territories of Canada close to a gold mine. It appears that arsenic fumes were emitted from the extraction plant which contaminated the surface runoff water. The trapper obtained his drinking water from a nearby lake and kept it in a kettle on top of the stove. The kettle was always hot and concentrated the arsenic by constant evaporation. Since the water had to be carried to the cabin, there was no desire to drain the contents of the kettle from time to time to reduce the arsenical concentration. Eventually, it exceeded the limits of tolerance and killed him.

Barium (Ba)

Barium in municipal waters should be absent but must not exceed 1.0 mg/ℓ. Until 1962, when it was first listed by the U.S. Department of Health in their Drinking Water Standards, barium was not considered a health hazard.[2] Many of the barium salts are soluble, but the carbonate and sulfate are highly insoluble. Consequently, it is to be expected that any barium ions discharged into natural waters will be precipitated and absorbed in the bottom mud. Barium ions are not normally present in natural surface or ground waters in measurable concentrations, although they have been detected in a few springs and in effluents from areas where baryte ($BaSO_4$) and whitherite ($BaCO_3$) are mined.

The fatal oral dose of barium for adult humans is reported to be 550–600 milligrams. Barium salts are considered to be muscle stimulants, especially for the heart muscle. By constricting blood vessels, barium may cause an increase in blood pressure. On the other hand, there is no evidence that barium accumulates in the bone, muscle, kidney, or other tissue. It is excreted more rapidly than calcium and, therefore, there is reported to be no danger from a cumulative effect.[5]

Boron (B)

The boron content of drinking waters was considered satisfactory if it was less than 5 mg/ℓ,[1] however this has now been reduced to 1 mg/ℓ.[3] The ingestion of excessive dosages of borates may cause nausea, cramps, convulsions, coma, and other symptoms of distress. The fatal dose for adult humans is reported between 5 and 45 grams. The wide difference between the lower and upper magnitudes of a fatal dose is probably because boron, in most humans, is promptly excreted in urine. Some investigators have suggested that an upper limit of 20 mg/ℓ for municipal waters would not constitute a health hazard.[2]

Cadmium (Cd)

Ideally cadmium should not be detected in drinking waters. The maximum permissible limit should not exceed 0.01 mg/ℓ.[1,6] Cadmium is used in metallurgy to alloy with copper, lead, silver, aluminum, and nickel. It is also used in electroplating, ceramics pigmentation, and photography. It is reported that ground water contamination by cadmium to the extent of 3.2 mg/ℓ occurred on Long Island, N.Y., as a result of effluents from electroplating industries. High concentrations of cadmium have been reported in Missouri mine waters where one spring is reported to contain 1000 mg/ℓ of cadmium.[5]

Several health departments have forbidden the use of cadmium-plated food containers and pipes. Cadmium contained in ice cubes in cold drinks have caused acute gastritis symptoms within one hour. Consumption of cadmium salts causes cramps, nausea, vomiting, and diarrhea. Since cadmium tends to concentrate in the liver, kidneys, pancreas, and thyroid of humans and animals, it has resulted in a number of deaths. Once it enters the body it is likely to remain, and there is no evidence that cadmium is biologically essential or beneficial.[5]

Chromium (Cr)

Chromium ions can be present in two principal forms, trivalent and hexavalent. The ingestion of trivalent chromium is reported to have no physiological effects in the quantities normally associated with drinking waters, but hexavalent chromium should not exceed 0.05 mg/ℓ.[1,6] There is no evidence that chromium salts are essential or beneficial to human nutrition. When administered orally, chromium salts are not retained in the body but are rapidly excreted. Although the salts of trivalent chromium are not considered to be physiologically harmful, there is evidence that large doses of hexavalent chromates lead to ulcerations in the intestinal tract. On the other hand, there is evidence that a family on Long Island, N.Y., drank well-water for 3 years with varying degrees of chromium contamination without apparent effect.[5]

Cyanide (CN$^-$)[5]

Cyanide is a chemical compound, not an element. It is reported that most of the cyanide in water supplies is in the form of hydrogen cyanide (HCN) which is largely undissociated at pH values of 8 or less, and it is therefore the hydrogen cyanide rather than the cyanide which is the major toxic chemical.

Cyanides occur in the effluents from coal-gas works and coke ovens, from the scrubbing of gases in steel plants, from metal cleaning and electroplating processes, and from chemical industries. Hence they are found in river waters receiving industrial wastes. Cyanide in small doses (2.9 to 4.7 milligrams of cyanide per day)[14] is normally not lethal to adult humans as the liver is capable of detoxifying the chemical. However, death will occur with larger dosages when the detoxification processes of the liver are overwhelmed. Fish are quite sensitive to cyanide in the water. It is reported that concentrations of 1 mg/ℓ will kill trout in 20 minutes, but even small concentrations as low as 0.05 mg/ℓ have killed brook trout in less than 6 days.[5] The toxicity of hydrogen cyanide to fish is a function of stream temperature where a rise of 10°C will produce a two to three-fold increase in the lethal action.

Cyanides in waters are usually decomposed by bacteria. The percolation of wastewaters through soil columns rich in organic matter will materially reduce the cyanide content. The biological degradation process is temperature-sensitive and is most effective between 10–35°C, proceeding at a slower rate at temperatures both above and below these limits. Cyanides and their related compounds, the cyanates, can be oxidized to non-toxic chemicals by the action of chlorine.[15] Drinking water standards for cyanide (CN) will usually accept 0.01 mg/ℓ with a maximum permissible limit of 0.20 mg/ℓ. However, the odor threshold for hydrogen cyanide in water is 0.001 mg/ℓ, and the taste and odors may become accentuated by chlorination with the danger of inducing the operator to reduce chlorine dosage in order to reduce tastes and prevent the oxidization of the cyanides into innocuous chemicals.

Lead (Pb)

Ideally, lead in drinking water supplies should not be detectable. An acceptable limit is considered to be less than 0.05 mg/ℓ, and the maximum permissible limit should not exceed 0.05 mg/ℓ. Some natural waters contain lead in solution, as much as 0.04 to 0.8 mg/ℓ where mountain limestone and galena are present.[5] Well waters have been recorded as high as 15 mg/ℓ. Waters of low pH, hardness, and alkalinity have greater plumbo-solvency than harder waters which have more alkaline characteristics. Lime-softened waters are said to be safe with regard to lead pipes in home plumbing installations.[18] Most water systems now use copper, galvanized, or plastic pipe, and very little lead is used even in paints compared to previous formulations.

The acceptable level of lead the body can safely tolerate has not been established. It is known that lead in excess of the quantity which can be excreted is accumulative and, in the advanced stages of lead poisoning, results in constipation, anemia, abdominal pains, and gradual paralysis, particularly in the arm muscles. Lead can, of course, enter the body through food, air, and tobacco smoke as well as from water and other beverages.[5] In some cases the lead in natural waters is combined with other constituents which can be removed by flocculation, clarification, and filtration.

Nitrates and Nitrites (NO_3 and NO_2)

Nitrate is oxidized nitrite, but in drinking water standards they are grouped together under one heading and should be less than the maximum permissible limits of 10 mg/ℓ expressed as nitrogen (N) or 45 mg/ℓ expressed as nitrate (NO_3).

Nitrogen enters the water system from many sources, primarily from the decomposition of organic matter, sewage, surface water runoff, and use of nitrogenous fertilizers. In some areas well waters consistently contain 4 to 5 mg/ℓ as nitrogen (N).

In domestic water supplies, if the nitrate and nitrite contents are appreciably above the acceptable limits, a disease known as infant methemoglobinemia (blue babies) may occur; but not all infants are susceptible.

Nitrates are not normally a problem in surface water supplies where they rarely exceed 5 mg/ℓ and are usually less than 1 mg/ℓ. However, ground water supplies can contain from 0 to 1000 mg/ℓ as nitrogen (N).[14] Depending on the topography of the terrain, an increase in the nitrogen content of well waters is sometimes an indication of sewage pollution of the aquifer.

Selenium (Se)

Due to the possibility that the presence of selenium may result in cancer, the maximum in drinking waters is set at 0.01 mg/ℓ as (Se) but preferably should not be detectable. Proof of human injury by the ingestion of selenium is scanty. The symptoms of selenium poisoning are similar to those of arsenical poisoning. Selenium in trace amounts appears to be essential for the nutrition of humans and animals, although very little is known about the mechanisms of its actions.[5]

Selenium occurs in some soils and in organic compounds derived from decayed plant tissue. In some areas in South Dakota and Wyoming, soils may contain up to 30 mg of the element per kilogram of soil. It is believed that there is more selenium ingested in foods than from water supplies. The element is used in many manufacturing processes such as pigmentation of paints, dyes, and glass production, as a component of rectifiers, semi-conductors, and photoelectric cells, and can therefore be expected to be present in their industrial wastes.

Silver (Ag)

Silver is found in a number of ores, but since many silver salts are insoluble, the silver ions are not normally present in significant quantities in natural waters but should not exceed 0.05 mg/ℓ. It is reported that nine municipal water supplies in the U.S. have shown the presence of 0.05 mg/ℓ of silver ion, and at Denver a concentration of 0.2 mg/ℓ has been recorded.[5]

Silver oxides are used for the disinfection of water in the Katadyn process.[18] However it is because of the proposed use of silver as a disinfectant, which discolors the skin, that a standard limiting the amount of silver in water has been set.

PESTICIDES AND HERBICIDES

The pesticides and herbicides are usually extremely toxic organic chemicals used by fruit growers and farmers to control diseases common to orchards and crops. *The Canadian Drinking Water Standards and Objectives* (*1968*) specify the acceptable and permissible limits in Table 6-7. However, when the U.S. Safe Drinking Water Act comes into effect, the limits will probably be as per Table 6-8. One of the major problems with the determination of biocides in waters is the sampling procedure that must be adopted if reasonably accurate results are to be obtained. It will be noted that chlordane must not exceed 0.003 mg/ℓ which is the same as 3 parts per billion, equivalent to 1 gram in 333 cubic meters, or 1 lb in 40 million U.S. gallons. One system, presently in use, is to fill a surgically clean 5 gallon glass bottle to which about 700 milliliters of analytical grade hexane has been added. On arrival at the laboratory the sample is then concentrated to a very small quantity suitable for analysis in a Gas Chromatograph. It can be appreciated that an upper limit of chlordane in a 5 gallon bottle will only amount to 0.000,000,125 lb and a quantity of organic chemical as minute as this can easily be absorbed on the sides of an unclean bottle or bung. The problem is to obtain a truly representative sample and then analyze it.

With the use of degradable biocides the problems in water supply systems should not become too severe providing a monitoring program is maintained. It is necessary, however, that a water engineer be aware of this problem when designing intakes and waterworks in areas where pesticides and biocides may become a problem, and to orient the intake with the intention of avoiding the surface and storm water runoff as much as possible.

It is generally stated that waters should contain no measurable quantities of pesticides or herbicides but where they are used the concentration should not exceed 1 % of the median tolerance limit (T.lm.).[16] Reference should be made to the 13th Edition of *Standard Methods*[7] for procedures for conducting bioassays. *The Canadian Drinking Water Standards and Objectives for 1968* require the following limitations.

TABLE 6-7 Biocides and Pesticides Limitations in Drinking Water[1]

Biocide*	Objective and Acceptable Limits	Max. Permissible Limits—mg/ℓ
Aldrin	Not Detectable**	0.017
Chlordane	Not Detectable	0.003
DDT	Not Detectable	0.042
Dieldrin	Not Detectable	0.017
Endrin	Not Detectable	0.001
Heptachlor	Not Detectable	0.018
Heptachlor Epoxide	Not Detectable	0.018
Lindane	Not Detectable	0.056
Methoxychlor	Not Detectable	0.035
Organic Phosphates + Carbamates***	Not Detectable	0.100
Toxaphene	Not Detectable	0.005
Herbicides (e.g. 2,4-D, 2, 4, 5-T, 2, 4, 5-TP)	Not Detectable	0.100

* Conventional water treatment has little effect on these dissolved biocides.
** Not detectable by an acceptable method of analysis as approved by the control agency.
*** Expressed as parathion equivalents in cholinesterase inhibition.
NOTE: Maximum Permissible Limits are adopted from the Report of the National Technical Advisory Committee on Water Quality Criteria to the Secretary of the Interior, U.S. Department of the Interior, 1968. These limits are to be regarded as *tentative standards* since they are still being evaluated.

When the new Safe Drinking Water Act comes into force in the U.S., the maximum contamination levels for organic chemicals in community public water supplies will probably be as follows:

TABLE 6-8 Maximum Contaminant Levels for Organic Chemicals in Community Water Supply Systems[4]

		Level, Milligrams Per Liter
(a)	Chlorinated hydrocarbons:	
	Endrin (1,2,3,4,10, 10-hexachloro-6, 7-cpoxy-1,4, 4a,5,6,7,8,8a-octahydro-1,4-endo endo-5,8—dimethano naphthalene).	0.0002
	Lindane (1,2,3,4,5,6-hexachloro-cyclohexane, gamma isomer).	0.004
	Methoxychlor (1,1,1-Trichloro-2, 2—bis [p-methoxypheny] ethane).	0.1
	Toxaphene ($C_{10}H_{10}Cl_8$-Technical chlorinated camphene, 67–69 percent chlorine).	0.005
(b)	Chlorophenoxys:	
	2,4-D, (2,4-Dichlorophenoxyacetic acid).	0.1
	2,4,5-TP Silvex (2,4,5,-Trichloro-phenoxypropionic acid).	0.01

NON-TOXIC CHEMICALS

Aluminum (Al)

Statutory limits have not as yet been established for aluminum in water supplies. Aluminum sulfate is widely used for coagulation and clarification of water and aluminum metal is used for the manufacture of cooking utensils.

Aluminum pipes are used for the transmission of water since they have good resistance to corrosion, particularly for waters with pH values near 7. Their advantage over other materials is their high strength to weight ratio. However, particular precautions must be taken to avoid copper in any of the alloys preceding aluminum, since particles of copper will form anodic cells on the aluminum surfaces resulting in severe pitting.

Ammonia

In times past the term "albuminoid" ammonia was often mentioned in chemical reports referring to an analytical method devised by Professor Wanklyn for estimating the free and saline ammonia in water which is usually derived from decaying organic matter. This term is no longer used in North America because its significance is difficult to define.[5] Ammonia is strongly ionized in water particularly under acidic conditions; nevertheless, it is always reported as nitrogen (N) and is a measure of possible organic pollution. It should be limited to 0.05 mg/ℓ because of the possibility that it may indicate organic pollution particularly of shallow-well waters. Ammonia may be derived from decaying vegetation without affecting the sanitary purity of the water, and with no hygienic significance; however, an increase in the ammonia content above its normal background level is an indication that pollution may have occurred. Although several drinking water standards suggest an acceptable upper limit of 0.5 mg/ℓ, it is also suggested[5] that ammonia concentrations in excess of 0.10 mg/ℓ render the water suspect of recent pollution.

Whenever it is necessary to form conclusions concerning the safety of a water supply, it is very difficult to form objective opinions from one or more samples. If reliable sampling and analytical techniques are conscientiously observed over a period of years, the background information which this type of program provides is invaluable in assessing the viability of a water when abnormalities suddenly appear. When ammonia is present in water it will react with the chlorine applied to disinfect it. The end-products of the reaction between chlorine and ammonia are known as chloramines. A number of investigators have reported that the disinfecting properties of chloramine are only one-twentieth to one-thirtieth the power of free residual chlorine. On the other hand, once the chlorine is compounded with ammonia to form a chloramine, it is more stable, or less reactive, than free chlorine and may show a chloramine residual at the end of the distribution system. Free chlorine under similar circumstances may be absent long before the water reaches the end of the system due to absorption by the organic deposits in the distribution system.

In some water supply systems where the introduction of chlorine results in accentuated taste problems, ammonia gas or ammonium sulfate is purposely added to the water ahead of the chlorine injection, specifically for the purpose of forming chloramines which do not have the same taste-provoking problems of free chlorine. While every water supply operator must be sympathetic with a system that provokes a taste and odor problem on the introduction of chlorine, there is still some doubt whether or not this approach is valid since the chloramines are less able to produce the desired end-result as compared to free chlorine and also trichloramine (NCl_3) causes a pronounced taste. In these circumstances, it may be advisable to use chlorine dioxide, or to apply breakpoint chlorination, to oxidize the offending taste producer, dechlorinate to remove the unpredictable quantities of free chlorine left in the water, and then rechlorinate to the desired residual level before the water is pumped into the distribution system.

Calcium (Ca)

Calcium is essential to human nutrition. The body requires 0.7 to 2.0 grams of calcium per day which is far in excess of the amount that could possibly be ingested from drinking even very hard waters.

Calcium in domestic water supplies causes problems due to hardness and scale formation. For these reasons, the objective limit is less than 75 mg/ℓ. Very soft waters are pleasant waters to use, but medical evidence indicates that insufficient calcium in water can have adverse physiological reactions.[5]

Chlorides (Cl)

Chlorides in water supply up to about 250 mg/ℓ. Chlorides have no adverse physiological effects, but, similar to ammonia and nitrates, a sudden increase above their background level may indicate pollution. Depending on the geographical location, the source of the pollution could be seawater, sewage, or industrial wastes.

The tolerance of chlorides by humans varies with the climate, exertion, and the chlorides lost through perspiration, which must be replaced in either the diet or the drinking water. It has been reported that waters containing as much as 1000 mg/ℓ of chloride have been consumed without adverse effect. Chlorides impart a salty taste, in some instances with a concentration as low as 100 mg/ℓ, whereas in other waters, with different chemical composition, a chloride content as high as 700 mg/ℓ may go almost unnoticed by most people.[5] The taste of water high in chloride may not be unpleasant; in fact some people prefer waters of this type compared to flat and tasteless waters. There are some indications that chlorides are injurious to people suffering from diseases of the heart, or kidneys.

Copper (Cu)

Copper and aluminum are antagonistic to each other, and much damage is done to the aluminum and zinc coating of the galvanized steel components of a water supply system by minute traces of dissolved copper salts in water. Metallic copper is insoluble in water, but many of its salts are highly soluble. It is the latter compounds which generally cause the corrosion.

Copper is essential for nutrition in small dosages. Most of it is excreted and is not considered a cumulative poison like lead or mercury. Even workers in the copper industries, who have absorbed enough copper to turn their skin and hair green, have shown no evidence of copper poisoning.[5]

Oral dosages of 60 to 100 mg have caused symptoms of gastroenteritis with nausea and intestinal irritation. An outbreak of 18 cases of gastroenteritis occurred at a factory canteen in England owing to the contamination by copper of tea made with water from a corroded drinking fountain. The water was found to contain 44 mg/ℓ of copper.[5]

Copper sulfate is widely used in lakes and reservoirs to control the algae content. Although it can be toxic to fish if fed in excess, it has not been considered a public health hazard when used in reasonable dosages. The nontoxicity of the metal is reflected in the wide range of limits between 0.01 mg/ℓ as the objective, up to 1.0 mg/ℓ as the acceptable limit.

Iron (as Fe)

Totally dissolved iron is present in most natural waters and particularly in well waters. Like copper, it is not normally a public health problem, but it is a nuisance in a water supply for many reasons. Iron hydroxide stains porcelain fixtures and laundered clothes. The iron is a source of food for the so-called iron bacteria, *crenothrix*, *leptothrix* and *gallionella* with their associated slime growths, taste, and odor problems.

In ground waters the iron is usually in the form of ferrous bicarbonate ($Fe(HCO_3)_2$); ferrous sulphate ($FeSO_4$); or ferrous chloride ($FeCl$). On first exposure the waters are clear and bright. If there is sufficient iron to give an iron taste, the waters are often referred to as "chalybeate" meaning that they taste of iron or "ferruginous" meaning rust-colored. Once the waters have been exposed to oxygen, the bicarbonates break down to hydroxides and carbon dioxide in accordance with the following equation:

$$4Fe(HCO_3)_2 + O_2 + 2H_2O = 4Fe(OH)_3 + 8CO_2$$

The water which is at first clear and bright turns cloudy with a reddish tinge, and eventually the precipitated ferric hydroxide ($Fe(OH)_3$) settles to the bottom, hence the term ferruginous or rusty waters.

There is, however, another form of iron which is frequently encountered in water supplies and for want of a better name is frequently called "organic iron." These waters are usually from surface runoff sources which contain organic acids, commonly derived from decomposing vegetation or muskeg, and appear to chelate the iron and prevent it from being readily oxidized. They are usually highly-colored and in the process of removing the color, by means of flocculation and clarification, the iron is also removed. This type of water may contain 6 ppm or more iron which is not always detected by some analytical field kits unless the samples are strongly acidified before testing. Colored shallow-well waters frequently contain organic or chelated iron and should be carefully investigated since the normal treatment for the removal of ferrous bicarbonate by oxidation is totally unsuitable for the removal of "organic" or chelated iron.

The acceptable limit for iron is generally acknowledged to be 0.3 ppm for esthetic and taste considerations since the body needs 1 to 2 mg per day and the normal dietary intake is 7 to 35 mg per day. This could not possibly be supplied by the drinking water since water with even 1.0 mg/ℓ of iron would be unpalatable and unesthetic in appearance.

It is not always advisable to condemn a water supply on the basis of one or more chemical analyses without a careful field investigation, since many well waters have shown high iron contents above the normal acceptable limits and, for some reason or other, have not caused any problems either from rust stains, slimes, or tastes. Iron can also be picked up from cast iron or steel pipes. Adjustment of pH and alkalinity will usually cure this problem, or, alternatively, the use of phosphates or silicates such as hexameta-phosphate, tri-sodium phosphate, or activated silica in very small dosages will gradually coat the inside surfaces of the pipes and prevent further attack, thus reducing the iron pick-up in water suplies to customers.

Precipitated iron hydroxide will often settle out in the distribution mains and will periodically cause considerable taste and odor problems in an otherwise perfectly satisfactory water supply system. Each case must therefore be examined and corrected according to the circumstances in which it occurs.

Magnesium (Mg)

Magnesium is a co-partner with calcium in causing hardness in water supplies. Magnesium is essential to nutrition and is not considered to be a public health hazard because the taste becomes unpalatable long before toxic concentrations are reached. The hydrate of magnesium sulfate ($MgSO_4 7H_2O$) is known as "Epsom Salts" and is used extensively as a bowel purgative. Magnesium is precipitated as the hydroxide at pH values of 9.6 to 10.4 and forms a gelatinous precipitate, which is sometimes difficult to settle in conventional treatment processes, resulting in shortened filter runs. The objective and recommended acceptable limits are less than 50 and 150 mg/ℓ, respectively, but apparently the upper limitations are for esthetic reasons rather than for health reasons.

Manganese (as Mn)

Manganese often appears in water supplies together with iron. Since both elements are necessary for nutrition, they are limited for esthetic and economic considerations rather than physiological hazards. Manganese, similar to iron, is usually present in well waters as manganous bicarbonate ($Mn(HCO_3)_2$), manganous chloride ($MnCl_2$), or manganous sulphate ($MnSO_4$). It is also found in stagnant areas at the bottom of many reservoirs where anaerobic conditions prevail due to the decomposition of vegetable and organic matter. Flentje[17] reports that manganese was present in 22 reservoirs in the United States and shows that the manganese content of the bottom samples ranged from less than 0.2 mg/ℓ in the winter months to an excess of 2.0 mg/ℓ in the summer. The samples taken from near the surface were normally less than 0.1 mg/ℓ. Similar to ferrous bicarbonate, waters containing manganous bicarbonate when first taken from a well are clear and bright but do not readily oxidize to the insoluble manganese without the addition of an oxidizing agent such as chlorine, chlorinedioxide, or potassium permanganate. Increasing the pH value of water between 9 to 10 will also cause the manganese to precipitate in the insoluble form. This often occurs with well waters containing manganese. When they are pumped to the surface the reduced pressure liberates the free carbon dioxide that holds the manganese in solution, and causes the pH of the water to increase. If the water is carried in concrete or cement mortar-lined pipes, the absorption of alkali from the mortar will often cause the pH to increase and the manganese to precipitate. When a hydrant is opened and the mains are flushed, the manganese deposits are disturbed and complaints of black stains in laundry, discolored tea and whiskey, are frequently received. Manganese in concentrations as low as 0.1 mg/ℓ is reported to cause laundry stains, and in excess of 0.16 mg/ℓ it also causes turbidity. The reactions of a large panel of tasters indicated that the threshold taste levels varied widely depending on the manganese compounds present and the type of water, whether spring or distilled.[5] For domestic waters, the acceptable limit has been set at 0.05 mg/ℓ with the objective less than 0.01 mg/ℓ.

Methylene Blue Active Substance (MBAS)

This substance is a means of measuring and reporting the quantity of synthetic detergents present in raw water. An acceptable limit of 0.5 mg/ℓ with an objective of less than 0.2 mg/ℓ has been established as suitable since the presence of detergents is indicative of wastewater pollution.

Phenolic Substances (as Phenol C_6H_5OH)

Phenols are widely used in the manufacture of disinfectants, synthetic resins, medical and industrial chemicals, and are often present in industrial and petroleum refinery waste-waters. Drinking waters are limited to 0.002 mg/ℓ or 2 parts per billion on the basis that phenol and chlorine react—producing chlorinated phenols,[5] and a very strong medicinal odor.

Phosphates (as PO_4)

Phosphates are used extensively as fertilizers and are present in natural waters from surface runoff. They are also used in water treatment and as builder compounds in some synthetic detergents, contributing to the problems of accelerated eutrophication. Many sewage treatment plants are designed to remove phosphate. The limit for phosphates is 0.2 mg/ℓ in domestic raw water supplies to minimize coagulating difficulties.

Total Dissolved Solids (T.D.S.)

Total dissolved solids are generally recommended to be less than 500 mg/ℓ with an acceptable upper limit of 1000 mg/ℓ. However, in the Canadian prairies where only highly-mineralized waters are available, a value of 1500 mg/ℓ has been established as an acceptable criteria.[16] The total dissolved solids refer only to solids in solution or, in other words, the solids remaining in the filtrate after all the suspended solids have been removed on the filter. One frequently used approximate method of determining the dissolved solids content is to measure the specific conductivity of the water. Such measurements indicate the capacity of a sample to conduct an electrical current, which in turn is related to the concentration of ionized substances in the water. Most dissolved inorganic substances in water supplies are in the ionized form and so contribute to the specific conductance. A notable exception is silica which may be as high as 60 mg/ℓ. Although this measurement is affected by the nature of the various ions, their relative concentrations, and the ionic strength of the water, it can give a practical estimate of the variations in dissolved mineral content of a given water supply. For each particular water there is a relationship between the total dissolved solids and the specific conductivity by an empirical factor varying from approximately 0.55 to 0.9.[11]

In addition to the total dissolved solids, the suspended settleable and volatile solids are also important particularly in raw or natural waters prior to treatment.

Total dissolved solids are usually determined by weighing the residue that remains after evaporation and drying at 103 to 105°C.[7] Even so, substances with appreciable vapor pressure below 103°C will have been lost during evaporation, and it is difficult to accurately determine the T.D.S. unless the sample is evaporated at low temperatures under vacuum. Loss on ignition to 600°C is a method of determining the volatile and the fixed solids in the sample, as 600°C is about the lowest temperature at which organic matter, particularly carbon residues resulting from the pyrolysis of carbohydrates, can be oxidized at reasonable speed and at this temperature the decomposition of inorganic salts is minimized. Any ammonium compounds not released during drying are volatilized, but most other inorganic salts are relatively stable, with the exception of magnesium carbonate ($MgCO_3$) which breaks down into magnesium oxide (MgO) and carbon dioxide (CO_2) at 350°C.

Organic Chemicals

Organic chemicals are more complex and varied in composition and characteristics than inorganic chemicals. They may include petroleum or other industrial wastes, decaying plants, and animals.

Most of the taste and odor problems in water supplies originate from the organics and most of them have an affinity for chlorine. A sudden influx of organic material into the intake of a water supply system can play havoc with the tastes and odor problems and may also destroy the free residual chlorine, leaving the water inadequately disinfected. Since most of the organic chemicals are extremely complex in structure and widely varied in chemical composition, they are most difficult to identify and quantitatively determine. However, Gas Chromatography has eliminated many difficult and tedious analytical procedures and considerably reduced the time necessary for testing. In water chemistry, apart from an occasional heavy contamination of organic compounds demanding identification, it is not usually necessary to specifically identify the compound. The organics are frequently measured by the amount of oxygen required to oxidize them, for example the B.O.D. (Biochemical Oxygen Demand) and the C.O.D. (Chemical Oxygen Demand). A third method known as T.O.C. (Total Organic Carbon) also indicates the magnitude of the organics present. For more detailed descriptions of these parameters, the reader is referred to Sawyer and McCarty,[11] and *Standard Methods*.[7]

Organic compounds are usually absorbed with activated carbon either by applying powdered carbon onto the top of the filters, if the problem is infrequent or seasonal, or in separate granulated carbon filters if the problems are continual.

Sulfates (SO_4)[5]

Sulfates occur in natural waters as a result of leaching from gypsum and other common minerals. They may also occur as the final oxidized stage of sulfides, sulfites, and thiosulfates, e.g., iron pyrites (FeS) may be leached from abandoned coal mines. They may also be derived from the oxidized state or organic matter in the sulfur cycle. Industrial wastes from pulp mills, tanneries, and chemical plants may contain appreciable quantities of inorganic and organic sulfur compounds.

For domestic supplies it is usually recommended that the sulfates do not exceed 250 mg/ℓ, but many water supplies in the area of alkali lakes frequently exceed this with no apparent health hazards. People not accustomed to drinking waters high in sulfates will probably notice a marked laxative effect but otherwise no apparent taste or other physiological reactions. In many parts of the world the sulfate content exceeds 1000 mg/ℓ, and some are even in excess of 3000 mg/ℓ.[5]

Sulfates can only be removed by ion exchange or by electrodialysis, reverse osmosis, or distillation.

Although sulfates are not normally considered a serious problem in most water supplies, they can nevertheless be troublesome if sulfate-reducing bacteria are present. These microorganisms known as De-sulfovibrio are capable of reducing the sulfate radical to hydrogen sulfide (H_2S) with the accompanying rotten-egg odors. The sulfate reducers are strictly anaerobes and may be found in large numbers where sulfates and other conditions associated with their habitat requirements suitable for their propagation are present. They have been known to grow profusely in the dead ends of distribution systems, in filter beds, and in carbon filter columns. Chlorination and flushing will often inactivate the organisms even in activated carbon filters.

Sulfides (SO_3)

Sulfides are present in many industrial wastes, but they are also generated in sewage and in some natural waters, particularly some ground waters, by the anaerobic decomposition of organic matter. Fortunately, the unpleasant taste and odor which accompany sulfides in water would prevent persons or animals from consuming a harmful dosage. The thresholds of taste and smell are reported to be 0.2 mg/ℓ of sulfides in pulp-mill wastes. For many industrial processes, sulfides are often detrimental but do not appear to cause serious consequences in irrigation waters. The sulfides dissociate into soluble ions which in turn react with the hydrogen ions to form HS^- or H_2S, the proportion of each depending upon the resulting pH value. The degree of toxicity appears to depend largely on the pH value and the predominant form of dissociation, increasing in toxicity to fish as the pH value decreases.

Uranyl Ion (UO_2^{--})

Uranium and its many salts may be capable of damaging kidneys.[1] The chemical may also produce objectionable taste and color in water. Fortunately, the threshold level for taste is approximately 10 mg/ℓ as UO_2 which is much less than the safe limit of ingestion from a physiological viewpoint. *The Canadian Drinking Water Standards*[1] have set the maximum permissible limit of 5.0 mg/ℓ as UO_2 based on color and taste considerations.

Zinc (Zn)[5,14]

Most health authorities limit the zinc content of drinking waters to 5.0 mg/ℓ with less than 1 mg/ℓ as the desirable objective. Zinc is an essential nutrient element beneficial to both human and animal metabolisms, and water containing as much as 40 mg/ℓ was reportedly consumed by 200 people without harmful effects. However, in other instances, 30 mg/ℓ caused nausea and fainting. Zinc in association with other elements and radicals such as the highly soluble chlorides and sulfates, will have a different threshold taste parameter than when it is associated with other elements or radicals such as nitrates. The acceptable limit of 5 mg/ℓ has been prescribed largely because of taste problems rather than health hazards since the water becomes almost unpalatable long before the concentrations become dangerous to public health. However, it is reported[5] that zinc-bearing waters should not be used in acid drinks like lemonade because zinc nitrate and organic zinc compounds can be poisonous.

Zinc is a common metal used in many industrial processes and therefore appears in many industrial waste effluents. When zinc is in the presence of copper the two metals appear to have a synergistic effect. Doudorff[5] observed that testfish in soft water could tolerate a concentration of 8 mg/ℓ of zinc alone for 8 hours; however, most of the fish died within 8 hours when exposed to a solution containing only 1 mg/ℓ of zinc plus 0.025 mg/ℓ of copper.

The removal of zinc from water supplies is relatively difficult except by ion exchange if it is associated with the soluble chlorides and sulfates; but on the other hand, if they are in the insoluble form of zinc carbonate, zinc oxide and zinc sulfide, these compounds are relatively insoluble and will precipitate out of solution in most natural waters. One method of keeping the concentration of zinc below 1 mg/ℓ is to keep the pH above 8.0 before the water is filtered.

REFERENCES

1. "Canadian Drinking Water Standards and Objectives (1968)." Publication of the Department of National Health and Welfare, Canada. Catalogue No. H48-1069. Printed by Queen's Printer for Canada, Ottawa, 1969.

2. Public Health Service, "Drinking Water Standards 1962." Published by U.S. Department of Health, Education and Welfare, Washington, 25 D.C.

3. *Water Quality and Treatment*, 3rd Ed. Published by the American Water Works Association Inc. New York: McGraw-Hill Book Company, 1971.

4. EPA Report—"National Interim Primary Drinking Water Regulations." *J.A.W.W.A.*, Vol. 68, p. 57, February 1976.

5. McKee, J. E. and H. W. Wolf, "Water Quality Criteria," 2nd Ed. Published by the Resources Agency of California, State Water Quality Control Board, Sacramento, California. Publication No. 3-A, 1963.

6. Reid, F. and J. H. McDermott, *Drinking Water Quality Regulations, Measuring, Monitoring and Managing*. Water and Sewage Works, Vol. 123 (6), p. 47. June 1976.

7. *Standard Methods for the Examination of Water and Wastewater*, 13th Ed. Published jointly by A.W.W.A., American Public Health Association, and Water Pollution Control Federation, 1971. Available from the Publications Sales Department. American Water Works Association, 2 Park Avenue, New York, N.Y.

8. Packham, R. F., "Studies of Organic Color in Natural Water." Published in Proceedings of the Society of Water Treatment and Examination, Vol. 13 (4), p. 316, 1974.

9. Ungar, J. and J. F. J. Thomas. "Further Studies on the Measurement of Organic (Coloring) Matter in Natural Waters." Published by Department of Mines and Technical Surveys, Ottawa. Technical Bulletin. TB39. 1962.

10. Cox, C. R. *Operation and Control of Water Treatment Processes*. Published by World Health Organization, Geneva. 1969.

11. Sawyer, C. N. and P. L. McCarty. *Chemistry for Sanitary Engineers*, 2nd Ed. New York: McGraw-Hill Publishing Company, 1967.

12. Behrman, A. S. *Water is Everybody's Business*. Anchor Books, New York: Doubleday & Company Inc., 1968.

13. *Handbook of Chemistry & Physics*, 51st Ed. Edited by R. C. Weast. Published by the Chemical Rubber Co.—Cleveland, Ohio, 1970.

14. Miller, A. P. "Water and Man's Health," published by the Office of Human Resources and Social Development Agency for International Development Washington, D.C. U.S.A. 1962. Technical Series No. 5.

15. White, G. C. *Handbook of Chlorination*. New York: Van Nostrand Reinhold Company, 1972.

16. "Water Quality Criteria." Published by the Saskatchewan Water Resources Commission, Regina, Saskatchewan, 1970.

17. Flentje, M. E. "How is Your Manganese?" Published in Water Works Engineering, Vol. 113, p. 228, April, 1960.

18. Taylor, E. W. *The Examination of Waters and Water Supplies*, 7th Ed. London: J. & A. Churchill Ltd., 1958.

7

RADIONUCLIDES

Surface and ground waters may acquire a small amount of radioactivity from the rocks and minerals in which they have been in contact. This is known as "background level" radiation. Industrial wastes may contain small quantities of radioactive materials, since radioactive chemicals are extensively used in the X-ray examination of welds and the structural soundness of materials. Small quantities of radioactive materials are used in medicine and in the watch industry. Atomic power plants use tonnage quantities of uranium to generate heat.

SOURCES OF RADIOACTIVITY

When seeking a new water supply where radioactive mineral deposits or industrial wastes may exist, it is advisable to determine the background level or radioactivity in the water if at all possible. Details for collecting samples and determining radiation are described in *Standard Methods*.[1]

The subject is somewhat complex since there are many variables involved. However, in the water supply industry, it is generally accepted[2] that water with a gross radioactivity of less than 10 picocuries per liter should be the objective, with a maximum permissible limit of less than 100 picocuries per liter.

Over 40 kinds of atoms are known to display the property of natural radioactivity and most have atomic weights greater than 200.[3] The heavy metal radioactive elements are uranium, thorium, and actinium. The uranium series has U^{238} as its parent substance and, after 14 successive transformations have occurred, the end product is Pb^{206}. Thorium (Th^{232}), after 10 transformations, remains relatively stable as Pb.[208]. The parent element of the actinium series is U[235] and, after 11 transformations, it remains as Pb.[207]. As the transformation proceeds the atomic weight of 238 is reduced to lead (Pb.[206]) with an atomic weight of 206. Ordinary lead has an atomic weight of 207.19.

The loss of atomic weight is due to the emission of alpha and beta and gamma rays. The alpha and beta rays have mass but the gamma rays are electro magnetic and have no mass. With Einstein's energy-mass equivalence formula

$$E = MC^2$$

it is possible for physicists to evaluate the radiation-energy. E is a unit expressed in g-cm/sec, or ergs; M = mass of the particle in grams; and C is the velocity of light (2.998×10^{10} cm/sec). Refer to Sawyer and McCarty for a further review of this subject.[3]

UNITS OF RADIOACTIVITY

In 1898, Pierre and Marie Curie concluded that the radiation emitted from uranium was an atomic phenomenon characteristic of the element, and they introduced the name radioactivity. The unit of measurement is known as the curie, named after the discoverers. The International Radium Standard Commission has recommended the use of a fixed value of 3.7×10^{10} or 37,000,000,000 disintegrations per second, which is known as a *Standard Curie* (c).

Some radioactive materials have a high rate of decay (short half-life) while others decay slowly (long half-life). Therefore, the mass of radioactive material which corresponds to an activity of 1 curie ranges from a fraction of a milligram up to many tons. It is therefore the activity of the element rather than the mass which is more important in water chemistry when assessing any possible hazard. The curie represents a large number of disintegrations per second and as a result small units have been defined (see Table 7-1).

TABLE 7-1 Units of Radioactivity

Unit	Magnitude	Disintegrations per Second
Curie	1 curie	3.7×10^{10}
Milli-curie	10^{-3} curie	3.7×10^7
Micro-curie	10^{-6} curie	3.7×10^4
Nano-curie	10^{-9} curie	3.7×10
Pico-curie	10^{-12} curie	3.7×10^{-2}

The proposed *objective limit* for water supplies of 0.1 ICRP* unit for constant 168 hour-per-week exposure[2] of gross radiation is less than 10 pico-curies per liter. This is equivalent to less than 3.7×10^{-1} or 0.37 disintegrations per

*ICRP—International Commission on Radiological Protection.

second per liter of water. It is also recommended[2] that levels of radiation in excess of this limit require further investigation to determine the nature of the activity and the desirability of systematic surveillance.

Considerable research has been done in recent years on the biological effects of radioactivity. The characteristics of the environment have considerable influence on the permissible levels for a wholesome water supply as well as on other physical and chemical characteristics of the water. Therefore, whenever a water supply is discovered to have a gross radioactivity in excess of the acceptable limit of 1/3 of the ICRP unit, or approximately 30 pico-curies per liter, an expert opinion as to the suitability of the water should be sought before developing the project. Not only is the problem of radioactivity of importance to the municipal or domestic water supply industry, it is also important in industrial water supplies; for example, a plant producing or developing photographic film has a very low tolerance to radioactive radiation.

FALLOUT FROM ATOMIC EXPLOSIONS

Many practical articles have been published on the possible problems and precautions to be taken in the event of nuclear warfare. The United States Department of Defense, Office of Civil Defense, published "Civil Defense Aspects of Waterworks Operations" (FG-F3.6, June 1966)[4] which describes in considerable detail the precautions to be taken in the event of nuclear attack. In Ottawa, Canada, the Department of National Health and Welfare published "Emergency Health Services—Control of Radioactive Fallout in Water Systems—A Manual for Water Engineers."[5]

WATER TREATMENT PROCESSES

The effectiveness of various treatment processes in removing radioactive material has been estimated. The ranges covered are generally as shown in Table 7-2.

TABLE 7-2 Removal of Radioactivity by Water Treatment Processes[6]

Process	Fraction of Radioactivity Remaining After Treatment (in percentages)	
	Soluble	Suspended
Alum coagulation and sand filter	15 to 75%	10 to 50%
Coagulation with iron salts sand filter	15 to 75%	10 to 50%
Lime-soda softening	10 to 70%	2 to 50%
Retreatment with clay slurries, coagulation and filter	20 to 50%	5 to 50%
Post-treatment ion exchange	less than 0.1 to 1%	less than 0.1 to 1%

Radioactive fallout particles will, in most cases, gradually settle to the bottom of a sedimentation tank. The reduction in radioactivity at the 10-foot depth is reported to be reduced as shown in Table 7-3.[6] This indicates one advantage of having intake gates at various levels in a reservoir. However, no reservoir is ever close to the ideal settling chamber, and there are many variables due to winds, thermal currents, algae, and other microorganisms which will tend to upset the ideal settling pattern.

TABLE 7-3 Radioactive Removal by Settling[6]

	Remaining Radioactivity from Initial Level
After 12 hours settling	20.0%
After 24 hours settling	11.0%
After 2 days settling	5.5%
After 3 days settling	3.5%
After 4 days settling	2.5%
After 5 days settling	2.0%
After 6 days settling	1.6%
After 7 days settling	1.4%
After 8 days settling	1.2%
After 9 days settling	1.0%
After 10 days settling	less than 1%

Whenever possible, only ground waters should be used immediately following a serious deposition of fallout until some initial settling and radioactive decay in the surface waters has occurred.

MONITORING MUNICIPAL WATER SUPPLY SYSTEMS

When the new USA Safe Drinking Water Act is enforced in 1977 all community water supply systems will be required to monitor radioactivity as follows:[7]

Community system must monitor for gross alpha within 3 years of June, 1977. If gross alpha does not exceed 5pCi/ℓ, monitor every 4 years or as approved by the state.

Surface supplies serving 100,000 population must monitor for gross beta, Sr-90 & H-3 within 2 years of June, 1977. If gross beta does not exceed 50 pCi/ℓ, Sr-90 ℓ 8 pCi/ℓ, H-3 20,000 pCi/ℓ and annual dose equivalent 20,000 pCi/ℓ does not exceed 4 millirem, monitor every 4 years.

Any community, as designated by the state, downstream from a nuclear facility must, within 2 years of June, 1977, begin quarterly monitoring for gross beta and I-131, and annual monitoring for Sr-90 and H-3. Non-community—as required by the state.

REFERENCES

1. *Standard Methods for the Examination of Water and Wastewater*. 13th Ed. Published jointly by A.W.W.A., American Public Health Association, and Water Pollution Control Federation, 1971. Available from the Publications Sales Department. American Water Works Association, 2 Park Avenue, New York, N.Y.

2. "Canadian Drinking Water Standards and Objectives (1968)." Publication of the Department of National Health and Welfare, Canada. Catalogue No. H48-1069. Printed by Queen's Printer for Canada, Ottawa, 1969.

3. SAWYER, C. N. and P. L. MCCARTY. *Chemistry for Sanitary Engineers*, 2nd Ed. New York: McGraw-Hill Publishing Company, 1967.

4. "Civil Defense Aspects of Waterworks Operations." FG-F 3.6. June 1966. Department of Defense, Office of Civil Defense, Washington, D.C. Printed by the U.S. Government Printing Office.

5. "Control of Radioactive Fallout in Water Systems—A Manual for Water Engineers." Published by the Emergency Health Services Department of National Health and Welfare. Government of Canada, Ottawa, 1965.

6. WHITE, S. N. "The Water Engineer and Radioactive Fallout." A paper presented to the Canadian Section American Waterworks Association. April 2, 1962.

7. REID, F. and J. H. MCDERMOTT, "Drinking Water Quality Regulations, Measuring, Monitoring and Managing" *Water and Sewage Works*, Vol. 123 (6), p. 47, June 1976.

8

WATER RESOURCES

HYDROLOGY All fresh water on the earth's surface has been precipitated, from snow, hail, rain, or mists, previously evaporated from the oceans and carried over the land by winds and air currents. It drains into creeks, rivers, and underground streams, and eventually back into the oceans from whence it came. This phenomenon, essential to all life on earth, is known as the hydrological cycle.

Rainfall irregularities, runoff diversifications, temperature variations, forestation, grasslands, land development, all affect the rate of available water from surface runoff.

WATER SHEDS All land masses, whether they be vast continents or small islands, consist, in the water engineer's eyes, as a series of water sheds. A water shed is an area contained by high ground which forms a divide between one catchment area and another. In water supply engineering this contained area, called a "water shed" sheds its water into a stream. In river control engineering, it would be called a drainage basin or catchment area, and is the land enclosed by ridges of mountains or higher ground, and water that falls or flows on to it would find its way into the stream, river, or lake on the valley floor.

The first thing to do in choosing a site for a large water supply project is to map the confines of the water shed on as large a scale as possible and estimate its area. Topographical maps showing 5 to 10-foot contours are necessary for the preliminary investigation. Closer contours at say 0.5 meters or 2-foot intervals, particularly at the higher and lower levels, will be required if it is necessary to construct dams to form an impoundment. Excellent contour maps can be prepared from aerial photogrammetry and this service is available in most countries.

Having obtained a good view of the contour of the water shed, the next problem is to determine the volume of water that will be precipitated annually. Few water sheds have adequate precipitation and creek runoff data to enable an accurate determination of the anticipated yield and possible ground water recharge. There are exceptions however, and for example, the Dallas, Oregon, water shed has 32 sampling stations strategically located within its area. Hunting is only permitted on the water shed if the gauging stations indicate a significant depletion of the undergrowth which would reduce the ability of the land to retain surface water and retard the runoff. Trees are only removed if they show signs of disease but this amounts to less than 5% per annum. The use of herbicides is not permitted unless authorized by the U.S. Department of Health, Education and Welfare. All chemicals used on the water shed must have prior approval. In general they prefer to use chemicals or mechanical equipment to control situations on the water shed rather than use people. Non-toxic fire retardant paints are also used to prevent forest fires.

Water shed management is a recent innovation, but it is relatively nonexistent in countries that need it most. In order to manage and control a water shed there must be rain and stream flow gauging stations in order to estimate its probable yield. As Lord Kelvin (1824–1907) once remarked "When you can measure what you are speaking about and express it in numbers, you know something about it, but when you cannot measure it in numbers, your knowledge is of a meager and unsatisfactory kind."

This comment is particularly relevant to water shed data. The annual data is required for many years before it is possible to know what can be anticipated during dry years when there is little precipitation or in wet years when there is too much. The rate and the duration of the runoff period is equally important before dams and spillways can be designed.

Water works managers responsible for large communities should install adequate gauging stations in order to determine dependable yield and flash flood characteristics. When the time comes to extend the water system, it is necessary to know whether or not the existing water sheds can satisfy the needs without obtaining further resources.

Water shed control is a long term commitment, and large water undertakings should have their aims and objectives clearly defined for at least 20 to 30 years into the future in order to ensure the reliability of their supply. Satellite telecommunications have made possible many important advances and their use can substantially increase the amount of data that can be economically obtained from a water shed area.

Stream gauging stations, rain gauges, and temperatures can be recorded and beamed to the overhead satellites, which in turn transmit the data to one of the central receiving stations such as Washington, D.C., or Ottawa, Ontario, for teletyping to any community on the North American continent. The more information that can be obtained over as long a period as possible, the more detailed and accurate will be the yield forecast. If a dam, spillway, and pipeline have to be constructed at a later date, complete hydrological data may substantially reduce the cost of future construction.

CHOICE OF WATER SUPPLIES

Smaller communities may be fortunate to have a choice of raw water supplies from wells, rivers, or lakes. An evaluation of all possible alternatives and a conceptual design report should be made before detailed engineering design commences, with reference to yield, reliability, water quality, treatment, piping, and pumping costs. The topography of the area with respect to access roads, rail, power and effluent disposal, should be studied. The cost of acquiring the necessary land and pipeline right-of-way is becoming increasingly more expensive. The municipality should have as many available alternatives as possible. Last but not least, determine the present legal status of the proposed water supply with reference to existing water licensing and riparian rights. Lakes and streams in Canada are usually controlled by the Provincial Water Rights Offices which issue licenses to municipalities, companies, and private individuals wishing to abstract water from the lakes or rivers. There are no licensing rights on ground water in Western Canada at the present time.

Water licenses are issued to prevent the resource from being over-abstracted and depleted. The licenses show the date and the annual quantity that can be taken out of the river or lake. It is essential to evaluate the status of the previous licensees together with the dates of their licenses and the quantities that they are permitted to abstract. A prolonged drought, with low river flows, will give the earliest licensees first preference. It could happen that later licensees would be prevented from taking the water.

Riparian rights were first formulated under French civil law and is law in both Canada and the U.S. The earliest concept gave the legal right to the owner of the land abutting on a stream or other natural body of water, the use of such water. It allowed each riparian owner, "to require the waters of a stream to reach his land undiminished in quantity and unaffected in quality," except for minor domestic uses. This interpretation has been annulled or abolished by repeal in a number of the western United States, and greatly modified in others. In general, each riparian owner is allowed to make a reasonable use of the water on his riparian land, the extent of such use being governed by the reasonable needs and requirements of other riparian owners and the quantity of water available.

This prevents a manufacturer or a municipality from buying a few acres of land on a stream or river bank for the purpose of building a water intake without other prior agreements.

There is also the question of plant effluents and in many cases they are much bigger headaches to the owners and their consulting engineers than the treatment of the water itself. Until recently, a water treatment plant could be located on the banks of a river, abstract the water it required, and return the clarifier sludge and filter backwash water back into the stream, usually a few feet downstream from the intake. This was considered reasonable practice on the grounds that the plant only returned into the stream dirt it took out of it in the first place. However this was not entirely true, as in many cases alum and other chemicals were added to the water in order to clarify the suspended matter, remove the color, and most of these added chemicals are discharged to the river.

It may now be necessary to treat the plant effluents and return only clear water to the stream from where it came. In some cases the cost of treating the effluent in order that it can be poured back into the stream will be more expensive than treating the water in the first case.

With all the verifications of legal requirements, licensing, capital, and operating costs, a detailed feasibility and conceptual design report is necessary in order to evaluate the possible alternatives before land acquisition and detailed engineering design should be authorized. If there are several alternatives, such as ground water, lake water, or river water, each scheme should be studied and cost-estimated for all locations, and for as many options in each category as possible. The following notes should be helptul in preparing such an evaluation.

GROUND WATER

Consult a good ground water geologist who is familiar with the area at the beginning of the investigation. He will obtain photographs from an aerial photograph library and study the surrounding rock formations to determine where gravel and sand layers may exist. Existing well-logs of wells drilled in the area are invaluable in determining the underlying subsurface material. If no previous wells have been drilled, it will probably be necessary to drill a number of 2 in. diameter test holes at various locations in order to determine what is beneath the surface. The geologist may also use geophysical methods of resistivity and electric logging to determine the characteristics of the subsurface geology. From this preliminary survey the possibilities of developing a ground water scheme can be assessed. A ground water geologist, from a study of the area, is usually able to indicate unfavorable areas where ground water is not likely to be found, and will indicate the most likely areas to drill the test holes. However he cannot assess the proposed scheme until the holes have been drilled, the core samples of the subsurface material analyzed, and a well-driller's log compiled which will show the depth of the various layers of soils, silts, clays, sand gravels, and rocks. If the underlying strata reveal the presence of an underground aquifer and the water is of suitable quality, the next step is to determine the extent of the aquifer and its mode of recharge. This will mean drilling a number of test holes around the proposed site of the well.

The extent of the preliminary investigation necessary to fully evaluate the capacity of the aquifer will be influenced by the requirements of the proposed scheme. An aquifer of good porous sands and gravels, several feet thick, with good recharge facilities, should be adequate for a small system of say, 100 to 150 U.S. gpm. (0.378 to 0.568 cubic meters per minute). If the project calls for a yield of 10 to 15 mgd (U.S.) (37,850 to 56,780 cubic meters per day) with an estimated capital expenditure of several millions of dollars, it is necessary to make as full an investigation as possible before embarking on a construction program. The extent of the aquifer must be determined together with its rate of recharge. It will also be necessary to drill one large diameter hole complete with screen and casing, together with a number of small observation wells, and pump-test the aquifer to determine the slope and characteristics of the "cone of influence." This will provide the draw-down data used to estimate the capacity of the aquifer. Having pumped water at a constant rate for several days and determined the static water levels under these conditions, the pump should be stopped. The rate at which the cone of influence returns to the static level of the water table provides much information about the characteristics of the aquifer. During the pump tests, care must be taken to pipe the pumped water into a creek, downstream of any possible recharge location, otherwise you may simply be recycling the water within the confines of the aquifer.

Having completed the ground water investigation, be careful to cap off all wells including the observation wells. Apart from the possibility of people dropping stones and other objects into the well, the owner is liable for any third-party damages that may result.

"If a man digs a well and doesn't cover it, and an ox or a donkey falls into it, the owner of the well shall pay full damages to the owner of the animal, and the dead animal shall belong to him." (EXODUS 21: 33–34.) This goes back nearly 3500 years to the time of Moses.

GROUND WATER RECHARGE

Much work has been done in recent years toward the understanding of ground water recharge. The American Society of Civil Engineers has published a number of reports on this subject. In the last 20 years, techniques have developed to claim some of the mountain runoff water for ground water recharge in parts of Africa, California and other locations where the entire year's rainfall, and surface runoff, occurs over a very short period of time, resulting in flash floods and loss of much of the water into the sea.

WATER DIVINING

Water divining, also known as dowsing or water-witching, has been known for centuries. A water diviner holds a vee-shaped hazel, willow, or hickory branch in both hands, close to the horizontal position. For some unknown reason, the pointed end of the branch exerts a strong tendency to pull down toward the ground when the operator walks over certain areas of the land. There is no doubt that this phenomenon does take place, and the force with which it does it can be quite substantial. Whatever this phenomenon is, it has not yet been defined in scientific terms. The fact that dowsing seems to indicate ground water in some instances and not in others is sufficient reason to doubt its validity.

Nordel[1] has some scathing comments about water diviners, but the subject has no application in engineering. A willow stick may find a source of ground water, but it has no means of determining the extent and depth of the aquifer, its porosity, and ability to recharge. Ground water exploration is expensive and frequently unsuccessful. If the art of water divining was as successful as some of its followers claim, then every ground water geological and well drilling company would employ them.

LOCATION OF WATER WELLS

The location of water wells with respect to septic tanks and disposal fields is obviously critical. Health regulations normally call for a minimum of 50 feet (15.24 meters) horizontal distance from the nearest septic tank to the well and a minimum of 100 feet (30.48 meters) from the nearest tiled disposal field to the well. It is obvious that the safe distance will depend on many other parameters as well as distance. This regulation primarily governs the installation of small private well systems for summer cottage or other occasional use. One hundred feet is insufficient for a large collection system where the cone of influence may be several thousand feet in diameter. Other factors which influence the minimum safe distance are the elevations of the aquifer, disposal field, source of recharge, and whether the well is "shallow" or "deep." A shallow well may still be several hundred feet deep but a "deep well" is one which penetrates an impervious layer of clay or other material, and it may not be as deep as a "shallow well." Sometimes the impervious layers are only thick lenses of clay or impervious silt, and may not be continuous over the whole area of the aquifer and are not to be regarded as impervious membranes protecting the aquifer from the surface water immediately above it.

What is thought to be a deep well may in fact be a only a shallow well since the impervious membranes do not completely cover the entire aquifer.

ADVANTAGES AND DISADVANTAGES OF GROUND WATER SUPPLIES

Approximately one-fifth of the fresh water withdrawals in the United States are from ground water resources. This figure would be higher if sufficient resources in the right place were available. When a new water supply system is required, ground water is the first place to look. The advantages may be listed as follows:

Advantages

— The water is free of suspended solids and normally it is "clear and bright," but it may be colored.

— It is usually free of pathogenic bacteria and providing the wells are properly designed, constructed, and maintained, it will continue this way.

— The water is uniform in temperature; ground water at a depth of 100 feet (30.48 meters) below the surface is approximately the same as the average annual air temperature. If adequate ground water is available, it is possible to develop ground water schemes at considerably less capital and operating costs than most surface water supplies.

Disadvantages

— Unless there is good geological data on the area, the exploration could be expensive and speculative.

— Some ground waters are highly mineralized, and contain large quantities of iron, manganese, sulfates, chlorides, calcium, magnesium, and other elements which are expensive to remove. Some ground waters are high in color due to organic acids from the topsoil vegetation.

— Elements such as iron and manganese are held in solution, at low pH values, in the aquifers by the presence of carbon dioxide. Once the water is pumped to the surface the free carbon dioxide is liberated and the ferrous and manganous ions precipitate out of solution when oxidized.

— Wells and aquifers can become contaminated by various types of nuisance bacteria such as *crenothrix, leptothrix, gallionella*, generally referred to as "iron bacteria," or the sulfate reducing bacteria such as desulphovibro-desulphuricans. Whenever these situations occur, the well should be "worked over" by removing the pump, and locating a drilling rig over the hole and surging it with solvents and disinfectants to remove the contamination. One scheme is to pour 12% sodium hypochlorite bleach (NaOCl) into the well with several pounds of dry ice (compressed solid carbon dioxide) and cap off the well. The gas pressure will push the hypochlorite into the formation and disinfect it. If the problem persists it may be cheaper to cap it off and start again.

— Occasionally a well water may deteriorate in quality because of earth movements, or changes in the migration of underground streams. When an aquifer is over-pumped, water from an entirely new and different geological formation may gain access into the original aquifer. Near the coast it could be salt water from the sea, or brackish water from swamps. Iron and manganese content can frequently change. Care must always be taken to insure that the dumping of scrap cans, old batteries, and other forms of garbage, does not take place where the seepage from these objects can contaminate the ground water, particularly in shallow wells. Even relatively small deposits of scrap metals can have serious consequences. An outbreak of brain disease in 1942, in a town near Tokyo,

Japan, originated from an area near a water well where 400 disused dry-battery cells had been buried.[2]

In most cases, problems can be avoided if the project is properly engineered, well constructed, adequately managed and maintained.

LAKES AND RIVER WATER SUPPLIES

The quality of lake water is not as consistent as the quality of ground water, but is more consistent than river water. The turbidity of river water may change rapidly during a heavy rain storm or flash runoff from melting snows. The lake water quality changes due to density modulations and from wind generated currents. Relatively shallow lakes with maximum depths of 30 to 50 feet (9–14 meters) with shelving shorelines, frequently contain substantial deposits of dead organic matter from decayed leaves, algae, and dead fish. Under the winter-ice the live fish and the remnants of the autumn algae blooms have first call on the dissolved oxygen with the result that the bottom deposits become anaerobic. The pH decreases from the surface down to the bottom. Many of the natural lakes and reservoirs have substantial quantities of iron and occasionally, manganese, in their bottom deposits. Normally they are of no particular importance to water engineers since they are oxidized and precipitated or chelated with organic compounds. However, when the bottom deposits in the relatively shallow areas of the lake become suspended due to wind induced currents, these deposits can enter the intakes.

Deep lakes are affected by changes in density which cause the water at the bottom of the lake to "turn over" and rise toward the surface. Since water is at its greatest density at 4°C (39.2°F), the colder water will migrate toward the bottom of the lake with the warmer, lighter water, nearer the surface. As the winter approaches, the surface water cools, increasing in density and displacing the warmer water immediately underneath it. During the winter, the water temperature becomes fairly constant at each interval of depth.

WATER SAMPLING

More can be learned about a proposed new surface water supply by walking along the shores of the lake or river, plus a trip in a boat to collect off shore water samples at various seasons of the year than any other form of investigation. Each water sample should be carefully labelled, and adequately documented. The information accompanying a sample should accurately pinpoint the location where it was taken, and whether it was collected from the surface or subsurface. Some notes of a general nature should be made as to whether the river was in full flood or a mere trickle, and whether there was much algae or floating debris in the lake or river when the sample was taken. Weather conditions influence the behavior of lakes and rivers considerably and the notes should indicate whether there had been light, medium, or heavy rainfall on the water shed prior to the sampling date. Any comments from local residents would be useful as to whether or not the conditions are normal for the time of the year. The temperature, color, and if possible, the pH of the water should be taken at the sample point when the sample is taken as on-site data.

On many occasions we have received unlabelled samples in pickle jars, or in plastic containers used for bleach, and even in old beer bottles plugged with a piece of twig, and asked to give an opinion on the use of the water for a municipal supply: In many cases the sampler has not bothered to wash the bottle first, and bleach bottles can never be adequately washed for use as sampling bottles. Even a well organized sampling program can only give a very sketchy appreciation of the water since the volume collected is infinitesimally small when compared to the amount of water in the lake or the river.

Water taken from the banks of a river may be almost undiluted ground water or surface runoff and may bear little resemblance to the water in mid-stream taken at the half depth (i.e., mid-way between the surface and the river bottom).

Photosynthesis of algae and phytoplankton convert the carbon dioxide in the water into sugars and oxygen.

If the samples are taken in the late afternoon the dissolved oxygen (D.O.) content will be much higher than it will be first thing in the morning, just after sunrise.

In many rivers the bottom deposits may have turned anaerobic and will often absorb the dissolved oxygen if stirred up by the passage of vessels through the navigable channels. Not only will there be a reduction in the dissolved oxygen content, commonly known as "oxygen sag," but there will also be an increase in the suspended solids, turbidity, and other chemical parameters.

The distance between a sampling point and a sewage or industrial waste outfall will materially influence the water quality. The purification processes that take place in rivers is a function of time and turbulence, so that the rate of flow becomes important together with the degree of dilution. All these changes are reflected in the analytical results of the sample which are only typical of a very small portion of the water at the point and moment of sampling. Where the sewerage system is combined, that is to say, the storm sewers collecting the surface runoff in the same sewerage collection system as the sanitary sewage, there is a considerable increase in sewage following a rainstorm, and the treatment plants are unable to handle the sudden increase and overflow diluted raw sewage directly into the streams.

In more modern installations, separate sewage collection systems are provided, whereby the sanitary sewage flows to the treatment plant and is eventually discharged to the river and is therefore unaffected by storms while the storm sewers discharge directly into the rivers. Even so, the first storm water runoff from the separate storm sewer system, after a period of dry weather, can have coliform counts higher than sanitary sewage.

If chemical and bacteriological samples were taken just after the start of the freshet, the results may indicate a grossly polluted river, whereas a sample taken an hour or two later may tell an entirely different story.

Before a project progresses beyond the feasibility stage, a suitable sampling and testing program must be initiated. In some of the older developed countries, records of water analysis of lakes and streams have been compiled over the years and are invaluable records. During the recent interest in mercury contamination of rivers and lakes, the investigators obtained stuffed exhibits, believed to be over 100 years old, from wildlife museums in order to assess what the background levels have been in the past. They found mercury levels were higher than had been anticipated.

An adequate sampling and testing program is time consuming and expensive, but it is absolutely essential to achieving satisfactory results. A properly coordinated and intelligently interpreted program should result in substantial savings.

RUNOFF

The volume of runoff and the various rates of flow are vital statistics to any water shed collecting system. Most lakes have one or more outlets into river systems which should be equipped with recording-level gauges and calibrated channels so that the flows can be calculated.

Once the equipment is installed and calibrated, it is necessary to accumulate several years readings, correlated with other meteorological data, in order to obtain meaningful results.

If day to day values are required, radio transmitting water-level indicating devices can be beamed to a satellite and back to a data collecting center for teleprinter dispatch to the interested party. In this way the necessity for numerous meter readers are avoided. If a dam is to be built across a stream to impound waters, it is essential to know the magnitude of the flood waters to size the spillway.

PROBABILITY METHOD

A convenient method of arranging river flow data consisting of spot checks taken at various intervals of time, is the probability method. This is a simple statistical tool which can be used to determine rates of flow likely to be exceeded for various percentages of the time, and can be done with or without the use of a computer. An example will serve to illustrate the system: the flows in Table 8–1 are taken from a river gauging station for the months of March to October. The remainder of the year the river is frozen over and ceases to flow. The measurements were recorded in acre-feet* but could be in any consistent units.

*An acre-foot of water is equivalent to 43,560 cubic feet. (1,233.4 cubic meters.)

TABLE 8-1 Stream Flows (Q) in Acre-Feet

Year	March	April	May	June	July	August	September	October	Annual Total
1971	[8] 1,150	[1] 26,100	[7] 4,150	[4] 10,540	[9] 2,520	[10] 730	[10] 600	[8] 870	[6] 46,660
1970	[5] 2,910	[6] 6,140	[5] 5,860	[1] 18,890	[4] 9,750	[7] 1,920	[7] 1,350	[5] 1,540	[5] 48,360
1969	[10] 480	[3] 19,320	[4] 7,210	[7] 7,290	[2] 19,070	[3] 3,260	[3] 2,050	[4] 2,010	[2] 60,690
1968	[6] 2,900	[10] 2,930	[9] 2,310	[9] 3,640	[10] 2,230	[4] 2,660	[4] 1,660	[3] 2,560	[10] 20,890
1967	[9] 910	[4] 12,070	[1] 17,360	[2] 15,990	[8] 3,550	[8] 1,730	[9] 760	[7] 1,090	[4] 53,460
1966	[1] 7,080	[5] 6,520	[3] 7,280	[6] 8,630	[3] 14,260	[2] 5,160	[2] 2,320	[2] 2,630	[3] 53,880
1965	[2] 6,200	[2] 24,550	[8] 3,270	[3] 12,730	[1] 25,360	[1] 7,920	[1] 9,830	[1] 5,960	[1] 95,820
1964	[3] 6,100	[9] 3,720	[2] 13,890	[5] 9,340	[5] 8,190	[9] 1,210	[6] 1,370	[6] 1,240	[7] 45,060
1963	[4] 2,930	[8] 4,220	[10] 1,800	[8] 4,230	[7] 4,250	[5] 2,360	[8] 1,190	[10] 600	[9] 21,580
1962	[7] 2,700	[7] 5,840	[6] 4,440	[10] 2,590	[6] 4,580	[6] 2,200	[5] 1,500	[9] 840	[8] 24,690
Totals	33,360	111,410	67,570	93,870	93,760	29,150	22,630	19,340	471,090
÷10 = Arith. Mean	3,336	11,141	6,757	9,387	9,376	2,915	2,263	1,934	47,109

The single number of 1 to 10 in the top left hand corner of each block indicates the descending order of magnitude for each month during the period. In this example, it is a period of 10 years and is designated by the symbol (m). The probability of recurrence in years can be calculated as follows:

$$\text{Recurrence interval (R.I.)} = \frac{n+1}{m}$$

Where:

n = number of years

m = order of magnitude
(small number in each block)

For records covering a 10 year period the occurrence interval in years for consecutive values of (m) are as follows:

$$\frac{n+1}{m} = \text{R.I.}$$

$$\frac{10+1}{1} = 11.0 \text{ years}$$

$$\frac{10+1}{2} = 5.5 \text{ years}$$

$$\frac{10+1}{3} = 3.67 \text{ years}$$

$$\frac{10+1}{4} = 2.75 \text{ years}$$

$$\frac{10+1}{5} = 2.20 \text{ years}$$

$$\frac{10+1}{6} = 1.83 \text{ years}$$

$$\frac{10+1}{7} = 1.57 \text{ years}$$

$$\frac{10+1}{8} = 1.375 \text{ years}$$

$$\frac{10+1}{9} = 1.222 \text{ years}$$

$$\frac{10+1}{10} = 1.1 \text{ years}$$

The percentage probability of recurrence is the reciprocal of the recurrence interval in years. For example in April ($m = 6$) a monthly flow of 6140 acre-feet or less will occur for 54.6% of the time. Whereas 6520 acre-feet or less will occur for 45.5% of the time, 12,070 acre-feet or less will occur for 36.4% of the time and so on, for the month of April.

This is shown in Table 8-2.

TABLE 8-2 Percentage Probability

Years (m)	1	2	3	4	5	6	7	8	9	10
R.I. = $\frac{n+1}{m}$	11.00	5.50	3.67	2.75	2.20	1.83	1.57	1.375	1.222	1.1
Percentage Probability (%)	9.1	18.2	27.2	36.4	45.5	54.6	63.7	72.7	81.8	91.0

From these data the monthly flows can be tabulated or plotted on log paper as shown in Figure 8-1 for the month of April. Similar graphs can be plotted for the other months and also for the annual total flow.

From the graph, Figure 8-1, the following assumptions for the months of April can be made as shown in Table 8-3, Probability of Flow, for the months of April (1962–1971).

Statistical data must always be judged with caution, and the computation can never be more accurate than the information from which it has been derived.

Stream flow measurements are rarely better than plus or minus 5% for a well instrumentated, carefully profiled, and calibrated stream. For important measurements the stream should be re-calibrated at least once per year since changes in stream profile can take place due to moving gravels and frost heave which can change the calibration. For these reasons, information from many gauging stations is probably not within plus or minus 30% of the actual flow.

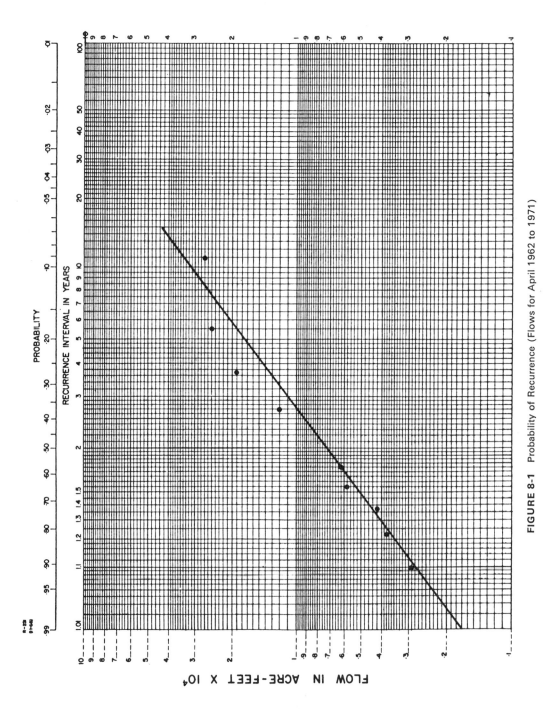

FIGURE 8-1 Probability of Recurrence (Flows for April 1962 to 1971)

TABLE 8-3 Probability of Flow for the Months of April (1962–1971)

Percentage of Time	Flow in Acre-Feet Will Be Equal to or Less Than
99%	1,650 acre-feet
95	2,300 acre-feet
90	2,800 acre-feet
80	3,800 acre-feet
70	4,800 acre-feet
60	6,000 acre-feet
50	7,200 acre-feet
40	9,200 acre-feet
30	12,500 acre-feet
20	18,000 acre-feet
10	32,000 acre-feet

There is always the potential for flash floods which can exceed all previous records. Land that has been stripped of most of its timber will have a faster runoff than well-timbered slopes when the snow starts to melt. Since changes can take place many times over during the life of a water shed, it is as well to monitor on a regular basis. Areas that have been extensively logged have larger and faster runoff rates than well-treed undisturbed slopes.

An engineer designing overflow spillways for dams is specifically interested in peak flows which can be obtained from recording-level gauges. If the normal level gauges are not capable of recording maximum flood levels then a maximum flow gauge should be installed.

REFERENCES

1. NORDELL, E. *Water Treatment for Industrial and Other Uses*, 2nd Ed. New York: Reinhold Publishing Corporation, 1961.
2. TAYLOR, E.W. *The Examination of Waters and Water Supplies*, 6th Ed. Philadelphia: The Blakiston Company, 1949.
3. *"Glossary—Water and Wastewater Control Engineering."* Published jointly by A.P.H.A., A.W.W.A., A.S.C.E., and W.P.C.F. and is available from any of these Associations, 1966.

9

GROUND WATER AND WELL PUMPS

WELLS The services of a competent ground water geologist to perform the preliminary survey, direct the drilling operations and completion of the well, is a worthwhile investment in the development of any large ground water collection system. His services will probably reduce the overall cost of the system as well as the risk of developing a well that produces sand to the detriment of the pumping equipment.

It is also important to establish coordination between the owner, geologist, well driller, engineer, and pump supplier. Unless the project has been correctly engineered and coordinated from the beginning it is possible that the diameter of the well casing will be too small to accommodate the most suitable pump for the job, and it may be necessary to install a smaller high-speed unit, that will wear out more frequently. The type and general characteristics of the well pump should be determined before the final production well has been drilled.

Particular care must be exercised in the selection of the well screen. Occasionally the well driller may "eyeball" the size of the material he is removing from the hole as he approaches the pay zone of the aquifer and size the screen accordingly. If the screen openings are not correctly chosen in accordance with a sieve analysis of the well cuttings, then serious problems are likely to develop. Either the screen will be too tight and the water yield will be reduced, or the openings will be too large and the well will continue to pump sand long after the normal development phase is over. Particular care must be taken when drilling in old glacier deposits where there may be little uniformity in the grain size of particulate materials in the water-bearing aquifer. If the well screen is properly sized, the larger particles of sand may bridge across the screen openings and retain the finely divided silts. But if the screen openings are too big, it may be impossible to bridge with the fine gravel and coarse sand of the aquifer with the result that the fine materials will continue to penetrate the screen.

The anticipated performance of a well can be envisaged during its initial development and pump test. The amount of sand that penetrates the screen determines the economic capacity and useful life of the well. It is unreasonable to consider 3600 rpm pumps in wells continuously producing sand since the pumps will wear rapidly and will have to be replaced frequently.

Most modern well screens are made of wedge-shaped stainless steel wire. The following Table 9-1 is representative of the open areas of Telescope-Size Johnson Well Screens. It is customary to determine the capacity of a well on the basis of 0.1 ft per second approach velocity.

The slot numbers refer to the size of the openings in a Johnson Well Screen, i.e., No. 60 slot openings has slots 0.060 inches wide. (1.524 mm.) For example, the open area for an 8-inch Johnson Well Screen with No. 60 (0.060 inch) slot openings is 113 square inches per linear foot of screen. With an approach velocity of 0.1 feet per second, this screen would have a capacity of

$$\frac{113}{144} \times 0.1 \times 60 \times 7.5 = 35.3 \text{ U.S. gpm/ft}$$

This calculation can be simplified by multiplying the open area of the screen in square inches by 0.3125 to give the U.S. gpm per linear foot of screen.

Increasing a #60 slot screen from 6 inch to 8 inch diameter only increases the area of the openings by 33 percent, but the large diameter well casing makes an enormous difference when selecting the pump. The larger well, i.e., an 8 inch diameter well is 77 percent bigger in cross-sectional area than a 6 inch diameter well, enables a larger diameter pump to be used. This could mean a reduction in the speed of the pump and/or the number of stages required to produce the same capacity and developed head.

Well drilling is a highly speculative business and nobody is anxious to drill any larger than he has to in case it proves to be a dry hole, or the water quality is unsuitable for municipal use, or the quantity of water is inadequate to meet the requirements of the water system. However, if the well turns out to be a good one, and since wells are not particularly cheap, then it is reasonable to complete it and put it into service. The alternative is to drill a larger production well. It is at this stage of the proceedings that the pump selection engineer runs into conflicting situations. He endeavors to make the best use of the well, to pump as much water as possible, and at the same time to select a pump capable of long and trouble-free life. These two concepts are diametrically opposite. Since pump capacity is directly proportional to shaft speed (rpm), and developed head is proportional to the speed squared $(rpm)^2$, it follows that the faster the shaft speed

TABLE 9-1 Representative Open Areas of[1] Telescope-Size Johnson Well Screens Telescope-Size Intake Areas (Square Inches per Lineal Foot of Screen)

Nominal Screen Size (in.)	10-Slot (0.010") (0.25mm)	20-Slot (0.020") (0.50mm)	40-Slot (0.040") (1.0mm)	60-Slot (0.060") (1.5mm)	80-Slot (0.080") (2.0mm)	100-Slot (0.100") (2.5mm)	150-Slot (0.150") (3.7mm)	250-Slot (0.250") (6.2mm)
3	15	26	41	52	59	65	65	76
$3\frac{1}{2}$	18	31	49	61	70	77	77	90
4	20	35	57	71	81	88	88	104
$4\frac{1}{2}$	23	40	64	80	92	100	100	118
5	26	45	72	90	102	94	112	132
$5\frac{5}{8}$	28	49	79	99	113	104	124	143
6	30	53	65	85	100	112	132	156
8	28	51	87	113	133	149	138	173
10	36	65	110	143	168	188	174	219
12	42	77	86	117	142	164	206	260
14	28	53	95	130	158	183	229	289
16	32	60	108	148	180	208	261	329
18	36	69	124	169	206	237	298	375
20	41	77	139	189	231	267	268	354
24	61	113	122	171	213	250	327	432
26	63	118	127	178	223	262	341	452
30	75	140	151	211	263	309	404	534
36	86	160	172	240	300	353	461	609

- Open areas may differ somewhat from these figures; extra strong construction for example, reduces open area in some cases because heavier material is used to increase strength.
- The maximum transmitting capacity of screens can be derived from these figures. To determine GPM per ft. of screen, multiply the intake area in square inches by 0.31. It must be remembered that this is the maximum capacity of the screen under ideal conditions with an entrance velocity of 0.1 ft per sec.

the larger the pump capacity, and the greater the head developed per stage. Since the available space inside a well casing is very limited, a high-speed pump has many advantages over a slower speed pump in terms of capacity and cost but with the exception of longevity. It has been said that wear is proportional to shaft speed squared (rpm)², that is to say that a pump of 3600 rpm would have four times the wear of an 1800 rpm $\left(\frac{3600}{1800}\right)^2 = 4$.

If, therefore, under normal conditions an 1800 rpm pump would be expected to last 10 years then a 3600 rpm pump would probably require replacement in less than 3 years. If there are abrasive grits present in the well water, then the life of either pump would be appreciably reduced, particularly the high-speed units since the tolerance on shaft and impeller clearances are very much less for higher shaft speeds.

ARTESIAN WELLS

A well that penetrates into an artesian aquifer in which the water is confined under pressure greater than atmospheric by an overlying, relatively impermeable layer, is known as an artesian well. The word is believed to have derived from Artois (France) where the first deep wells into pressurized aquifers were drilled.

The general profile of an artesian aquifer is shown in Figure 9-1. There are two types of artesian wells:

(a) Nonflowing artesian

(b) Flowing artesian

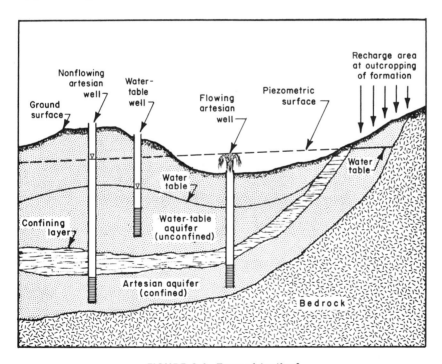

FIGURE 9-1 Types of Aquifers[2]

Nonflowing Artesian Wells

Nonflowing wells have the advantage that the water is higher in the casing and there is less pumping head required to bring the water to the surface. In some instances it may be possible to use shorter pump columns than would be required with an ordinary well.

Flowing Artesian Wells

A ground water geologist can often predict that a well will be artesian and if it is believed to be a flowing artesian, special precautions in drilling and completing the well must be observed. The wells vary in depth from a few feet to several thousand feet. Controlling the flow, particularly if the water pressure at the surface is appreciable, is sometimes difficult and can be extremely dangerous if not competently controlled. An account of some of the problems that can occur is given in the following excerpt.[3]

Control of Flow

> The main problem with artesian wells is control of flow. The flow should be controlled to prevent waste of water which may, over a long period, lower the artesian pressure in the aquifer so that the wells no longer flow, or flow at a reduced rate. Another reason for controlling the flow is to prevent flooding and waterlogging of land. If the well is properly constructed, shutting off the flow may cause water to flow up outside the casing. When this happens, all control of the flow may be lost and damage may be extensive. In some cases, improper well construction under artesian conditions may permit water to flow from an aquifer under artesian pressure into overlying aquifers. This, of course, wastes water and may cause waterlogging and other undesirable conditions.
>
> Some artesian waters, particularly those from certain sandstones, may be of very poor quality. When such is the case, the poor quality water may cause serious damage on surface or may leak into an overlying good aquifer contaminating it with bad water. Artesian wells encountering bad water should be permanently sealed off.

Construction

> Where flowing artesian conditions are known or suspected, a method of construction using casing should be used. The accepted method under artesian conditions is to seat the casing with cement in the formation overlying the artesian aquifer. This is most easily done before drilling through the confining layer by forcing cement grout down the casing in such a way that it flows up around the outside of the casing as far as possible. After the grout has set, the drilling is continued into the aquifer.

This, of course, cannot be done where a hole unexpectedly encounters artesian conditions. In these cases, it may be possible to shut off the water until the hole has been grouted as described above, or grout may be injected around the outside of the casing by some other means. With the casing properly grouted, it is quite safe to shut off the water by means of a valve on the surface.

Corrosion Problems

A very troublesome problem sometimes encountered with artesian wells is corrosion of the casing. The constant flow of water, particularly if it is corrosive, will corrode the casing, causing it to leak. Where artesian water is known to be corrosive, a smaller diameter flow pipe may be installed in the well. This may be made of corrosion resistant material or may be replaced if it is destroyed by corrosion. Where the casing has become corroded, a slightly smaller diameter casing may be installed inside the old casing and sealed in place with cement grout. In any case, it is important to maintain an artesian well in good condition as it is quite difficult to repair an artesian well when the casing has been destroyed. Obviously thin-wall or sheet metal casing should not be used in artesian wells.

Screens

A screen is just as important in an artesian well as in a nonflowing well. Such wells are just as prone to yielding water with sand or becoming plugged with sand, as are nonflowing wells. There is no reason to believe that artesian flow of a properly constructed well will decrease if it is shut off, although such an idea is prevalent. In fact, the opposite is true as waste of artesian water will often eventually cause a decrease in artesian pressure. This idea has probably arisen from experience with poorly constructed, small diameter, open-end artesian wells in fine-grained aquifers where constant flow keeps the casing free of sand and silt. When such a well is shut off, the sand may settle in around the bottom of the casing, causing it to become clogged.

Artesian wells are often pumped for increased capacity. In this case, proper well construction, as far as use of a screen is concerned, is just as important as with a nonflowing well.

Pumps for Artesian Wells

The installation of vertical turbine or submersible pumps into flowing artesian wells should be avoided. If the well has a shut-in pressure of more than a few feet of head then the installation of a pump inside the well casing is impractical if not impossible. The pumps should be either canned-vertical turbine, end-suction, or split-case centrifugals installed in a pump house on the surface. Ample space must be available around and over the well to provide access to a drilling rig if subsequent work-overs are necessary.

Flowing artesian wells should be avoided if possible, since they can be difficult to control and maintain and they can be a potential hazard and liability, particularly in developed areas.

WELL PUMP SELECTION

There are usually two choices, vertical turbine pumps and submersible pumps. The vertical turbine (see Figure 9-2a) has the bowl assembly adjacent to the top of the well screen and the motor is located on the surface, immediately above the discharge head. Between the motor and the bowl assembly is a long length of drive shafting centered in guide bearings spaced at approximately 10 foot

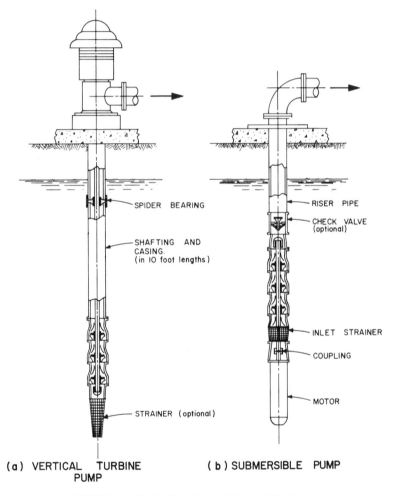

FIGURE 9-2 Vertical Turbine and Submersible Pumps

intervals. The submersible pump (see Figure 9-2b) has its motor below the bowl assembly, thus eliminating the long length of drive shafting and guide bearings.

The total capital cost of the submersible pump installation is likely to be less than the vertical turbine. However, if the well casing is small in diameter, the annulus space between the outer casing of the motor and the inside diameter of the well casing may cause an appreciable restriction to the flow of water from the well screen to the pump inlet (see Figure 9-3b). Since centrifugal pumps are

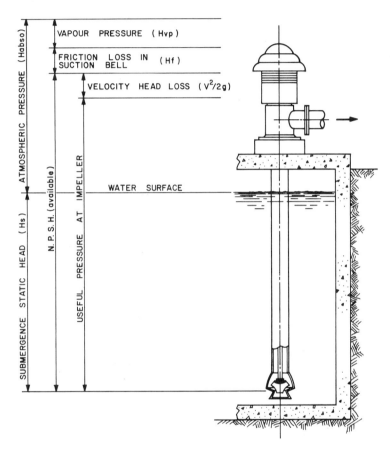

FIGURE 9-3 Well Pumps

essentially kinetic energy machines, converting velocity energy into pressure energy, they must have an adequate flow of water into the pump suction. They are unable to suck water in the same fashion as a plunger pump. If the intake to a centrifugal pump is restricted there will be insufficient flow to the pump and it will tend to cavitate and not function efficiently. This is due to insufficient net positive suction head (NPSH).

NET POSITIVE SUCTION HEAD (NPSH)[5]

It is important to realize that there are two values of NPSH; the available NPSH which depends on the location and design of the intake system and can be calculated; and the required NPSH, determined by manufacturers bench scale tests. The required NPSH is the head required at the inlet of the impeller to insure that the liquid will not boil inside the impeller under the reduced pressure conditions, and the water flow to the impeller will flow smoothly without cavitation. It is essential that the available NPSH exceed the required NPSH with a reasonable margin of safety, at least 2 to 3 feet or more if possible. See Figure 9-4.

$$\text{NPSH}_{(available)} = H_{abso} + H_s - H_f - H_{vp}$$

where:

$\text{NPSH}_{(available)}$ is Net Positive Suction Head measured in feet.

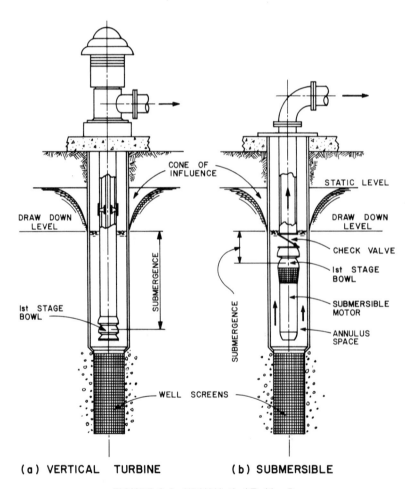

(a) VERTICAL TURBINE   (b) SUBMERSIBLE

FIGURE 9-4 NPSH Vertical Turbine Pump

H_{abso} is the absolute pressure on the surface of the liquid in the well measured in feet.

H_s is the static elevation of the liquid above the centerline of the pump (on vertical turbine pumps to the entrance eye of the first stage impeller) expressed in feet. If the liquid level is below the pump centerline, H_s is a minus quantity.

H_f is the friction head and entrance losses in the suction piping expressed in feet, or around the annulus (see Figure 9-4b).

H_{vp} is the absolute vapour pressure of fluid at the pumping temperature expressed in feet of liquid. Figure 9-3 shows a typical vertical turbine pump installed in a larger pump chamber. The same concepts apply to well pumps but with added complications.

Figure 9-4 illustrates a vertical turbine pump and a submersible pump in two exactly similar wells with the same static water level, draw-down, and well screen capacity.

VERTICAL TURBINE PUMPS

Submergence

The vertical turbine has its suction close to the top of the well screen and has therefore maximum submergence.

Well Alignment

The straightness of the well is important if the shaft is to have a long trouble-free life. A gentle curvature is not serious, and in some cases has advantages since the shaft will run smoothly in its intermediate bearings without chatter. However, if the well is seriously out of line with an abrupt change in direction then shaft fatigue and excessive bearing wear will result.

Well alignment should be checked before the screen is installed in case the well has to be abandoned.

Practical advice on the construction, drilling, testing, screening, and disinfection of wells is available in the "American Water Works Association Standard for Deep Wells."[4] Municipal water wells should be drilled in accordance with these specifications. The standard describes the acceptable procedure for determining the plumbness and alignment of the finished well. This should preferably be measured and plotted to scale before the screen is installed in case the well has to be abandoned due to malalignment. In a borderline case, a 40-foot long spool or dummy should be made from a rigid spindle with three rings, each ring to be not more than 1/2-inch diameter smaller than the inside diameter of the casing and not less than 12 inches in width. One of the rings is welded at the center and one ring at either end. The dummy or spool is lowered into the well and must be able to move freely to ensure that the well is straight enough to install a vertical turbine pump and shafting.

In the selection of suitable alternative pumps for well installation, reference should be made to the "American National Standard for Deep Well Vertical and Submersible Types. A.W.W.A. Standard E101-71."[6]

SUBMERSIBLE PUMPS

Submergence

With the submersible pump the depth of submergence will be less than with the vertical turbine pump by the length of the motor. With large horsepower pumps the motor can be 10 feet or more in length. In some cases two motors may be coupled together in series which will appreciably increase the pump length. The submersible motors can sometimes be installed down into the well screen in order to increase the depth of submergence and NPSH. However, if the annulus space between the outside diameter of the motor and the inside diameter of the well screen is restricted, the flow pattern of the water through the screen and past the outer periphery of the motor will be distorted. The velocity of approach normally 0.1 ft/sec into the screen may be considerably increased near the top of the screen at the expense of the lower section near the bottom of the motor. If this occurs there may not be sufficient water flowing over the entire surface of the motor to keep it uniformly cool. Although the motor is immersed in water, it relies on the velocity of water over its outer surface to dissipate the heat and failure to do this can result in the motor becoming overheated.

Annulus Space

Between the outer perimeter of the motor and the inner surface of the casing there is an annulus space through which the water must pass to go from the well screen to the pump suction. The velocity and head loss through this annulus should be calculated if there is any possibility that it will cause a serious obstruction to flow, and result in an appreciable loss of available NPSH. On the other hand, the velocity of flow past the motor should not be less than 0.5 feet per second in order to dissipate the heat generated within the motor.

Well Alignment

Normally, a crooked well is not as serious with submersible pumps as it is with vertical turbines, since there are no long lengths of shafting. However, if there are serious deflections in the well alignment that could prevent the pump from being installed, or if it was installed and was too tight to be easily removed, then it would be as well to use a full size dummy equivalent to the length and diameter of the pump and motor. The dummy can be lowered into the well casing to insure that it can be moved in and out of the well without difficulty before installing the well screen and an expensive pump.

SUCTION LOCATION

It is obviously advisable to get the pump suction as low down the well as possible but it is not considered good practice to have the suction lower than the top of the well screen. If the hydraulic grade line of the water bearing aquifer drops below the top of the well screen, it will change the screen environment with deleterious results. This is known as "drawdown" and is the magnitude of the lowering of the water surface in a well, and of the water table in the vicinity of the well screen resulting from the withdrawal of water by the pump.

The environment of the aquifer is normally anaerobic, nonoxidizing, due to the absence of air and the presence of free carbon dioxide. If the waters contain iron, as ferrous bicarbonate, exposing the screen may cause the soluble ferrous iron to oxidize to insoluble (rust) ferric hydroxide and plug the screen. Once partial plugging occurs, the drawdown in the well casing is aggravated, and the situation rapidly deteriorates until the screen is completely plugged.

WELL PUMP OPERATION

Well pumps discharging directly into the distribution system, thereby maintaining the water level in an elevated storage reservoir, are operated between high and low reservoir water levels. Usually they are set to allow the pump to operate continuously for several hours without stopping. If, however, instead of a large reservoir, the buffer storage of relatively small water supply systems is a hydropneumatic tank, then the sequence of the well pump starting and stopping is more frequent. Since most motors cannot tolerate too many starts in any one period of time, because of overheating resulting from frequent surges of large inrush currents when starting (up to 6 times full load current), then consideration must be given to the selection of equipment suitable for these conditions. Alternatively, it may be possible to redesign the system using variable-speed drives to reduce the number of starts.

Each well pump should have a low water level cut-out probe located inside the well casing to stop the pump when the water level is drawn down close to the bottom.

When a well pump starts, the water in the well casing is the first available supply for the pump suction and in consequence the level in the casing drops rapidly creating a differential head between the water in the aquifer and the water in the well. This differential head enables the water in the aquifer to flow through the screen into the well casing. Once equilibrium conditions prevail, the difference in water level between the water in the aquifer and the well casing is sufficient to maintain an adequate flow through the screen. The water level in the aquifer in the immediate vicinity of the well is lowered, and a cone of influence is formed. It is important that these levels are recorded at periodic intervals in order that the performance of the well screen and the aquifer can be monitored. A water level probe consisting of an air bubbler tube extending to the bottom of the inside of the well casing or a battery operated electrode which deflects the needle of a galvanometer when the probe touches the water is used to measure the static and drawdown levels in the well casing. In order to monitor the cone of influence, small diameter observation wells are sometimes drilled at various locations adjacent to the well, but they are only provided in relatively large ground water installations.

The drawdown level in the well casing must be correlated to the static water level and the rate of pumping. If the drawdown is increasing, or if the pump frequently cuts out on low water level in the casing when the static water level remains unchanged, then this is an indication that the screen is becoming plugged.

Well screens can become plugged for a variety of reasons as follows:

Carbonate Scale

Hard water will often cause encrustations of calcium to restrict the free flow. Often these problems can be solved by acidizing with an inhibited acid. Suitable formulations in tablet form are available for this purpose.

Iron Deposits

Waters high in ferrous bicarbonate may result in deposits of insoluble ferric hydroxide on the well screen resulting from aeration of the ground water. Acidization and surging will help to remove the deposit.

Iron Bacteria

Slimes of iron bacteria can also effectively restrict the flow through a well screen. Dosing with strong disinfectants and detergents together with surging may eliminate the problem providing it is caught early. However, if there are extensive slime deposits they may be difficult or impossible to remove without the use of a drilling rig to surge and redevelop the well.

Sand

Migrations of finely divided silt and other granular materials may be the cause of a reduced yield and about the only solution to this problem is surging with a plunger attachment on a drilling rig. The use of phosphate compounds will help to break down the conglomerates and loosen the build-up on the outside of the screen. If there are appreciable deposits of sand inside the screen they must be removed by a bailing attachment on the drilling rig, otherwise the screen capacity will be seriously reduced.

In order to adequately maintain a water well it is desirable to be able to remove the pump and place a drilling rig over the top of the well to surge and bail it. If it is not possible to do this, then there is less likelihood of correcting a severe case of screen plugging.

HORIZONTAL COLLECTORS

Horizontal collectors consist of vertical hollow concrete caissons usually 13 to 16 feet in internal diameter (3.96 to 4.87 meters) by approximately 2 feet thick (0.609 meters). They extend from the surface to just below the bottom of the aquifer. The concrete caisson is slip-formed on the ground and falls into place as the top soil and gravel is excavated from the center. When the caisson is down to its final depth the bottom is plugged with tremie concrete. Holes are formed in the bottom section of the caisson in one or more circumferential rows several feet above the bottom concrete plug. Through these holes the horizontal collectors are power-jacked through the caisson openings into the aquifer. Due to the limited internal diameter of the caisson, the collectors are made up from short lengths approximately a meter long, about 8.5 inches in diameter, and projected to a distance of approximately 150 feet. There are two basic systems, one uses slotted steel pipe. The other projects plain steel pipe into the formation and plastic screens are inserted into the steel pipe. When the plastic screens are in place, the steel pipe outer casing is withdrawn leaving the screen in direct contact with the aquifer. As in the case of the vertical wells, the screens are selected on the basis of their slot openings with respect to the sieve analysis of the aquifer material.

Blanks can also be used if fine sands are encountered which could cause plugging of the laterals. The diameter and length of the laterals are fixed by the equipment necessary to jack them into the formation. There are limitations on the number of laterals that can be effectively installed in each situation. It is customary design practice with one system of horizontal collectors to assume a flow of 1 liter per second per meter of screen, irrespective of slot size. This is equivalent to 4.84 U.S. gpm per foot of screen. Based on the free area of slot opening for the smallest 1 mm slots, this is equivalent to a velocity of approach of 0.06 feet per second, and for the larger 5 mm slots it is equivalent to 0.016 feet per second which is well below the normally accepted velocity of 0.1 feet per second for vertical well screens. The screen sections are 1 meter in length with a choice of 1, 3 or 5 mm slot openings, and are solvent-welded to each other in order to form a complete screen. Any lateral could have all 3 sizes of slot openings if the sieve analysis indicated that this was necessary. A lateral 150 feet long could have a designed capacity of approximately 725 U.S. gpm. If 15 laterals are installed the entire collector would have a capacity of approximately 10,850 U.S. gpm or 15.7 mgd (U.S.), providing, of course, the aquifer can maintain this withdrawal rate. If the aquifer is adequate and the recharge rate will maintain a larger flow, it is possible to install additional collectors located at suitable intervals.

There is proven technology and equipment available to abstract ground water up to many millions of gallons per day as required.

DEVELOPING THE WELL

When a vertical well has been completed with the screen in place, it is usual to develop the well using contractor's tools. The primary objective is to repair the damage done to the formation by the drilling process and to remove the fine material which will pass through the screen and leave the coarser material on the outside of the screen. This process can be done by the use of surging tools and bailers without damaging or plugging the screen. Once the developing procedure has been completed, (see *Ground Water and Wells*,[1] also A.W.W.A. A100-66[4] for details), the well should be tested using, if possible, a contractor's or well driller's pump. There may be some sand in the well or attached to the well screen, and for these reasons it is preferable not to use the regular pump since the sand may rapidly reduce its efficiency.

DISINFECTION

Once the regular well pump has been installed, it is necessary to thoroughly disinfect the well, screen, and pump with sufficient chlorine solution to insure a minimum of 50 mg/ℓ of free chlorine on all surfaces. There is always the possibility that the aquifer in the immediate vicinity of the well screen could have been contaminated by nuisance as well as pathogenic organisms which could have gained access to the well from the surface during its construction. These organisms can cause taste and odor problems as well as slimes and other forms of growth on the screen and on the pump. To avoid this, the chlorine solution should be forced into the aquifer during the surging operation when the well is being developed, or by forcing the chlorine solution through the screen under pressure when the pump is in place. This can be done by applying compressed air on the outer casing, assuming that all the openings on the top of the well can be effectively plugged. An alternative method is to inject sufficient chlorine solution down into the well to insure that it passes into the aquifer. The chlorine solution at not less than 50 mg/ℓ should be in contact with the well for not less than 2 hours, and then pumped to waste. In order to still have not less than 50 mg/ℓ at the end of the 2 hour period, it is necessary to use not less than 0.1% solution strength (1000 mg/ℓ) initially.

Bacteriological sampling and analysis should be performed after disinfection.

REFERENCES

1. *Ground Water and Wells*, 1st Ed. Saint Paul, Minnesota: Edward E. Johnson Inc., 1966.
2. GIBSON, U. P. and R. D. SINGER, "Water Well Manual" Published by Premier Press, P. O. Box 4428, Berkeley, California 94704, 1971.
3. "Practical Information on Groundwater Development" Published by Department of Environment Water Resources Service, Water Investigation Branch, British Columbia Water Resources Services, Victoria, B. C., Canada 1976.
4. "Standard for Deep Wells (A100-66)", *AWWA*. Available from the American Water Works Association Inc., Denver, Colorado.
5. WALKER, R., *Pump Selection*. Ann Arbor, Michigan: Ann Arbor Science Publishers Inc., 1972.
6. "Standard for Deep Well Vertical Turbine Pumps—Line Shaft and Submersible Types," *AWWA*, E101-71. Available from the American Water Works Association Inc., Denver, Colorado.

10

STRAINING AND SCREENING

TRASH RACKS River intakes should always be fitted with trash racks. They usually consist of vertical flat steel bars 3 in. × 1 in. on edge, with 2 to 3 inch open spaces between them.

Frazil ice on the trash racks can be a problem, and electrically heated bars can eliminate some of the difficulties.[1]

FISH SCREENS Fish screens are mandatory on many rivers where there are commercial salmon and trout. Government regulations specify the type, size of openings, and the allowable approach velocities permissible for each location. They are designed to protect the fish from leaving their habitat. There are two types, stationary and travelling. In each case, the size of the openings are normally not to exceed 1/10 inch (2.54 mm).

— Stationary screens can be designed for an approach velocity not exceeding 0.1 ft per second or 45 U.S. gpm per square foot (30.56 liters per second per meter^2).

— Travelling screens can be designed for an approach velocity not exceeding 0.4 ft per second or 180 U.S. gpm per square foot (122.24 liters per second per meter^2).

The reason for the four-fold increase in capacity for travelling as compared to stationary screens is that they will be kept appreciably cleaner and will not have the same restrictions to flow.

Whenever there is a large intake and substantial quantities of debris, the use of travelling screens is preferred. While the water passing through the screens will be devoid of large suspended material including fish, the use of a fish screen is a primary form of water treatment, and can be an integral part of the treatment process.

MICROSTRAINERS

Microstrainers for the removal of plankton from impounded waters were first developed by Glenfield and Kennedy in Scotland.[2]

A microstrainer consists of a cylindrical frame covered with stainless steel wire mesh. As the cylinder rotates, suspended particles adhere to the inside surface of the cylinder and are washed off by water jets. The axle supporting the cylinder is hollow and carries a hopper into which the particles are washed from the fabric. The drum axle, serving as a conduit, conveys the washings to waste. The head loss across the microstrainer is usually less than 6 inches. This can be controlled (assuming the volume and the load are within the capacity of the unit) by varying the speed of rotation and by controlling the pressure on the water jets. A toothed ring is fitted to the periphery of the drum at the driving end and is driven by the motor through a variable-speed gear.

Approximately 75% of the diameter of the microstrainer drum is submerged at any one time. This is equivalent to 67% of the total strainer area continuously in operation while the remaining 33% of the fabric is being washed and drained. There are many standard microfabrics available made from stainless steel wire.

Water for backwashing is drawn from the downstream side of the unit. Drum rotation and backwash are continuous operations, adjustable either manually or automatically with the flow rate.

Flow capacities may vary from 5 to 45 U.S. gpm per square foot, but the lower rates of flow are normally recommended. Each application has to be investigated individually by making on-site filterability tests on representative water samples with the specific fabric to be used. The desirable finished water quality determines the choice of fabric, and the seasonal variation in raw water quality determines the flow rate for the fabric selected.

Microstrainers are available in capacities up to 20,800 U.S. gpm (30 mgd U.S.). The washwater consumption varies according to the microfabric used and the algae loading but is usually between 1 and 3% of the microstrainer capacity. The pressure at the wash nozzle is recommended to be 60 psi.

They are essentially designed to remove plankton and algae, and will not handle sand, silt, or other abrasive material. The slower the drum revolves and the thicker the layer of debris that accumulates on the inside of the microfabric, the better the finished water quality. The only real problem in microstrainer operation is an occasional build-up of slime on the fabric. If it only occurs infrequently, it can be corrected by stopping the rotating drum and washing the fabric with sodium hypochlorite. If it is more than an occasional problem the use of ultraviolet irradiation equipment over the drum will inhibit bacterial and slime growth. Inorganic deposits on the microfabric can be cleaned with an inhibited acid or a special formulation of sulfamic acid (NH_2SO_3H) similar to that used for cleaning encrusted well screens.

MICRO-STRAINING SURFACE WATER SUPPLIES[3]

The clarification of waters by microstraining is particularly effective with particle sizes of 20 μ (micron) and above. The build-up of a layer of debris on the inside of the drum increases the effectiveness of the microfabric so that smaller particles, less than the smallest aperture, can be strained out.

Some of the organisms that are reported to be removed by microstraining are:

Algae	Protozoa
Coelastrum	Ceratium
Coelosphaerium	Euglena
Cosmarium	Mallonmamonas
Pediastrum	Trachelomonas
Scenedesmus	Vorticella
Straurastrum	

Crustacea	Cyanophyceal
Boamina	Anabaena
Cyclops	Aphanizomemon
Daphnia	Microsystis
Nauplius	Rotifera
	Anuarea

Chlorine injection should be after the microstraining rather than before it for the following reasons:

1. Live algae are easier to clean off the microfabric than dead algae.

2. Ferrous bicarbonate that may be present in the water, particularly under anaerobic conditions from icebound reservoirs, will tend to be oxidized to a sticky gelatinous ferric hydroxide in the fabric.

3. Prechlorination is likely to cause taste problems, particularly if certain species of taste and odor algae are present.

4. Water that has been microstrained is relatively free of suspended organic matter, and will require less chlorine.
5. Free chlorine can result in crevice corrosion in the microfabric.

Disinfection with ozone[4,5] after microstraining has been successful.

PRETREATMENT BY MICROSTRAINERS[6,7]

Microstrainers are installed for the primary clarification of waters before chemical coagulation. There are a number of good reasons for this: a few of them are discussed as follows:

1. Microorganisms and organic matter are difficult to coagulate since they do not retain surface active charges compared to inorganic substances. Furthermore, they usually have a specific gravity less than 1.0 and will float rather than settle to the bottom of a clarifier.
2. Living microscopic organisms become food for the higher forms of life, and if these aquatic organisms are removed at the first stage of the process many other problems are eliminated. With certain species of algae in the raw water, prechlorination can cause taste and odor problems by forming substitution products. Once these taste and odor compounds are formed, they are difficult to eliminate. This problem also concerns the pulp and paper industry in the manufacture of special papers.

Microorganisms can get between the surface-charged particles of coagulated floc and reduce the forces of cohesion, with the result that the fragile flocs may break apart into smaller pieces. Once the particles of floc disintegrate they will not recoagulate since their electric charges are neutralized one to another. Considerable improvements in clarifier performance are experienced due to eliminating the microorganisms before the coagulation processes.

PRESSURE STRAINERS

One disadvantage of a microstrainer is its gravity operation. There are strainers that operate under pressure and back flush automatically. They are effective for removing wood chips, most algae, and the larger microorganisms. Equipment of this type is capable of handling up to 30 mgd (U.S.) per unit with pipeline pressure of 150 psi and a head loss of 1 to 10 psi across the fabric.

REFERENCES

1. CREAGER, W. P. and J. D. JUSTIN, *Hydroelectric Handbook*, 2nd Ed. New York: John Wiley & Sons, Inc., 1949.

2. SKEAT, W. O. (Editor). "Manual of British Water Engineering Practice," 4th Ed. Published for the Institution of Water Engineers, 11, Pall Mall, London, by W. Heffer & Sons Ltd., Cambridge, England.

3. WALKER, R. "Microstraining Clarification of Water." *Tappi*, Vol. 48 (12), p. 129A, 1965.

4. CAMPBELL, R. M. "The Use of Ozone in the Treatment of Lock Turret Water." Journal of the Institution of Water Engineers, Vol. 17, p. 333, 1963.

5. CAMPBELL, R. M. and M. B. PESCOD, "The Ozonization of Turret and Other Scottish Waters." Journal of the Institution of Water Engineers, Vol. 19, p. 101, 1965.

6. BERRY, A. E. "Removal of Algae by Microstrainers." *J.A.W.W.A.*, Vol. 53 (12), p. 1503, 1961.

7. EVANS, G. R. "Use of Microstrainer Unit at Denver." *J.A.W.W.A.*, Vol. 51 (3), p. 354, 1959.

11

AERATION AND DEAERATION

AERATION

The aeration of municipal water supplies either by natural or artificial means has many advantages:

(a) *Removal of Dissolved Gases* Gases such as hydrogen sulfide (H_2S) and carbon dioxide (CO_2), resulting from the decomposition of organic matter, can be largely removed by aeration. Raw water impounded in large reservoirs during the winter months under the ice tends toward anaerobic conditions, particularly at the bottom. If a hole is made in the ice and a pH electrode is lowered into the water, the pH is highest immediately under the ice and drops in value as the electrode is lowered toward the bottom. At lower depths, the oxygen content is also reduced to much lower values than near the surface. Forced draft aerators are usually installed as the first stage in the treatment of these waters and the value of this process can be judged by the obnoxious odors that are emitted from the aerator vent.

(b) *Microorganisms* There are two broad classifications of microorganisms of interest to sanitary engineers: the aerobic which depend on free oxygen in the water for their survival, and the anaerobic which can survive on chemical oxygen such as that in the sulfate radical ($SO_4^=$) reducing it to free hydrogen sulfide (H_2S) which has a repulsive rotten egg odor. Some of these organisms are said to be facultative, that is to say, they can adapt to changing conditions from aerobic to anaerobic environments. At the beginning of the winter the pond life may be actively aerobic, and by spring almost entirely anaerobic.

The fish population requires free oxygen and some fish have trouble if the oxygen content goes much below 5 mg/ℓ. Decaying algae also have a high oxygen demand. If there is insufficient oxygen to satisfy all the various life forms in the water, then anaerobic conditions will prevail and the water quality will deteriorate.

(c) *Oxidation* Aeration, particularly of well waters, will oxidize any soluble ferrous iron to the insoluble ferric iron in order that it can be removed by settling and filtration. Some well waters contain large quantities of natural gas (CH_3) that must be removed before piping the water to the distribution system. It is not unusual to see in the Canadian prairie newspapers, photographs of natural gas flames burning from a water tap where a well had been drilled through a natural gas-capped aquifer, and piped directly into a private water supply system.

Theoretical Concepts

Gases are dissolved or liberated from aerated water until equilibrium is reached between the content of each specific gas in the atmosphere and its partial pressure in the water. Henry's Law states that "the solubility of a gas in a liquid at constant temperature is proportional to the partial pressure of the gas" and can be used to calculate the amounts of oxygen and nitrogen present at saturation at any given temperature. The solubility of both nitrogen and oxygen varies with temperature and salinity. Both are less soluble at high temperatures and higher total dissolved solids. At saturation the dissolved oxygen varies from 14.7 mg/ℓ at 0°C in fresh water to 11.3 mg/ℓ at the same temperature in saline water containing 20,000 mg/ℓ of chloride. As the temperature increases the oxygen saturation value diminishes (see Table 11-1).

The gas transfer rate either into or out of solution follows the laws of gas absorption and mass transfer. Downing[2] has presented a paper on the subject of aeration in water treatment. Since aeration is of vital importance in the activated sludge treatment of sewage, the subject is extensively discussed in sanitary engineering textbooks.[1,3,4,5,6]

Design of Aerators

The objective is to expose the maximum area of water surface to the atmosphere. In order to have the most efficient transfer from one medium to the other, it is essential that there be enough turbulence between the water and air so that there will be no stagnant air-water interface films which retard the transfer rates.

TABLE 11-1 Solubility of Oxygen in Water[1] (in equilibrium with air at a pressure of 1 atmosphere.)

Temperature		Oxygen mg/liter
°C	°F	
0	32.0	14.7
1	33.8	14.3
2	35.6	13.9
3	37.4	13.5
4	39.2	13.1
5	41.0	12.8
6	42.8	12.5
7	44.6	12.1
8	46.4	11.8
9	48.2	11.6
10	50.0	11.3
11	51.8	11.0
12	53.6	10.8
13	55.4	10.5
14	57.2	10.3
15	59.0	10.0
16	60.8	9.8
17	62.6	9.6
18	64.4	9.4
19	66.2	9.2
20	68.0	9.0
25	77.0	8.2

Danckwerts[6] has suggested that one method of approaching ideal conditions is to have the liquid phase in very thin films moving over an inclined surface until it reaches an edge where it will roll itself into a droplet, fall a short distance, and break into many smaller droplets equivalent to a fine mist. The fine droplets would then settle onto another surface forming a moving film until a new droplet is formed. The whole process, in an atmosphere of forced air, is one of forming and reforming droplets in rapid succession. There are several types of equipment which come close to this concept.

Natural and Forced Draft Aerators

Each design must be considered in respect to the application, pumping cost, and blower horsepower. If a water has a high odor problem or contains other dissolved gases, the first phase of the aeration process will be the elimination of these gases before oxygen saturation commences.

What is rarely appreciated with forced or natural draft aerators or cooling towers is that they not only pick up oxygen, but they also scrub the air from all forms of airborne algae, dust, insects, and other particulate matter. A new aerator design has been pilot-plant tested at the South Tahoe, California, Public Utility Advanced Wastewater Treatment Plant. It consists of a packed tower with three corner-shaped slats fairly close together which follows very closely with Dr. Danckwert's concept.

Spray Aerators

Spray aerators are effective and providing they can be economically designed, as a decorative fountain they can be attractive; however, they have severe limitations. To produce an atomizing jet, a large amount of horsepower is required. The losses from wind carry-over and the nuisance problems can be considerable. Climatic conditions, particularly in cold regions, limit their usefulness. Information on spray aerators is given by C. R. Cox.[5]

Weirs and Waterfalls

Many natural water courses provide excellent aeration, particularly with fast-flowing shallow rivers cascading over waterfalls and weirs. Much data has been accumulated over the years on this subject.[7] Pollution on many rivers, including the Rhine in Germany and the Thames in England, has exceeded their capability to maintain a reasonable degree of oxygen saturation, and additional artificial aerators have been installed. Large surface agitators such as those used in activated sludge plants have been effective on the River Thames. Dr. Bouthillier, of the University of Alberta, developed an ingenious device for injecting air into the Red Deer River in Central Alberta.

Forced Aeration

Bubbling compressed air through water is not necessarily the most economical and efficient form of aeration. On the other hand, lifting the water 20 feet or more to flow down a tray type aerator can also be an expensive operation. Nevertheless, the forced draft tower is probably the best solution for small flows. If the quantities of water are large, surface agitators or combined surface and air jet agitators should be evaluated.

Aerating Reservoir

Depending on the depth and degree of oxygen saturation required, an air-barrier could be useful not only for the purposes of aeration but to help mix the lower water levels which are usually low in oxygen with the oxygen-rich upper layers. In this way the lower stagnant waters can be exposed to a free air-water interface at the surface and also the benefits of sunlight. These systems are sometimes used to raise the warmer water from the bottom during the winter and prevent the surface from freezing. Since very little aeration can take place once the body of water is iced over, this system has a number of side benefits which compensate for the low absorption efficiency of the air bubbles.

Disadvantages of Aeration

Occasionally, aeration immediately prior to flocculation can cause problems with the production of floc and its settling characteristics. This is often true with highly-colored waters which may contain chelated iron. In the first case, aeration will not oxidize the chelated iron, and the very light floc tends toward buoyancy due to the presence of air bubbles on the floc, preventing it from settling. In these circumstances it is better to rely on chlorine oxidation rather than aeration.

DEAERATION

Water containing dissolved oxygen, unless it is chemically inhibited, is corrosive to unprotected carbon steel and cast iron. Water is injected underground into the oil-bearing formations of oil fields to increase the pressure and to force the oil toward the production wells and up to the surface. To keep the corrosion of high pressure pipelines and injection well casings to a minimum, it is necessary to deaerate the water and to chemically scavenge the last traces of oxygen that may still be present. Not only is the wastage of metal and leakage a serious problem, but the products of corrosion can plug the oil-bearing formation. There are many other industrial applications where deaerated and degassified waters are needed.

Steam Deaeration

The deaeration of boiler feed water is also a necessary operation; otherwise corrosion of the boiler and condensate return systems becomes a serious problem. Steam deaeration is an effective process and, since hot water is required for boiler feed, steam deaeration and chemical scavenge is the obvious choice for this application.

Vacuum Tower Deaerators

For cold-water systems the vertical vacuum tower is in use in many installations where deaeration and degassification is required. A vacuum deaerator consists of a tall tower packed with rings to spread the water evenly into as many thin films as possible similar to aeration. Preferably two vacuum pumps are used to exhaust the air and dissolved gases. Because the tower is under full vacuum, the water has to be pumped from the tower. This arrangement is shown in Figure 11.1.

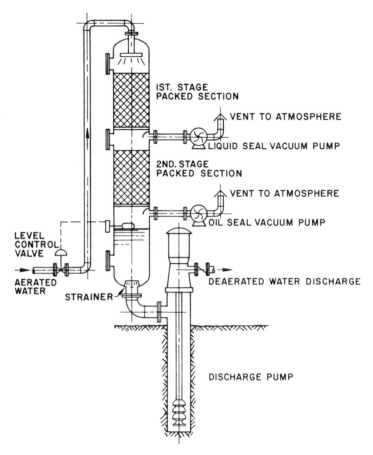

FIGURE 11-1 Vacuum Tower Deaerator

Water is pumped to the top of the tower where it falls onto a distribution plate. The rate of flow is controlled by the level of water in the bottom of the tower. The first stage packed section is filled with "Pall" rings, or other patented devices, which expose as large a surface area as possible. Ceramic Raschig rings in $1\frac{1}{2}$ to 2 inch sizes are also used, but the thick walls of the rings take up too much free volume and are heavy. Plastic rings of various designs, particularly the BASF Pall ring, have been much more effective then Raschig rings. Other similar packing designs are commonly used with equally good results. The second stage packed column is basically identical to the first. In vacuum technology the unit of vacuum is known as a Torr which is equivalent to 1 mm of mercury (Hg). See Tables 11-2, 11-3, and 11-4. This data has been provided by the courtesy of Kinney Vacuum Company.[9,10]

TABLE 11-2 Pressure Conversions[9]

Pressure To Obtain → Multiply ↓ By Factor	atm.	bar	psi	kg/cm^2	in. Hg (32°F)	in. Hg (60°F)	in. H_2O (39.2°F)	in. H_2O (60°F)	torr	newton/m^2
Atmosphere	1.0	1.01325	14.696	1.0332	29.921	30.005	406.79	407.19	760.00	101325
Bar	0.98692	1.0	14.504	1.0197	29.510	29.613	401.47	401.87	750.00	100000
Pounds Per Sq. In. (PSI)	0.06805	0.06895	1.0	0.07031	2.0360	2.0418	27.679	27.708	51.715	6894.8
Kilogram Per Sq. CM KG/CM^2	0.96784	0.98067	14.223	1.0	28.959	29.041	393.71	394.09	735.56	98067
Inches of Mercury in. Hg (32°F)	0.03342	0.03386	0.49116	0.03453	1.0	1.0028	13.596	13.609	25.400	3386.4
Inches of Mercury in. Hg (60°F)	0.03333	0.03377	0.48977	0.03443	0.99718	1.0	14.022	13.570	25.329	3376.9
Inches of Water in. H_2O (39.2°F)	0.00245	0.00241	0.03613	0.00246	0.07355	0.07132	1.0	0.9678	1.8063	249.08
Inches of Water in. H_2O (60°F)	0.00246	0.00249	0.03609	0.00254	0.07348	0.07369	1.0333	1.0	1.8655	248.84
Torr mm Hg (32°F)	0.00158	0.00133	0.01934	0.00136	0.03937	0.03948	0.5536	0.5358	1.0	133.32
Pascal newton/meter^2	98.6×10^{-5}	1.00×10^{-5}	14.5×10^{-5}	10.1×10^{-6}	29.5×10^{-5}	29.6×10^{-5}	4.03×10^{-3}	4.02×10^{-3}	7.50×10^{-3}	1.0

1 torr = 1 mm Hg = 1000 millitorr = 1000 micron(μ)Hg = 1333.2236 dyne/cm^2 : 1 dyne/cm^2 (bayre) = 1 newton/m^2 (pascal)

TABLE 11-3 Volume Conversions[9]

	Volume				
To Obtain → Multiply ↓ by Factor	$in.^3$	$ft.^3$	m^3	Li.	gal.
Cubic Inch	1.0	5.787×10^{-4}	1.6387×10^{-5}	1.6387×10^{-2}	4.329×10^{-3}
Cubic Feet	1728	1.0	2.8317×10^{-2}	28.317	7.481
Cubic Meter	6.1023×10^{-4}	35.314	1.0	1000	264.17
Liter	61.02	3.5316×10^{-2}	0.001	1.0	.26418
Gallon (U.S. Liq.)	231	13.368×10^{-2}	3.7854×10^{-3}	3.7854	1.0

TABLE 11-4 Flow Conversions[9]

Rate of Flow To Obtain → Multiply ↓ by Factor	torr- cfm	torr- l/sec.	torr- m^3/hr
Torr—Cubic Foot Per Minute	1.0	0.472	1.699
Torr—Liters Per Second	212	1.0	3.60
Torr—Cubic Meter Per Hour	0.5886	0.2777	1.0

1 atm-cm^3/sec = 1 sec = 760 millitorr-l/sec
1 millitoor-CFM = 1 micron-cfm = .001 torr-cfm
1 millitorr-l/sec = .001 torr-l/sec
1 millitorr-m^3/hr = .001 torr-m^3/hr

REFERENCES

1. FAIR, G. M., J. C. GEYER, and D. A. OKUN, *Water and Wastewater Engineering*, Vol. 2. New York: John Wiley & Sons Inc., 1968.

2. DOWNING, A. L. "Aeration in Relation to Water Treatment." Proceedings of the Society for Water Treatment and Examination, Vol. 7, part 2, p. 66, 1958.

3. SAWYER, C. N. and P. L. MCCARTY, *Chemistry for Sanitary Engineers*, 2nd Ed. New York: McGraw-Hill Book Company, 1967.

4. CAMP, T. R. *Water and Its Impurities.* New York: Reinhold Publishing Corp., 1963.
5. COX, C. R. "Operation and Control of Water Treatment Processes." World Health Organization, Geneva, 1969.
6. COULSON, J. M. and J. F. RICHARDSON, *Chemical Engineering*, Vol. 2. London: Pergamon Press, 1955.
7. GAMESON, A. L. H. "Weirs and the Aeration of Rivers." Institute of Water Engineers, London, Vol. 11, (6), p. 477, 1957.
8. OSTROFF, A. G. *Introduction to Oil Field Water Technology.* Englewood Cliffs, N.J.: Prentice-Hall, Inc., 1965.
9. "Selector Manual for Rotary Vacuum Pumps." Kinney Vacuum Company, Boston, Mass., U.S.
10. SPINKS, W. S. *Vacuum Technology.* Palisade, N.J. Franklin Publishing Company, Inc., 1962.

12

MIXING, FLOCCULATION AND CLARIFICATION

MIXING Chemical solutions for flocculation and coagulation are added prior to the settling and clarification basins. The designers of earlier water treatment plants had little appreciation of the importance of rapid mixing. Chemicals were added to the influent water to mix naturally with no attempt to mechanically accelerate the process. When chemical reaction kinetics were studied, it was realized that flocs form almost instantaneously when the chemicals first come into contact with the water, and the importance of rapid mixing was appreciated. Much of the research has been done by Camp.[1] Some of his conclusions are as follows:

(a) The floc volume concentration and size distribution is determined during flocculation by the mean velocity gradient and time, and both the concentration and the size of the floc may be varied over a wide range by changes in velocity gradient and time.

(b) After flocculation is complete, at a particular velocity gradient, continued mixing at the same velocity gradient results in little change in floc volume concentration and size distribution.

(c) Rapid mixing at sufficiently high velocity gradients of floc already formed will dispense such floc into colloidal particles.

The problems inherent in mixing a small quantity of chemical solution into a large quantity of water are discussed in Chapter 18. Complete mixing of the chemical into the water before the two react is strongly emphasized. If the mixing is too slow some parts of the water will be subjected to chemical overdose while other parts, the larger majority of the flow, will have no chemical dosage at all. If agitation is too violent, the flocs initially formed will be broken into colloids. This can be worse than the initial condition, since colloids may not recoagulate and are too small to settle out. No clear-cut guidelines appear to exist for determining the optimum power dissipation, design of mixer, or detention time required to disperse chemicals into a flow of water. However, detention times of 20 seconds or less are not uncommon, and mixing units providing 1 to 2 hp for each cubic foot per second of flow rate (450 U.S. gpm or 375 igpm) are not unreasonable.[2]

VELOCITY GRADIENTS

Much of the modern design data is in terms of velocity gradients. To illustrate the meaning of this term, if two particles of fluid in a tank are 0.1 feet apart and one is moving with a speed of 1 fps relative to the other, the velocity gradient between them is 10. The velocity gradient is therefore the relative velocities of two particles divided by their distance apart. Velocity gradients can be calculated from the following equations.[3]

For baffled basins:

$$G = \sqrt{\frac{62.4\,H}{\mu T}}$$

For mechanical agitation:

$$G = \sqrt{\frac{550\,P}{V \mu}}$$

where:

G = velocity gradient has the dimensions of $\frac{\text{Velocity}}{\text{Distance}}$ which is ft/sec ÷ ft and is equal to (sec^{-1}).

H = head loss due to friction in feet.

μ = viscosity in $\frac{\text{lb sec}}{\text{ft}^2}$ units which is equivalent to centipoise × 2.088 × 10^{-5} (see Table 12-1 on following page).

T = detention time in seconds.

V = volume of basin in cubic feet, and

P = water horsepower.

Camp[1] studied the impact of rapid mixing on floc formation. He reported that mixing at G values of 500 sec^{-1} to 1000 sec^{-1} for about 2 minutes produced essentially complete flocculation and that prolonged rapid mixing at this intensity accomplished practically nothing more. For rapid mixing with small G values,

TABLE 12-1 Viscosity of Water

Temperature		Viscosity	
32°F	0°C	1.79	Centipoise
40	4.4	1.54	
50	10.0	1.31	
60	15.6	1.12	
70	21.1	0.98	
80	26.7	0.86	
90	32.2	0.81	
100	38.0	0.77	

a 2 minute period was insufficient to complete the flocculation; on the other hand, mixing at excessively high G values (10,000 sec^{-1} or more) for as long a time as 2 minutes seemed to substantially retard floc formation.

The total number of particle collisions is proportional to GT. Mixing and flocculation are more efficient when imparted as controlled turbulence, caused by propeller mixers and baffles, rather than as a general rotation of the mass of water or a flow along a conduit. Reference 1 contains a graph showing the increase in mean velocity gradient (G sec^{-1}) for baffled beakers as opposed to unbaffled beakers at equal revolutions per minute of the agitator.

Other authorities[4] have found that the degree of agitation produced is of greater importance than the time over which the mixing continues and, if a special mixing chamber is provided, adequate dispersal of chemicals can be obtained in a period as short as 30 seconds to 1 minute if sufficient power is provided for the purpose. The power required for efficient mixing is given as 9 to 18 inches (228 to 457 mm) head of water which is approximately equivalent to 0.25 to 0.5 hp per 1000 igpm (1200 U.S. gpm) (76 l/sec).

Parameters found to be suitable in the United Kingdom are considerably lower than those recommended in the United States. Since viscosity is an important criterion in mixing processes and is inversely proportional to water temperature, data from one source cannot be compared with that of another unless this factor is considered. Furthermore, the water treatment process appears to have some bearing on the optimum velocity gradient to use.[3] It is suggested that the following ranges of G (sec^{-1}) should be considered:

Process	G (sec^{-1})
Turbidity and color removal (no solids recirculation)	50–100
Turbidity and color removal (Solids contact reactor) (5%—volume in suspension)	75–175
Softening (Solids contact reactor) (10%—volume in suspension)	130–200

Process	G (sec^{-1})
Softening	250–400
(Ultra high solids contact)	
(20% to 40% volume in suspension)	
Rapid Mix (Blending)	
20 sec contact	1000
30 sec contact	900
40 sec contact	790
Longer contact	700

Variable-speed driven agitations capable of a wide variety of velocity gradients should be installed. The enormously improved results in treated water quality, and the reduction in chemical dosages when optimum mixing is achieved, justifies the additional costs. It is not too difficult to design a suitable process to mix chemicals into natural water. The difficult problem is to add additional chemicals to modify already formed flocs, for example, the addition of polyelectrolytes as filter aids after the flocculation and clarifier processes. In this area of technology we do not have very good answers, as rapid mixing for dispersion of the secondary flocculant or filter aid would break the already formed floc into colloids. A considerable amount of research has been devoted to the subject of flash mixing and velocity gradient.[5,6]

FLOCCULATION

Coagulation, the destabilization and initial aggregation of colloidal and finely divided suspended matter by the addition of floc-forming chemicals such as alum or one of the iron coagulants, takes place in the flash mixing zone or immediately afterwards. The flocculation process is the agglomeration of colloidal and finely divided suspended matter after coagulation by gentle stirring through either mechanical or hydraulic means.

In other words, the flocculating is where the small pin-point flocs grow into larger settleable flocs. In the days of the large square or rectangular cross-flow clarifiers it was important to develop good settling flocs that would fall to the bottom of the clarifiers, not passing over the launders. Coupled with the poor chemical mixing ability of the older plants, the flocculation section covered a substantial area and had 30 minutes or more retention time.

Where softening reactions occur the flocculating chambers provide the retention facilities and, since chemical reactions are slower in cold waters, surface waters need longer retention during the winter than in the summer. The period of retention and the degree of agitation are critical factors in the operation of flocculators for the same reasons that too much mixing can break the floc into unsettleable colloids. One authority[4] suggests a retention period of 5 to 30 minutes with paddle-tip speed velocities of $\frac{1}{2}$ to 2 feet per second. Higher velocities are required where the suspended matter is relatively heavy as, for example, in a lime-softening reaction or where a turbid river water is treated. Various designs of horizontal or vertical revolving paddles are used. In one design, reciprocating paddles move up and down in the flocculating chamber. The objective is to cause as many gentle collisions as possible between the already formed flocs with the intention of making them adhere together to form large conglomerates. Earlier designs used large wooden slats revolving at 5 to 6 rpm consuming a fair amount of horsepower in stirring and moving the water. It is possible that better results could be obtained with less power if expanded metal or heavy gauge wire mesh were used instead of the large wooden paddles. The suspended flocs would be drawn together, tending to coagulate each time they pass through an opening in the rotating wire mesh.

THEORY OF COAGULATION

The present theory of coagulation and flocculation assumes that there are surface-active electrical charges on the outside of each particle of floc.[7,8] The flocs come together because of the differences in their surface-active charges and hold together partly by this electrical attraction and partly by the cohesion between the two irregular rough surfaces. It is reasonable to assume that the differences in their electrical charges become partially neutralized as they remain in contact with each other and they assume a common electrical potential. When this occurs, the largest forces holding them together will be the cohesion between them. The larger the floc grows, the more delicate it becomes, and, although it may settle more readily because of its size, it will break up into smaller pieces more easily than the smaller, stronger flocs. Once it breaks it will not recoagulate since the surface active charges are equal. The only way to reform the floc is to introduce an entirely different coagulant to the first one and start all over again. It is therefore of the utmost importance that once a floc has been formed and allowed to grow it must be handled with increasing respect the larger it grows. It is necessary to consider the effect the flocculator paddles will have on the floc as the floc develops. Variable speed ranges of 1 to 4 should be provided. Several authorities on flocculation have suggested that there should be three or four separate flocculation compartments in series with independent flocculator drives, so that succeeding flocculators will rotate at slower shaft speeds to avoid breaking the floc.

The flocculation basins should be provided with sloping floors to enable the floc that may settle out of suspension to be drained off in the same way as clarifier sludge. The sediment should not be allowed to accumulate but should be removed from the chamber. Microorganisms will collect on the floc and populate in the settled sludge which can become septic and float to the surface on the gas bubbles they generate.

Another serious difficulty is the turbulence in the pipes, channels, or conduits from the flocculators to the clarifiers. Streamline or viscous fluid flow cannot exist under these conditions; the flow is always in the turbulent flow region. The question then arises, how should the floc be handled to prevent it from disintegrating into smaller unsettleable particles? Average flow velocities of 1 ft/second are often recommended but unless the hydraulic design of the conduits is particularly good, velocities greatly in excess of 1 ft/second would occur when the flow is diverted or changes direction. In an attempt to improve flow patterns in older existing plants, many model studies and use of tracers have been described.[9,10]

The upflow clarifiers known as solids-contact units, or sludge blanket clarifiers, to a large extent overcome many of the problems with the older flocculators and cross-flow clarifiers. By the 1950's the principle of design had been reasonably well established.[11] The influent water having received the chemical dosage passes through a flash mixer, enters a central flocculation chamber, and then moves hydraulically downward into the sludge blanket zone. The particles tend to attach themselves to the sludge blanket and the clear supernatant water rises upward and over the launder.

Flocculation mechanism is still a subject for research. Although the principles of flocculation have been well defined in many published articles,[2,3,12,13] the method used to determine the best treatment for a particular water, under the specific conditions in which treatment is required, is still a subject for experimentation not only as a research project but at the application level for each specific water treatment project.

A review of current technology of coagulation has been compiled by Te Kippe and Ham.[14,15] They have evaluated the techniques used in the design and control of the coagulation process for turbidity removal including the conventional jar test, modified jar test, speed of floc formation, visual floc size comparisons, floc density, settled floc volume, floc volume concentration, residual coagulant concentrations, silting index, filterability number, membrane refiltration, inverted gauze filtration, cation exchange capacity, surface area concentration, conductivity, zeta potential, streaming current detection, colloid titration, pilot column filtration, filtration parameters, cotton-plug filtration, and electronic particle counting.

In the conclusion to their study Te Kippe and Ham reaffirmed that the effects and interactions of the chemical variables affecting colloidal particle destabilization in water treatment are very complex. For example, the optimum values of pH and alum concentration are sufficiently dependent upon each other so that a general statement of optimum of one such variable without qualification about the other may be of little value.

"... Of the coagualtion control technique tested, the settled turbidity jar test technique was found to be of the most practical value for evaluating the settleability of a coagulated suspension. However, the electronic counting technique does give a nearly comparable indication of settleability and can measure other floc properties simultaneously...." It is unlikely that the electronic particle-counting control technique will experience widespread practical use in the near future for several reasons:

(a) The equipment is expensive.

(b) Careful operational procedures must be followed to obtain meaningful results, and

(c) Large orifices, which are not easily clogged, do not accurately measure the turbidity fraction consisting of very small particles.

In all practical studies, the really meaningful results are those which are obtained on very fresh samples under on-site conditions. This means that complex and sensitive testing apparatus is unlikely to be of much value unless a large pilot study is envisaged and the equipment can be moved to the site of the new water treatment facility.

The old jar-stirrer has also been given an uplift to a much more sophisticated model designed by Dr. A. P. Black.[16,17] The new model eliminates many of the faults with the old multiple jar-stirrer, but it is still a fairly crude approach to coagualation and flocculation problems.

COAGULATION CONTROL

Coagulation is the most important and difficult step in the operation of a water treatment plant where the raw water conditions change frequently.[18] In many cases, with modern mixed media filters and to some lesser degree with dual media filters, preclarification before filtration is not always essential. Provided that fairly accurate raw water and filtered water turbidity measurements can be obtained, it is possible to control the coagulant dosage to suit the variable demands of the natural water.

Whether there is preclarification or not, water treatment plants located on fast changing rivers must have the capability of being able to adjust their chemical dosage rates as soon as the first signs of change occur. In order to do this a measurement to determine whether or not coagulation has been achieved is a necessary first step in coagulation control. Of all the methods in use today, jar tests, electrophoresis (zeta potential), streaming current, and pilot filters, the pilot filter system has the advantage of direct measurement of filtered water turbidity. Since turbidity is affected by every variable discussed, this is the only system that has an overall relationship to the entire concept of water treatment. In practical terms, it is the quality of the end product that matters most.

In some modern water treatment plants a Coagulation Control Center (CCC) is built into the main control panel of the plant. Raw water is pumped into the CCC from the vicinity of the plant intake or further upstream if possible. The important thing is to be able to get as representative a sample of influent water as possible and as early in the process of treatment as possible. A continuous sample of 0.5 to 1.0 U.S. gpm is all that is required. The pipeline from the sampling pump to the CCC should be sized to insure that the velocity through the line is in excess of scouring velocity, so that the water arriving at the CCC is truly representative of the water entering the plant. The turbidity, pH, temperature, and any other essential parameters of the raw water are measured and recorded on a circular or strip chart. Small chemical dosage pumps for each of the chemicals used in the process are incorporated in the CCC and operate in proportion to the metered water flow. Each chemical feed has a rotameter or other flow-measuring device which can be calibrated to read directly in terms of milligrams per liter (mg/ℓ) if required. Lengths of coiled tubing are used to provide the necessary retention time equivalent to the flocculation zone of the plant. The flow is then piped into either a tube settler or directly to a filter. The filter consists of a 4, 6, or 8 inch diameter clear plastic or glass vertical cylinder of up to 8 feet in length. The filter material is of the same composition and proportions as the filter beds in the plant. It also has backwash and surface wash facilities. Lights behind the filter permit observation of the floc formation and its filtering characteristics. The turbidity and other parameters, if necessary, are measured and recorded on circular or strip charts.

When changes in the river water quality occur, the CCC immediately indicates and records the change. Since water treatment processes only reduce the amount of turbidity passing through the plant, change in the influent water is quickly manifested in the quality of the filter effluent. The plant operator then adjusts the chemical dosage to suit the new conditions by observing the effects of the change in turbidity of the filtered water.

It is rather like adjusting a television set to produce the best picture from a number of variables. It is not necessary to have a degree in electronics to adjust the controls on the television set, and likewise, specialized knowledge in chemistry or chemical engineering is not required to adjust the chemical dosage rates in the CCC. Once the new dosage rates have been "tuned in," the chemical dosages can be transferred to the main plant which will then perform in a similar fashion.

The concept of the pilot filter or Coagulation Control Center (CCC) has many practical possibilities. It can be completely automated so that it will adjust its own chemical dosages in response to changes in the quality of the influent water, which in turn can be fed into the computerized control system of the main plant once the pattern of dosage rates for known changes is formulated. Normally, however, this degree of automation is rarely justified or even practical since in most cases every change in the river is unique with regard to the optimum chemical dosage. During periods of rapid change the operator is usually needed to make the final adjustments, but with the CCC one operator can do a much better job and much more quickly than a whole team of chemists and technicians with alternative methods.

CLARIFICATION

The objective of the clarification stage is to separate the clear supernatant water from the floc. The clear water goes to the filters and the sludge settles to the bottom of the clarifier and is collected for disposal.

The clarifier is therefore a chamber for separating the solids from the liquid. All chemical reactions with the possible exception of cold-lime softening will be completed before the water enters the separation zone of the clarifier. Any opalescence or milky appearance of the water, which should have been removed in the flocculation zone, will pass over the clarifier launder and on to the filters.

In earlier designs of upflow clarifiers it was usual to design on the basis of 1.0 U.S. gpm per square foot of surface area or less for light flocs from colored waters with a small amount of turbidity, and up 1.5 U.S. gpm for lime softening or highly turbid water with good settling floc characteristics. Rise rates were often expressed in terms of feet per hour.

Crossflow and upflow clarifiers are always sensitive to temperature changes, speed of rotation of the sludge removal mechanisms, and changing flow rates. Many manufacturers of upflow clarifiers may not guarantee their product if the influent water temperature varies substantially.

TUBE SETTLERS

Until the advent of tube settlers during the mid 1960's the whole aspect of both upflow and crossflow clarification was in a state of impasse. In order to allow the particulate suspended matter in the clarifier to settle out, it was necessary to reduce the forward velocities to less than settling velocity (particularly with an upflow clarifier where the particulate matter is trying to settle out of the water by gravity while the water is flowing upward).

Tube settlers consist of a matrix of square tubes about 2 inches square and 24 to 30 inches long inclined at 60° to the horizontal. The tube modules are below the surface of the water in the clarifier, under the clarifier launder. A cross-section of a tube settler module is shown in Figure 12-1 and the general arrangement in both upflow and crossflow clarifiers is shown in Figure 12-2 and 12-3 respectively.

The larger clarifiers use steeply inclined tubes (60°), but the smaller package water treatment plants use a different configuration with tubes inclined as little as 5° to the horizontal. Much literature has been published on the theory of high rate settling applying shallow depth sedimentation by Neptune Micro-FLOC.[19,20,21,22] A number of manufacturers, as well as providing new installations, can provide equipment for converting existing clarifiers to tube settlers. See Figs. 12-4, 12-5, 12-6 and 12-7 for examples of existing installations. The principle of tube settling is not confined only to water treatment applications but also extends to waste treatment, hydraulic mining, and other sedimentation processes. Typical diagrams of small and large treatment plants are shown in Figs. 12-8 and 12-9.

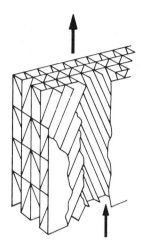

FIGURE 12-1 Tube Settler Modules[19].

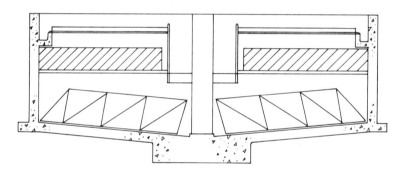

FIGURE 12-2 Tube Modules Fitted[19] Inside an Upflow Clarifier.

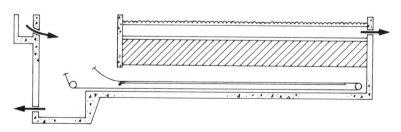

FIGURE 12-3 Tube Modules Fitted Inside[19] a Crossflow Clarifier.

FIGURE 12-4 Tube Settlers Installed in a 60-foot Diameter Circular Clarifier; Original Capacity 3 mgd, Increased to 6 mgd, at Buffalo Pound, Saskatchewan, Canada (Courtesy, NEPTUNE MicroFLOC, Inc.).

FIGURE 12-5 Floc Barriers Installed in a Graver Reactivator at Pickering Water Treatment Plant, Ontario, Canada (Courtesy, Ecodyne Limited, Graver Water Division).

FIGURE 12-6 Chevron-shaped Tube Settlers in a Package Precipitator (Steel Construction) (Courtesy, Permutit Company, Paramus, N.J.).

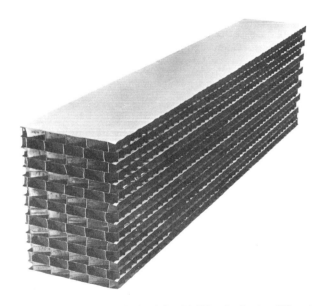

FIGURE 12-7 Tube Settler Module with Tubes Inclined at 60° to the Horizontal (Courtesy, NEPTUNE MicroFLOC, Inc.).

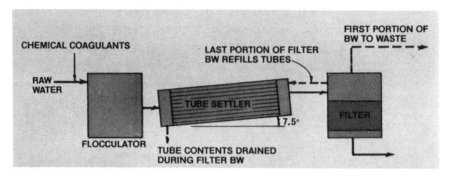

FIGURE 12.8 Diagram of Flocculation—Tube Settler—Filter Plant with 7.5° Tubes (Courtesy, NEPTUNE MicroFLOC, Inc.).

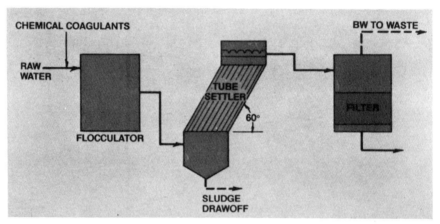

FIGURE 12-9 Diagram of Flocculation—Tube Settler—Filter Plant with 60° Tubes (Courtesy, NEPTUNE MicroFLOC, Inc.).

THEORY OF TUBE SETTLERS It was recognized by Hazen[23], in 1904, that settling basin efficiency is primarily dependent upon basin depth and overflow rate and independent of detention time. He proposed basin depths of as little as 1 inch. Over 25 years ago Camp[24] proposed settling basin depths of 6 inches and total settling basin detention time of 10 minutes. For example, a suspended particle capable of settling in perfectly calm and quiescent waters completely devoid of movement, may have an average settling velocity of 1 inch per minute. If the settling chamber is 10 feet deep and the particle enters the chamber close to the surface near the top of the chamber, it would theoretically require 2 hours for the particle to reach the bottom of the chamber. However, the conditions prescribe quiescent water which is completely devoid of any movement whatsoever and can never exist. Since there is always movement in the water the particles may never reach the bottom of the tank at all. However, if the water is slowly flowing upward through an inclined tube in viscous or streamline flow devoid of turbulence, the suspended particle can fall by gravity, unhindered by eddy currents and other disturbances. Considering a

2 inch square tube held at 60° to the horizontal, the greatest distance a particle moving up to the top side of the tube would have to fall in order to settle on the bottom of the tube would be 4 inches. If it is assumed that the particle has a settling rate of 1 inch per minute, then the particle entering the tube would settle within 4 minutes or less. If the tube is 24 inches long and the rise rate through the clarifier is 8 feet per hour (1.5 U.S. gpm per square foot) then the water will be inside the 2 foot long tube for 15 minutes. Since the particles only require 4 minutes to settle, it can be assumed that the particle will settle out about $\frac{1}{4}$ of the way up the tube. See Figure 12-10.

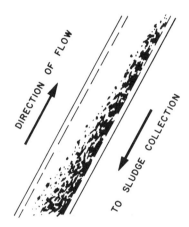

FIGURE 12-10 Inclined Tubes Permit[19] the Solids to Separate Quickly and Slide to the Bottom of the Clarifier as Sludge.

When the particle has come to rest on the inclined surface of the tube it will form a sludge with other particles and gently slide down the tube, falling to the bottom of the clarifier. Since the once separate discrete particles in suspension have joined together with other similar particles to form a heavier-than-water sludge, they will fall rapidly through the turbulent zone of the clarifier into the sludge blanket at the bottom of the tank. The sludge will then be raked across the floor of the clarifier and be discharged to waste.

REYNOLD'S NUMBER (Re)

With the older clarifiers, before the introduction of tube settlers, viscous flow conditions were impossible to achieve. The Reynolds number is important in hydraulic calculations and it has an interesting origin. Reynolds was engaged in research on blood and its conditions of flow through the arteries and veins. From his studies the dimensionless number known as the Reynolds number (Re) was derived. In its initial form it was expressed in grams, centimeters, and seconds (CGS) of the metric system, but since it is dimensionless it can be expressed in any consistent system of measurement. For engineering problems it is convenient to use the foot, pound, second system (fps). The Reynolds number is defined as:

$$Re = \frac{Dup}{\mu}$$

Where:

Re = Reynolds number.

D = Diameter of the pipe in feet.

u = Velocity in feet per second.

p = Density of the fluid in pounds per cubic foot, and

μ = Viscosity of the fluids in foot-pound-second units (fps) which is equivalent to centipoise $\times$ 0.000672.

The fact that the Reynolds number is dimensionless can be demonstrated as follows:

$$Re = \frac{Dup}{\mu} = \frac{L \times \frac{L}{T} \times \frac{M}{L^3}}{\frac{M}{LT}}$$

Where:

L = Length (i.e., diameter).

T = Time in seconds.

M = Mass, and

$$Re = \frac{L \times L \times M \times L \times T}{T \times L^3 \times M} = \frac{L^3 MT}{L^3 MT}$$

which cancels, leaving a number without a unit.

Equivalent values for diameter (D) in the Reynolds number, for several cross sections, are as follows:[25]

Shape or Cross Section	Equivalent Diameter
Pipes running full	D
Annulus (Where D = outer internal diameter) (Where d = inner external diameter)	$D - d$
Square, side = D	D
Rectangle, sides a and b	$\frac{2ab}{a+b}$
Open channels or partly filled ducts (Rectangle Depth = D. Width = W)	$\frac{4WD}{W+2D}$
Semicircle (Free surface on diameter D)	D

If the Reynolds number, no matter whether it is calculated in cgs or fps units is less than 2100, the flow is always viscous. If it is 4000 or greater, the flow is always turbulent depending on the configuration of the tube. However, in the design of tube settlers, the flow is always in the viscous flow range. The following example will serve to demonstrate why a clarifier without tube settlers will always be in a state of turbulence and with tube settlers the flow will be viscous.

With reference to Figure 12-2 consider a circular upflow clarifier 60 feet in diameter with an inner flocculation zone of 10 feet diameter. The surface area through which the water must rise in order to overflow the launders is as follows:

$$\text{Area of annulus} = \frac{(D^2 - d^2)\pi}{4}$$

$$= \frac{(60^2 - 10^2)\pi}{4} = 2750 \text{ ft}^2$$

If the clarifier is operating at a rise rate of 8 feet per hour, or 1.0 U.S. gpm per square foot, the capacity of the clarifier would be 2750 U.S. gpm equivalent to 3.96 mgd (U.S.). Assuming the water is at 50°F, has a density of 62.41 pound per cubic foot, and an absolute viscosity of 1.31 centipose, the Reynolds number for the upflow clarifier without tubes would be:

$$\text{Re} = \frac{(D-d) \times u \times p}{\mu} = \frac{(60-10)\left(\frac{8}{60 \times 60}\right)(62.41)}{(1.31)(0.000672)} = 7880$$

Since 7880 is greater than 4000, the upward flow must therefore always be turbulent. The water density (p) and its viscosity (μ) are constants for any particular set of conditions; in order to maintain a constant capacity of 2750 U.S. gpm, the only changes that can be made to reduce the Reynolds number are in the rise rate. If it is decided to reduce the rise from 8 ft/hour (1.0 U.S. gpm per square foot) to 6 ft/hour (0.75 U.S. gpm per square foot), it will be necessary to increase the cross sectional area of the clarifier by increasing the outer diameter (D) to 69 feet and the effective surface area in the annulus to 3660 square feet. The Reynolds number under these conditions would be 7000 which is still well within the turbulent range. No significant improvement in clarifier operation can be anticipated even though the clarifier has increased in total area by 33.3% and the capital costs will have increased proportionately.

If the tube modules are installed in the 60 foot diameter clarifier using the 2 inch square tubes and the original rise rate of 8 feet per hour, the diameter (D) component of the Reynolds number is reduced from 60 feet to 2 inches or 0.1667 feet and the new Reynolds number will be:

$$\text{Re} = \frac{(0.1667)\left(\frac{8}{60 \times 60}\right)(62.41)}{(1.31)(0.000672)} = 26.26$$

well within the viscous flow range of less than 2100.

If the particulate matter is capable of settling at the rate of 1 inch per minute or better, it should be possible to triple the capacity of the 60 foot diameter clarifier from 2750 U.S. gpm to 8250 U.S. gpm and still be able to settle the particles in the tube settler. The rise rate in the clarifier will have increased from 8 ft per hour (1.0 U.S. gpm per square foot) to 24 ft per hour (3.0 U.S. gpm per square foot). The retention time in the 2 ft long tube will now be 5 minutes instead of the previous 15 minutes, but since the flow up the tube will still be in viscous flow, it should still be able to settle out in 4 minutes and should therefore come to rest $\frac{4}{5}$ or 80% of the distance up the tube. The Reynolds number will be as follows:

$$\mathrm{Re} = \frac{(0.1667)\left(\frac{24}{60 \times 60}\right)(62.41)}{(1.31)(0.000672)} = 78.8$$

still well below any possibility of turbulent flow.

Not only is this a theoretical concept but it is a practical one which has been used on numerous occasions. It is not always possible to triple the former capacity of an old clarifier by the use of tube settlers since in many cases the piping and valves are incapable of carrying the additional flow, but it is frequently possible to double the capacity without too much difficulty.

With new construction and the use of tube settlers, the flow rates can be increased from the former conservative value of 1.0 U.S. gpm per square foot to 3.0 and even 5.0 U.S. gpm per square foot, but each project must be carefully engineered, particularly at the higher flow rates.

FUTURE DEVELOP-MENTS With the advent of tube settlers, new polyelectrolyte coagulant aids, developments using magnesium carbonates,[17] and other technological advances, there has been a trend toward a new concept in clarification. Essentially the process of coagulation and clarification is simply one of reacting two or more chemicals together and settling the resulting precipitate from the clear liquid. In the past it was believed that large tanks with several hours of contact and retention time were necessary for these reactions to go to completion and for the precipitates to settle. However it has now been demonstrated that the chemical reactions in water treatment do take place fairly rapidly providing the water is sufficiently above freezing to allow adequate mobility of ions. If the flash mixing processes are as efficient as possible, together with new coagulant aids, small pin-point flocs can be formed in several minutes. With tube settlers the need for producing large heavy flocs is no longer a necessity since the settling zones are not turbulent and in a state of random flow. New plants are being built using tube settlers of complicated design compared to the original devices. Some of the new plants will not have conventional clarifiers.

REFERENCES

1. CAMP, T. R. "Floc Volume Concentration." *J.A.W.W.A.*, Vol. 60 (6), p. 656, 1968.
2. *Water Quality and Treatment.* Prepared by *AWWA*, 3rd Ed. New York: McGraw-Hill Book Company, 1971.
3. "Water Treatment Plant Design." Published by AWWA Inc., 1969.
4. SKEAT, W. O. *Manual of British Water Engineering Practice.* Published for the Institution of Water Engineers, London. Cambridge: W. Heffer & Sons Ltd., 1969.
5. MORROW, J. J. and E. G. RAUSCH, "Colloidal Destabilization with Cationic Polyelectrolytes as Affected by Velocity Gradients." Presented at the AWWA 93rd annual conference, Las Vegas, Nevada, May 13–18, 1973.
6. VRALE, L. and R. M. JORDEN, "Rapid Mixing in Water Treatment." *J.A.W.W.A.*, Vol. 63, p. 52 (1), 1971.
7. RIDDICK, T. M. "Zeta Potential: New Tool for Water Treatment." (Part I and II), *Chemical Engineering*, June 26, 1961, (pages 121–126) and July 10, 1961, (pages 141–146).
8. SAWYER, C. N. and P. L. MCCARTY, *Chemistry for Sanitary Engineers.* 2nd Ed. New York: McGraw-Hill Book Company, 1967.
9. HUMPHREYS, H. W. "Hydraulic Model Study of a Settling Basin." Presented at the AWWA 93rd annual conference, Las Vegas, Nevada, May 13–18, 1973.
10. HUDSON, H. E. "Use of Tracers in Water Treatment Plants." Presented at the AWWA 93rd annual conference, Las Vegas, Nevada, May 16, 1973.
11. PRAGER, F. D. "The Sludge Blanket Clarifier." *Water and Sewage Works*, Vol. 97 (4), p. 143, April, 1950.
12. JORDEN, R. M. "Coagulation-Flocculation." Part I, II, and III, *Water and Sewage Works*, Vol. 118, Part I—p. 26 1971, Part II—p. 431 1971, and Part III—p. 77 1972.
13. HUDSON, H. E. and J. P. WOLFNER, "Design of Mixing and Flocculating Basins." *J.A.W.W.A.*, Vol. 59, p. 1257, 1967.
14. TE KIPPE, R. J. and R. K. HAM, "Apparatus to Examine Floc-Forming Process." *J.A.W.W.A.*, Vol. 62, p. 260, April, 1970.
15. TE KIPPE, R. J. and R. K. HAM, "Coagulation Testing: A Comparison of Techniques." Part I—*J.A.W.W.A.*, Vol. 62, p. 594, September, 1970, and Part II—*J.A.W.W.A.*, Vol. 62, p. 620, October, 1970.
16. BLACK, A. P. and R. H. HARRIS, "New Dimensions for the Old Jar Test." *Water and Wastes Engineering*, Vol. 6, p. 49, December, 1969.
17. THOMPSON, C. G., J. E. SINGLEY, and A. P. BLACK, "Magnesium Carbonate—a Recycled Coagulant." Part I—*J.A.W.W.A.*, Vol. 64, p. 11, January, 1972 and Part II—*J.A.W.W.A.*, Vol. 64, p. 93, February, 1972.
18. CONLEY, W. R. and R. H. EVERS, *Coagulation Control.* Published by Corvallis, Oregon: NEPTUNE MicroFLOC, Inc., 1967.
19. CULP, G. L. "A Better Settling Basin." *The American City*, Vol. 84 (1), p. 82, January, 1969.

20. Hansen, S. P., G. L. Culp, and J. R. Stukenberg, "Practical Application of Idealized Sedimentation Theory." Corvallis, Oregon: NEPTUNE MicroFLOC, Inc.

21. Culp, G. L., S. P. Hansen, and G. Richardson, "High Rate Sedimentation in Water Treatment Works." Corvallis, Oregon: NEPTUNE MicroFLOC, Inc.

22. Hansen, S. P. and G. L. Culp, "Applying Shallow Depth Sedimentation Theory." Corvallis, Oregon: NEPTUNE MicroFLOC, Inc.

23. Hazen, A. "On Sedimentation." *Transactions of American Society of Civil Engineers*, Vol. 53, p. 45, 1904.

24. Camp, T. R. "Sedimentation and Design of Settling Tanks." *Transactions of American Society of Civil Engineers*, Vol. III. (2285), p. 895, 1946.

25. Clarke, L. *Manual for Process Engineering Calculations*. New York: McGraw-Hill Book Company, 1947.

13

COLOR REMOVAL

DEFINITION OF COLOR

Natural waters are often colored in various shades of brown from very pale straw to dark chocolate. Some waters appear to be a bluish-green but this is due to algae and possibly other microorganisms which may be strained and filtered out of the water leaving the filtrate clear and bright. True color cannot be removed by filtration alone. Colored waters are frequently found in areas where there is muskeg, decomposing vegetation, spring runoff from fields of stubble, organic acids, and other sources.

ORGANIC ACIDS

Much research remains before a complete understanding of color in natural water is reached. It is understood that there are usually three types of organic acids present: humic acid, hymatomelanic acid, and fulvic acid.

R. F. Packham[1] studied samples of water from seven sources in England and Scotland. His results are contained in Table 13-1.

There is some doubt as to whether or not humic acid is a true solution or colloidal, but fulvic acid is believed to have true color and hymatomelanic acid is somewhere between the two. It seems likely, however, that humic acids may

TABLE 13-1 Proportions of Various Fractions in Humic Substances Isolated from Natural Waters

Source and Water Authority	Sample No.	Humic Acid %	Hymatomelanic Acid %	Fulvic Acid %
River Thames, Medmenham	1	3.9	13.1	83.0
Loch Humphrey (Dumbarton B.C.)	2	4.4	10.2	85.4
Loch Turret (Loch Turret Water Board)	3	3.8	10.2	86.0
Longdendale Reservoir (Manchester C.B.C.)	4	16.2	18.3	65.5
Stocks Reservoir (Fylde Water Board)	5	5.0	18.5	76.5
River Hull, Hempholme (Kingston upon Hull C.B.C.)	6	5.2	13.9	80.9
Burnhope Reservoir (Sunderland and South Shields Water Co.)	7	10.0	32.2	57.8

be in true solution but that a portion of the molecules is sufficiently large to exhibit colloidal properties.

In the interpretation of the chemical and physical analysis of a highly colored water, notice must be taken of the pH at the time the analysis is made. Iron and manganese are often present in highly colored waters since the water is usually low in pH and alkalinity and therefore able to dissolve the iron and manganese that may be present. One of the problems is the ability of the humic substances to chelate the iron and possibly manganese so that they do not readily come out of solution. They may be completely missed by some field water-testing kits. One source of water having approximately 100 color units also had 5 mg/ℓ of iron which was not apparent when field tested but registered when analyzed in accordance with *Standard Methods*.[2] When iron is chelated with color, the percentage of iron removal corresponds closely to the amount of color removed. Ungar and Thomas studied the measurement of coloring matter in natural waters[3] and cite a significant number of literature references.

COLOR REMOVAL PROCESSES

Although colored public water supplies are not known to be harmful to health, they are objectionable for esthetic reasons. Furthermore, they are natural, slow, self-coagulators. For example, colored waters supplied to a water distribution system will often gradually lose some color by depositing organic precipitates in the distribution system. The sediments are disturbed by scouring velocities during a fire or other heavy draw-off and become suspended in the water to the detriment of the consumers. A series of samples of a highly colored muskeg water were collected and shipped 600 miles. When the samples were first collected the color was over 500 APHA units but on arrival at the laboratory it was less than 20 APHA with a speck of dark brown dirt at the bottom of the bottle.

To remove color, some form of chemical oxidation followed by coagulation and filtration appears to be the best approach. Colored waters are the most difficult of all waters to coagulate, particularly those that are free from turbidity and low in mineral content. Preaeration is usually disappointing and often does more harm by supersaturating the water with air thus making the floc lighter and even more difficult to settle. On the other hand, prechlorination up to 10 mg/ℓ of chlorine usually works well followed by coagulation with alum. Ozone is said to be better for color treatment than chlorine since it is a powerful oxidizing agent and does not depress the pH as much as chlorine.[4] Ozone, together with microstrainers, has provided satisfactory treatment and color removal for Lock Turret water in Scotland.[5,6]

Potassium permanganate is also used to remove color, particularly if the color is partly due to iron and manganese. It is also interesting to learn that there is a close relationship between zeta potential (measured and frequently expressed as electrophoretic mobility) and residual color concentration.[7]

The effect of pH is also extremely important in the removal of color. Alum seems to work as well as any other coagulants, but in some cases, a pH of 5 or less is essential for good color removal. The addition of lime or soda ash to increase alkalinity is sometimes an essential prerequisite to coagulation with alum.

The mechanism by which color removal is aided by the addition of chlorine and/or ozone is not fully understood and is difficult to simulate in a jar stirrer unless there is an available supply of chlorine water. The use of sodium hypochlorite as a substitute for chlorine water is misleading in jar test studies since the caustic soda content of the hypochlorite will raise the pH of the water and intensify the color while the chlorine depresses it and reduces the color.

In cases where the water has contained an appreciable amount of mineral matter or alkalinity, sulfuric acid has been used to depress the pH, thereby reducing the alum dosage, but this approach is not always successful since the sulfuric acid is unable to produce a floc. Furthermore, if there is any suspended turbidity the reduction in pH becomes erratic.

The mechanism of color removal is really one of chemical precipitation followed by coagulation. The color is largely in the form of a true solution in its natural state and does not become a subject for coagulation until it has been destabilized. Not only is pH a significant factor which can range from as low as 3.35 to as high as 6.9 for optimum color removal, but the water temperature is also critical. For example, the optimum coagulation pH for a highly colored water decreases from pH 4.10 at 3°C (37.4°F) to pH 3.35 at 42°C (107.6°F).[8]

Color removal by the use of demineralizing resins is disappointing because the resins quickly become ineffective due to coating of organics on the active surfaces. The porous surfaces of activated carbon undergo a similar sealing of the surface, which quickly inactivates the carbon.

In some cases a plant can be controlled by adding sufficient alum to maintain the pH at the optimum isoelectric point. However, with highly colored surface waters the optimum pH and other parameters will change and it is necessary to frequently check the optimum pH for color removal for any particular state of river flow; this can sometimes be accomplished by the use of zeta potential. Riddick designed the water treatment plant at Waterford, New York, to be controlled with the use of zeta potential measurement in an electrophoresis cell.[9]

Once color has been destabilized by chemical oxidation, coagulation and flocculation with alum, aided by a polyelectrolyte, will usually produce a stable lightweight floc which sometimes prefers to float rather than settle. Some measure of flocculation by a mechanical agitator is often necessary to insure adequate color destabilization. The bulk of the floc can usually be removed in a tube settler, and it appears that the tube settlers approaching the horizontal are the most effective. With this type of tube settler floating flocs can be retained almost as effectively as settling, heavier-than-water flocs, and the bulk of the copious floc can be retained in the tubes which are eventually cleaned by backwashing. From the tube settler the water flows to a filter. Mixed media filters are preferred since they are better equipped to use polyelectrolyte filter aids to "polish" the water. In view of the sticky nature of colored floc, filter rates should be limited to 3 U.S. gpm per square foot for suitable mixed media and less for other types of filters in order to prevent short filter runs. Surface wash equipment, or facilities to manually rake the filter bed, are essential to avoid accumulations of mud balls.

REFERENCES

1. PACKHAM, R. F. "Studies of Organic Color in Natural Water." Proceedings of the Society of Water Treatment and Examination, Vol. 13, part 4, page 316, 1964.

2. *Standard Methods for the Examination of Water and Wastewater*, 13th Ed. Published jointly by *AWWA*, *APHA* and *WPCF*, 1971.

3. UNGAR, J. and J. F. J. THOMAS, "Further Studies on the Measurement of Organic (Coloring) Matter in Natural Waters." Department of Mines and Technical Surveys, (Bulletin TB 39) Mines Branch, Ottawa, Canada, August, 1962.

4. SKEAT, W. O. *Manual of British Water Engineering Practice*, 4th Ed., Vol. III, page 246. Published for the Institution of Water Engineers, London. Cambridge, England: W. Heffer & Sons Ltd., 1969.

5. CAMPBELL, R. M. "The Use of Ozone in the Treatment of Loch Turret Water." *Journal of the Institution of Water Engineers*, Vol. 17, page 333, 1963.

6. CAMPBELL, R. M. and M. B. PESCOD, "The Ozonization of Turret and Other Scottish Waters." *Journal of the Institution of Water Engineers*, Vol. 19, page 101, 1965.

7. *Water Quality and Treatment*. 3rd Ed. Prepared by the American Water Works Association. New York: McGraw-Hill Book Company, 1971.
8. "Coagulation and Color Problems." Joint Report, *J.A.W.W.A.*, Vol. 62, page 311, 1970.
9. RIDDICK, T. M. "Zeta Potential, New Tool for Water Treatment." *Chemical Engineering*, June 26th, 1961, (pages 121–126) and July 10, 1961 (pages 141–146).

14

LIME SODA SOFTENING

LIME SODA SOFTENING Water hardness results almost entirely from the presence of calcium and magnesium ions. Calcium can be reduced to its lowest saturation level by adjusting the pH to the range 9.0 to 9.4. Magnesium as well can also be reduced by adjusting the pH to approximately 10.2 to 10.6. To precipitate magnesium it is usual to compromise between pH 9.4 and 10.2. See equilibrium concentrations of calcium and magnesium ions in Figure 14-1.[1]

Water can also be softened by base exchange zeolites where the calcium and magnesium ions are exchanged for sodium ions. There are other softening processes including complete demineralization (see Chapter 22) where the cations (Ca, Mg, Na, etc.) are exchanged for hydrogen ions (H) and the anions (HCO_3, SO_4, Cl, etc.) are exchanged for hydroxide ions (OH).

SHOULD THE WATER BE SOFTENED? Whenever a new source of water supply is developed, particularly if the water is hard, the question always arises as to whether it should be softened or not and, if so, to what degree and which process should be used?

There can be no doubt that soft waters are pleasant to use for washing, bathing, and cleaning. However, with the use of detergents and sequestering

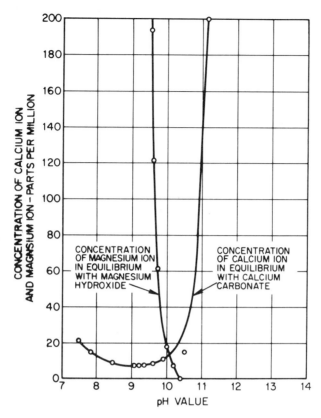

FIGURE 14-1 Equilibrium Concentration of Calcium and Magnesium Ions Plotted Against pH (Used with Permission of McGraw-Hill Book Company).

agents the insistence to soften water to improve laundry and to prevent "bath tub rings" is not as critical as it was 10 to 15 years ago. Industrial consumers have their own water conditioning equipment designed to treat water to their specific requirements and are not dependent on municipal softeners.

One of the biggest problems in assessing the whys and wherefores of water softening is the disposal of treatment plant wastes. Formerly it was acceptable to return the clarifier sludge and filter backwash into the river. Alum sludge is one of the difficult disposal problems since it cannot, at this time, be economically and satisfactorily reclaimed. Lime softening using magnesium carbonate[2] as a recycling coagulant instead of alum, has only recently been developed and could be a possible solution to some of the present water treatment problems. Lime is used to precipitate the calcium and magnesium at a high pH (above 11.0) which should also provide disinfection where adequate contact time is available. This may eliminate the need for prechlorination and essentially provide complete removal of any iron or manganese that may be present. The sludge from the clarifier contains calcium carbonate and magnesium hydroxide at a pH of 11 or above. Carbon dioxide is bubbled into the sludge to selectively solubilize the remaining magnesium as magnesium bicarbonate. The magnesium bicarbonate can be recovered by vacuum filtration, and the filtrate can be recycled and reused as the coagulant in place of alum. The filter cake from the vacuum filter, composed of calcium carbonate and other impurities, can then be recalcined, if in sufficiently large quantities to justify the cost of a kiln, or it can be disposed of as landfill. The treated water will also be carbonated with carbon dioxide to a pH of approximately 9.0 according to the requirements of the stability index.

LIMITATIONS OF LIME SODA SOFTENING

Lime soda softening will theoretically reduce the calcium and magnesium to approximately 30—35 mg/ℓ as $CaCO_3$, but in practice this degree of softening is rarely achieved. Normally, water of 80 to 100 mg/ℓ hardness as $CaCO_3$ is reasonable, providing the pH and alkalinity have been suitably adjusted by recarbonation or acid addition to insure that the water will be stable, neither scale forming or corrosive. Any further reduction in hardness should be left to the consumer.

HEALTH ASPECTS

There is a growing concern among public health authorities over epidemiological studies in several countries that appear to indicate higher death rates from cardiovascular diseases in areas using soft drinking water compared to areas with hard drinking water.[3] There seems to be some doubt as to why there should be less heart disease in areas high in calcium and magnesium compared to soft water areas, and the possibility of plumbo-solvency of soft waters in districts using lead pipes should not be overlooked. Certain metals and minerals are more soluble in soft waters which are generally more corrosive than hard waters. It is possible that the real causes for the higher disease rate will not be known until it is possible to obtain more detailed chemical analyses with particular reference to trace elements. There have been no directives from the medical profession to put limitations on the softness of drinking waters. Similarly, there is no program to add hardness to those natural waters which have very low calcium and magnesium content.

Sodium is not considered to be a serious health hazard in drinking waters within the limitations of the taste tolerance of 200 to 450 mg/ℓ as sodium chloride.[4] Softening water by means of sodium zeolite exchangers increases the sodium content since the calcium and magnesium ions are exchanged within the zeolite for sodium ions. People assigned to a sodium-free diet should avoid using waters softened by base exchange.

DISINFECTION AND VIRUS INACTIVATION

There has been growing concern in the water industry with the increasing evidence that chlorination on its own is not as effective in the inactivation of viruses as it is with bacteria. Fortunately, the excess lime-soda-ash process has resulted in 99.45 percent virus reduction.[5] High pH levels, lime flocculation, and rapid sand filtration through 8-inch deep filters have achieved a maximum of 99.997 percent removal.

QUALITY OF LIME SOFTENED WATER

The quality of a well treated lime-softened water supply is superior to that of waters softened with most other treatment processes. The water is low in color and turbidity and is often described as clear and bright. If it has been adequately recarbonated so that there is little or no caustic alkalinity, the water will not have a harsh feeling or leave the skin coarse and rough after washing.

Lime is used to precipitate the calcium and magnesium carbonate hardness, also called temporary hardness, i.e., $Ca(HCO_3)$ and $Mg(HCO_3)_2$. It will also react with the CO_2 and precipitate it as $CaCO_3$. Soda ash (sodium carbonate, Na_2CO_3) is used to precipitate the noncarbonate hardness also called permanent hardness, i.e., $CaCl_2$, $CaSO_4$, $MgCl_2$, and $MgSO_4$. Alum is used as a coagulant, but in some cases it has been possible to substitute one of the polyelectrolytes. In time it is possible that the magnesium carbonate process[2] may be established with less reliance on alum as a coagulant.

TEMPORARY AND PERMANENT HARDNESS

The terms "temporary hardness," referring to carbonate hardness, and "permanent," referring to noncarbonate hardness, have a descriptive meaning which helps us to understand the meaning of hardness in water supplies. Temporary hardness refers to the bicarbonate alkalinity associated with the calcium and magnesium. If the water is boiled the bicarbonate alkalinity tends to become unstable and liberates the half-bound carbon dioxide which raises the pH slightly and converts the soluble bicarbonate radical to an insoluble one resulting in scale deposition. Boiling does not result in any significant change in the permanent hardness.

EQUIPMENT USED FOR SOFTENING

The softening of hard ground waters free of organic contaminants is a relatively simple process. Aeration to remove the free carbon dioxide is frequently used prior to reacting with lime and soda ash. See Fig. 14-2 for a typical lime feeder and slaker. It is not always necessary to use soda ash particularly if the noncarbonate hardness is low. The softening reaction usually takes place in a

FIGURE 14-2 Lime Feeder and Paste-type Slaker for Water Softening and pH Adjustment; Capacities 1000 to 8000 pounds per hour (Courtesy, Penwatt Corp.—Wallace and Tiernan Div.).

clarifier fitted with a sludge removal mechanism. The pH of the treated water is usually above pH 10, and with the magnesium carbonate process, it is above pH 11. In order to reduce the pH to 8.0 to 9.0, carbon dioxide, from scrubbed flue gas, or liquified CO_2, is bubbled into the water in a recarbonation chamber immediately following the clarifier.

Reducing the pH from approximately 11 down to pH 8 to 9 will cause some further reduction in hardness by precipitating more calcium carbonate, because hydroxide alkalinity is converted to carbonate alkalinity by the carbon dioxide. By the use of carbon dioxide in the recarbonation process the pH is reduced from pH 10+ to pH 9.0–9.5, the point of minimum calcium solubility. The finely-divided precipitated calcium carbonate will turn the water from a clear solution to a turbid one and it is desirable to install a second clarifier to flocculate, coagulate, and remove the newly precipitated calcium carbonate. A second recarbonation chamber should then be installed following the second clarifier to reduce the pH from 9.0–9.5 down to pH 8.0–8.5. At pH 8.0 the equilibrium concentration of calcium ion is approximately 12–14 ppm. However, since the water, having passed through the second clarifier at a pH of 9.0–9.5, will have an equilibrium concentration of calcium ion of only 4–6 ppm, at a pH of 8 or less the water has a greater capacity for holding calcium ion in solution than the calcium available. The water may therefore be regarded as relatively stable and not likely to deposit calcium carbonate scale in the filters or the distribution system.

It is not suggested that the concentrations of calcium ion and the pH values shown in Figure 14-1 are those that will be achieved under water treatment plant operating conditions, as these values have been determined under laboratory conditions. However, it is believed that the pattern of calcium and magnesium ion equilibrium versus pH is realistic and leads to a better understanding of the process.

River and other hard surface waters can be softened in the same way as ground waters but the process is complicated by organic and other contaminants together with the frequent and often rapid changes in the raw water quality resulting from flash floods, thunderstorms, and other seasonal disturbances. In order to reduce this problem to a minimum a preclarifier is needed, thus making the plant a three-clarifier system. The primary clarifier serves to remove color and other contaminants that tend to inhibit the softening reaction. The removal of humic substances that constitute color is particularly important since colors are pH-sensitive and tend to increase in intensity as the pH values are increased in order to precipitate the calcium and magnesium ion. Fortunately, highly colored waters rarely require softening. On the other hand, river waters are often hard during the winter months and relatively soft, colored, and highly turbid during the spring and summer. It is difficult to supply a municipality with a consistently uniform year-round water supply under these conditions without the use of the three clarifiers. Edmonton, Alberta, has a treatment plant employing three clarifiers and two recarbonation chambers for treating North Saskatchewan river water. The primary clarifier may use only alum and possibly polyelectrolyte coagulant aids.

EFFECT OF INHIBITORS

The problems resulting from the presence of surface water contaminants in a lime-softening process should not be underestimated. Powell[1] writes: "One of the most important causes for the difference between the commercial performance of water-softening plants and the theoretical solubility data is the presence of contaminants in the water which retard precipitation or distort crystal formation." (The minimum solubility of calcium in water is dependent on pH but is often quoted to be about 30 mg/ℓ.) "In their presence, the apparent solubility of calcium carbonate may be increased and these same substances interfere with coagulation in the clarification of water. There are many substances in water which reduce the efficiency of chemical softening processes. Tannins, organic wastes, sewage, and other contaminants are objectionable in this respect. If they are present in a water supply, it is difficult to predict accurately the results of softening by chemicals without making laboratory jar tests and interpreting these data to determine adjustment of the treatment necessary in full scale operation."

"Chlorination will often remove the effect of these contaminants by the oxidation of the organic compounds. Such pretreatment is essential in many cases to insure good softening results. The failure of many water softening plants to produce the desired residual hardness in the treated water is traceable to failure to remove organic contaminant before adding the softening reagents."

FILTRATION OF LIME-SOFTENED WATERS

In the past, many operating problems have been experienced with the filtration of lime and lime-soda-softened waters. Most of these problems have resulted from insufficient pretreatment in the chemical reaction and clarification stages. If organic contaminants are present in the water during the chemical precipitation phase, it is possible that many of the coagulated conglomerates will not remain together but will break apart into finely divided, almost colloidal, particles. These very small discrete particles of calcium carbonate will pass through a recarbonation chamber without change, except for a possible slight reduction in their dimensions due to their contact with water of lower pH in the reaction zone of the chamber. The very small particles may adhere firmly to the sand grains of the rapid gravity filters to the extent that the filter beds grow in depth and in the sizes of the sand grains so that the beds have to be changed from time to time and replaced with new filter media.

The filter underdrains can become plugged with precipitated calcium carbonate which frequently gets into the clear-water well and even the distribution system. Some of the older rapid gravity filters used sintered porous ceramic filter underdrain systems which rapidly became plugged and had to be replaced. One of the advantages of the porous ceramic plates is the elimination of the four or more layers of graded gravel needed to support the sand. But the porous ceramic system of filter underdrains is not suitable for lime-softening plants unless there is a well controlled and adequate pretreatment of the softening and recarbonation processes ahead of the filters. On the other hand, if the pretreatment facilities are adequate and well operated, the filters do not have very much to do. Although it is believed that mixed media filters are better than dual media (i.e., anthracite and sand), they are also more expensive and may not be justified if adequate pretreatment is provided. Dual media filters can also be replaced for less money than mixed media if a process upset occurs in the softening process calling for the replacement of the filter media.

REFERENCES

1. POWELL, S. T. *Water Conditioning for Industry*. 1st Ed. New York: McGraw-Hill Book Company, 1954.
2. THOMPSON, C. G.; J. E. SINGLEY, and A. P. BLACK, "Magnesium Carbonate A Recycled Coagulant." Part I, *J.A.W.W.A.*, Vol. 64, page 11, January 1972. Part II, *J.A.W.W.A.*, Vol. 64, page 93, February, 1972.
3. CRAWFORD, M. D.; GARDNER, M. J. and MORRIS, J. N. "Mortality and Hardness of Local Water Supplies." *The Lancet*, p. 827, April 20, 1968.

4. MILLER, A. P. "Water and Man's Health." *A.I.D. Community Water Supply Technical Series*, No. 5. Office of Human Resources and Social Development. Washington, D.C., 1967.

5. SHELTON, S. P. and W. A. DREWRY, "Tests of Coagulants for the Reduction of Viruses, Turbidity and Chemical Oxygen Demand." *J.A.W.W.A.*, Vol. 65, p. 627, October, 1973.

15

RECARBONATION

RECARBONA-TION

The term recarbonation is often shortened to carbonation. In water treatment they mean the same thing and refer to the diffusion of carbon dioxide gas into a lime-softened water to convert the hydroxide and carbonate alkalinity to the more soluble bicarbonate alkalinity while at the same time reducing the pH. Lime-softened water is a supersaturated solution of calcium carbonate which will form crystals of calcium carbonate in the filters, clear-water wells, and distribution system unless the pH is lowered and the caustic alkalinity (OH) is converted to the more soluble bicarbonate (HCO_3) alkalinity.

Carbon dioxide is bubbled into the water after it leaves the lime-softening clarifier. It neutralizes the free excess lime and also converts the carbonate alkalinity to bicarbonate. Apart from the elimination of the unstable supersaturated calcium hydroxide and carbonate content which would result in after-treatment deposition and scale formation, the harsh treatment of skin from washing and bathing in waters high in caustic alkalinity is also considerably reduced. The chemical reactions are as follows:

1. Neutralization of the free excess lime to calcium carbonate which will precipitate out of solution in the last clarifier.

$$Ca(OH)_2 + CO_2 \longrightarrow CaCO_3\downarrow + H_2O$$
$$(74) \quad\quad (44) \quad\quad\quad (100) \quad\quad (18)$$

Since all forms of alkalinity are expressed in terms of calcium carbonate (molecular weight = 100), the calculations for the theoretical quantity of carbon dioxide required are:

CO_2 in pounds per million U.S. gallons of water treated

$$= \frac{44}{100} \times 8.33 \times (OH^- \text{ alkalinity in mg}/\ell \text{ as } CaCO_3)$$

$$= 3.66 \times (OH^- \text{ alkalinity in mg}/\ell \text{ as } CaCO_3)$$

The calcium carbonate produced from this reaction will settle in the next clarifier and reduce the calcium hardness of the water by that amount. It will also produce a reusable byproduct in recovered lime if a recalcining plant is installed.

2. Conversion of the relatively insoluble calcium carbonate to the soluble bicarbonate. This is done in the second recarbonation chamber following the last clarifier and immediately prior to filtration.

$$\underset{(100)}{CaCO_3} + \underset{(44)}{CO_2} + \underset{(18)}{H_2O} \longrightarrow \underset{(162)}{Ca(HCO_3)_2}$$

Using the same calculation as before, a similar amount of carbon dioxide is required per million U.S. gallons for each mg/ℓ of carbonate ($CO_3^=$ alkalinity):

CO_2 in pounds per million U.S. gallons of water treated

$$= 3.66 \, (CO_3^= \text{ alkalinity in mg}/\ell \text{ as } CaCO_3).$$

If, however, only one stage of recarbonation is used, then by combining the two equations:

$$Ca(OH)_2 + CO_2 \longrightarrow CaCO_3 + H_2O + CO_2 \longrightarrow Ca(HCO_3)_2$$

the carbon dioxide required will be:

CO_2 in pounds per million U.S. gallons of water treated

$$= 7.32 \, (OH^- \text{ alkalinity in mg}/\ell \text{ as } CaCO_3)$$

It must be appreciated that these values are based on stoichiometric relationships and pure CO_2. In actual practice with submerged combustion burners, twice the calculated values are used on the basis that the efficiency of the process of extracting carbon dioxide from the flue gases is only 50% efficient.

LIQUID CARBON DIOXIDE

The use of evaporated liquid carbon dioxide (99.5% CO_2) is much more efficient. The absorption of CO_2 into water forms weak carbonic acid (H_2CO_3) which is not hindered by quantities of inert gases which are contained in flue gas. The price of liquid carbon dioxide delivered in bulk depends on the freight and location of source and demand and whether or not it is in short supply. Table 15-1 indicates that carbon dioxide has a low critical temperature of 87.8°F (31°C), i.e., above this temperature CO_2 cannot exist as a liquid but only as a gas. Large storage tanks must therefore be refrigerated and well insulated. It is usual to store liquid CO_2 at 0°F (−17.8°C) and 300 psi.

TABLE 15-1 Vapor Pressure of Carbon Dioxide[1]

Temperature		Pressure
°F	°C	(psig)
0°	−17.8°	290.8
20°	− 6.7°	407.1
40°	4.4°	553.1
60°	15.6°	733.9
80°	26.7°	954.0
87.8°*	31.0°*	1,054.7*

*Critical temperature and pressure of carbon dioxide

Small containers (30–50# cylinders) are available, also road tankers of 10 to 20 tons and rail tankers of up to 40 tons. However, because of the high vapor pressure of carbon dioxide, heavy, thick-walled vessels of special steel, suitable for low temperatures, are required.

Large water treatment plants using the lime-softening process can sometimes justify the installation of a lime-recalcining plant for the recovery of quick lime from the precipitated calcium carbonate sludge. Carbon dioxide is produced in the recalcine process in accordance with the equation:

$$CaCO_3 \xrightarrow{\text{Heat}} CaO + CO_2$$
$$(100) \qquad (56) \quad (44)$$

The carbon dioxide in the flue gas can be and is used for recarbonation.

SUBMERGED COMBUSTION BURNERS

Natural gas or propane, mixed with the theoretical amount of air required for complete combustion, is compressed and piped to a submerged burner located approximately 15 feet or more below the water surface. A typical submenged combustion burner and control assembly are shown in Figs. 15-1 and 15-2. The submerged combustion burners range from 400 to 10,000 lb of CO_2 per day with a turndown ratio of 33% of full capacity. A large unit of 10,000 lb of CO_2 per day capacity would require 60 standard cubic feet per minute (SCFM) of 1000 BTU

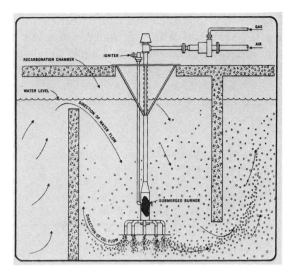

FIGURE 15-1 Submerged Combustion Burner Located in a Municipal Water Treatment Plant Recarbonation Chamber (Courtesy, Ozark-Mahoning Company).

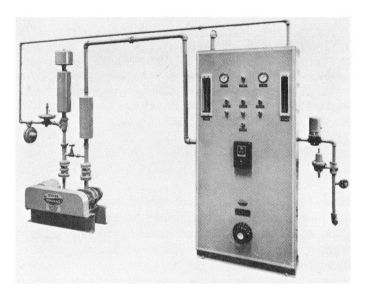

FIGURE 15-2 Motor Driven Blower and Panel for Controlling the Air-Gas Mixture for Submerged Combustion (Courtesy, Ozark-Mahoning Company).

per cubic foot natural gas and 629 SCFM of air. The CO_2 output can be controlled automatically from a pH meter with the electrodes located at the water outlet end of the recarbonation chamber. The recarbonation chamber must be sized and baffled to provide between 15 and 30 minutes retention time. A depth of 15 to 20 feet is necessary to insure good contact between the water and the combustion gases. The solubility of CO_2 in water is inversely proportional to temperature. Table 15-2 is based on evaporated liquid carbon dioxide gas (i.e., pure CO_2). However, Henry's Law of partial pressures applies and the combustion burners, producing a mixture of carbon dioxide, possibly some carbon monoxide (CO), excess air, nitrogen, and other inert gases will seriously disrupt the absorption process.

TABLE 15-2 Solubility of Carbon Dioxide

Temperature		Solubility	
		mℓ of CO_2/100 mℓ of Water	% by Weight
0°C	32°F	179.7 mℓ/100 mℓ	0.35%
20°C	68°F	90.1 mℓ/100 mℓ	0.177%

It is essential that the products of combustion be emitted from the burner in the smallest size bubbles possible in order to present as large a surface area of gas-to-water interface as possible. A recarbonation chamber using combustion gases must be much larger than would be required if evaporated liquid carbon dioxide were used. The design will be based on gas absorption criteria.

It is advisable to seek advice from the equipment manufacturer before completing the design of recarbonation chambers. Normally they are installed outside; however, if due to weather conditions it is necessary to install them indoors, precautions must be taken to insure that the building is thoroughly ventilated to prevent accumulation of exhaust gases.

SURFACE COMBUSTION

A number of manufacturers have developed surface units capable of using natural gas, propane, and oil for the production of carbon dioxide. The products of combustion are compressed and piped through corrosion-resistant piping into the recarbonation chamber. When flue gases from boilers and recalcining kilns are used, scrubbing devices may be necessary to remove particulate matter from the gas stream. The gases may then be compressed and piped to the recarbonation chamber. Culp[2] discusses various processes and suitable equipment for recarbonation that are becoming as important in wastewater treatment as in water softening.

In order to optimize the amount of carbon dioxide necessary to produce a satisfactory effluent, carry-over of suspended calcium carbonate must be minimized. Wide ranges of turbidity entering the recalcination chamber will vary carbon dioxide demand.

PRODUCTION OF CARBON DIOXIDE

With suitable equipment, CO_2 can be produced from a wide variety of fuels,[3] e.g., 3 lb of CO_2 from 1 lb coke; 115 lb of CO_2 from 1000 cu ft natural gas; and 20 lb of CO_2 from 1 U.S. gallon of kerosene. With good combustion control the flue gases would contain 10 to 12% CO_2. It must be appreciated that the choice of fuels used could be critical since phenols and other taste-producing compounds may be added to the water via the recarbonation chamber. Haney and Hamann[4] reviewed the subject of recarbonation with particular reference to the use of liquid carbon dioxide as a possible alternative to combustion processes.

ADJUSTMENT OF pH

Small lime-softening water treatment plants may find that recarbonation equipment cannot be economically justified. Under these circumstances the high pH of the water after softening can be reduced by the addition of sulfuric or hydrochloric acid. The products of the reaction will be calcium sulfate, calcium chloride, and CO_2, but the process should not be called recarbonation. Due consideration must be given to the safety of operating personnel who must be trained and equipped to safely handle acids.

THRESHOLD TREATMENT

Another possibility for preventing after-treatment deposition of calcium carbonate following a lime-soda-softening process is to chelate the calcium carbonate with a phosphate compound. With the use of chelating agents the pH of the water is not changed; although the deposition of carbonate is inhibited in cold water, it will not completely prevent deposition if the water is boiled.

REFERENCES

1. STANIAR, W., *Plant Engineering Handbook*. 1st Ed. New York: McGraw-Hill Book Co. Inc. 1950.
2. CULP, R. L. and G. L. CULP, *Advanced Wastewater Treatment*. New York: Van Nostrand Reinhold Company, 1971.
3. "Water Quality and Treatment." A manual prepared by the *AWWA*, 2nd Ed. Published by *AWWA*, 1951, New York.
4. HANEY, P. D. and C. L. HAMANN, "Recarbonation and Liquid Carbon Dioxide." *J.A.W.W.A.*, Vol. 61, p. 512, 1969.

16

FILTRATION

FILTRATION Water filtration is probably the most important single unit operation of all the treatment processes, and it would appear that more technical papers have been written on filtration than on any other water treatment subject. Baker[1] has well documented the history of filtration from ancient times to the middle of the twentieth century. In the last 20 years many advances have been made, and in the discussion of this subject the problem is not what to include but what can be left out.

FILTRATION THEORY Research into the various parameters of filtration has been conducted during the last quarter of a century, namely by Stanley,[2] Ives,[3] Miller,[4] O'Melia and Stumm,[5] Edwards and Monke,[6] Hudson,[7] Foess and Borchardt,[8] Cleasby,[9] Mohanka,[10] and many others. The correlation of filtration parameters becomes somewhat complex and highly mathematical. A number of relatively recent papers have been published in an attempt to rationalize the various theoretical concepts; Amirtharajah Cleasby,[11] Sakthivadivel, et al.[12] In October, 1972, an AWWA committee consisting of the most prominent filtration authorities published a report entitled "State of the Art of Water Filtration."[13] The title of this report and the use of the word art is in keeping with the present practical application of filtration technology. There are a number of undetermined factors which affect filter performance that are not completely understood. The reason why this subject has not yet been reduced to a few relatively simple concepts is covered by Dr. Ives.[3] He stressed that the authors did not regard the filter as a strainer. Particles considerably smaller than the pores (voids in the filter) are removed by the filter. As the range of action-of-surface forces was so small, the removal mechanism had to be considered in two stages; (1) a transport mechanism which brought particles to the surface of the filter media, or existing deposit; (2) an attachment mechanism to cause the particle to adhere to that surface.

Mintz[14] in his report to the Barcelona International Water Supply Congress made the following observation:

"There are complicated interrelations among many factors affecting the performance of rapid filters. These interrelationships vary with seasonal changes in the quality of raw water, chemical treatment, output, and load changes. Therefore, it is apparent that an attempt to work out an exact mathematical description, with theoretical constants, of the filtration process to hold for any conditions of filter operation is bound to fail. Obviously, it will always be necessary to determine the parameters of the process experimentally. The task of the theory is to provide a rational experimental procedure and a rational method of working out the experimental data so as to get the results required for engineering practice."

Particulate suspended matter penetrates into the pores of the filter bed and adheres to the grains of the filter media. As more and more particles are trapped in this fashion the free passage through the filter is diminished. If the same flow rate is maintained, the velocity of flow through the reduced areas is increased. With the high velocities the particles of floc are subjected to greater shear forces which will cause them to fragment into small particles that break off from the "parent floc" and pass deeper into the filter bed. At the end of the filter run, the filter either plugs and the run is terminated due to high head loss, or, the particulate matter breaks through the filter media and the run is terminated due to high effluent turbidity.

The capacity of a filter is the quantity of particulate matter it can retain before it requires backwashing. Conventional rapid sand filters demonstrated poor performance in their capacity to retain large quantities of dirt and to wash it out effectively during the backwash cycle. The old mono-media rapid gravity filters were designed for approximately 2—3 U.S. gpm per square foot and were backwashed at 10—15 U.S. gpm per square foot. During the backwash cycle the grains of sand graded themselves with the smallest grains on the top of the bed and the coarser heavier grains underneath. When the filter is returned to service after backwashing, the finest sand is at the top of the filter and as the voids become filled, the downward streaming velocity increases in inverse proportion to the reduction in the free flowing area. Once the deposited floc fragmented due to the increased shear forces, the particles would pass deeper into the filter bed where the voids were larger and, consequently, the finer particulate matter would often pass through the filter media into the underdrain system. Several ingenious filter designs are available to overcome the problem of natural gradation of filter porosity.

UPFLOW AND BIFLOW FILTERS[13]

Although these filters have not been widely used in North America, upflow and biflow filters have been used in Russia and The Netherlands. Russia's contact filter is an upflow filter that is used in treating colored waters. Coagulant chemicals are added immediately ahead of the filter and coagulation takes place within the filter. Sand, approximately 70 to 80 inches deep (1778–2032 mm) with a size range of from 0.5—2.0 mm, is used for the filter media. Filtration rates are generally about 2 U.S. gpm/ft^2 upward through the filter. At higher rates there exists the hazard of expanding the media as it becomes clogged. The Dutch upflow filter helps to avoid this by placing a grid of metal bars just below the sand surface at the top of the filter.

The biflow filters overcome the tendency toward expansion of the clogged bed by placing the filter effluent lines below the surface of the sand and introducing water to both the top and bottom of the filter.

DUAL AND MIXED OR MULTI MEDIA FILTERS

By far the greatest filter technology advance in the last 3 decades has been the introduction of dual media and particularly multi media, rapid gravity downflow filters, coupled with the use of polyelectrolyte filter aids.

The use of the dual media was refined in the late 1940's by W. R. Conley and R. M. Pitman. The dual media filter used only 6 inches (152.4 mm) of silica sand as opposed to the 30 inches (762 mm) of the typical rapid sand filter. Some 24 inches (609.6 mm) of anthracite coal, with a grain size of approximately 1 mm, overlies the sand layer. This provides a roughing filter on top of the sand, and a two-layer effect is accomplished equivalent to having two filters together in series, a coarse filter followed by a finer filter. Although dual media filters retain more material than the sand filters, much of the material is rather loosely held in the coal layer. If the rate of water flow applied to the filter changes, or a disturbance occurs in the flow pattern through the media, the accumulated material may be dislodged and is likely to penetrate to the sand layer, thereby rapidly increasing the head loss across the filter and may even appear in the filter effluent.

In the ideal filter the size of the particles making up the media should decrease in size uniformly in the direction of flow. The general characteristics of the three types of filter beds are shown in Figure 16-1 (a, b and c).

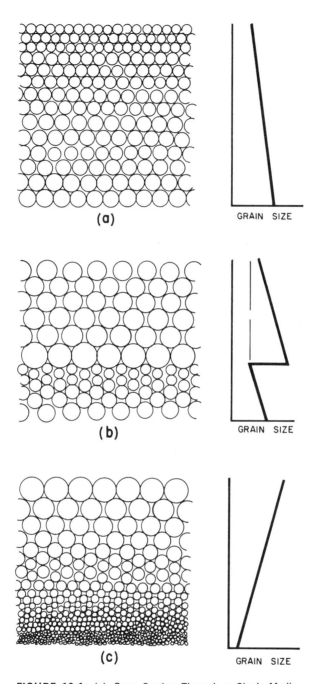

FIGURE 16-1 (a) Cross-Section Through a Single-Media Bed; (b) Cross-Section Through a Dual-Media Bed; (c) Cross-Section Through an Ideal Filter, Uniformly Graded from Coarse at Top to Fine at the Bottom.

MULTI MEDIA FILTERS

The use of multi media is patented by Neptune MicroFLOC and consists of a number of different types of media of different specific gravities. The density and size are so controlled that when the bed is backwashed, the grains hydraulically arrange themselves so that the coarsest media is on top and the finest at the bottom. The filter underdrains consist of layers of graded gravel necessary to support the mixed media on Wheeler or other patented filter bottoms. Examples of typical filter media support systems are shown in Figs 16-2 and 16-3. Above the supporting gravel the various media are carefully placed in the filter bed one at a time and backwashed to insure even distribution. Fig. 16-4 shows tube samples of multi media as installed and after backwashing. After backwashing each layer, the top $\frac{1}{2}$ to 1 inch (12.7—25.4 mm) is skimmed off to remove the "fines." This process is repeated until all media are in place. A typical bed will contain 60 percent coal which has a specific gravity on 1.65. This is the lightest of the materials and supplies the 1.0 mm material at the top of the bed. Thirty percent is silica sand which has a specific gravity of 2.6. Ten percent is garnet with a specific gravity of 4.2. The garnet supplies the finest material, 0.15 mm at the bottom of the bed.

A dual media bed consisting of approximately 80 percent coal and 20 percent silica is also available. This bed does not contain garnet but, while the elimination of the fine media results in a less effective filter as compared to the mixed or multi media, it is nevertheless a superior filter when compared to the mono-media.

In both the multi and dual media, once the filter has been finally backwashed, the various components do mix together and hydraulically regrade themselves

FIGURE 16-2 Wheeler Filter Bottom, Monolithic Type (Courtesy, Roberts Filter Manufacturing Company).

FIGURE 16-3 Filter Bottom Installation Using Plastic Strainers in Concrete Inserts (Courtesy, Ecodyne Limited, Graver Water Division).

into a reasonably close semblance of the ideal filter, uniformly graded with the coarse media at the top, becoming gradually finer toward the bottom.

There are many different formulations for multi and dual media filters depending on the service requirements, whether it be for color removal, lime softening, algae, iron, or manganese removal, or simply the removal of finely divided silt. It is usual to submit samples of water for laboratory evaluation followed by on-site pilot plant study using several 4, 6, or 8 inch diameter filter columns each filled with a different medium. In some cases the performance of a filter using a different grain size or proportion of one material compared to another has resulted in doubling the length of the regular filter run indicating the importance of the detailed preliminary investigation.

Mono-media filters were conservatively rated at 2-3 U.S. gpm per square foot to keep the streaming velocities in clogged media down to a reasonable figure. Dual and multi media filters can operate satisfactorily from 3 to 10 U.S. gpm per square foot without difficulty depending on the water and the filter application. To achieve a suitably low turbidity filter effluent at these high rates of flow, small dosages of polyelectrolyte filter aids are added to the water as it enters the filter. This does not necessarily increase the size of the floc, but it does increase the shear strength of the existing floc and causes the finely divided coagulated material to adhere to the filter media. Dosage rates may be as much as 2 mg/ℓ but they will usually be less than 0.1 mg/ℓ.

FIGURE 16-4 Tube on the Left Shows a Four-Component Sample of "Multi media" as Initially Installed. Tube on the Right Shows the same Media Hydraulically Graded after Backwashing (Courtesy, NEPTUNE MicroFLOC, Inc.).

Caution must be exercised in the use of excessive amounts of polyelectrolyte, otherwise the filter media will become too sticky and may tend to conglomerate the media grains into clumps with subsequent loss of performance. For this reason, it is always advisable to install facilities for surface washing of dual and mixed media beds. This technique will also help to remove mud balls. The decreasing voids from top to bottom in mixed media and dual media beds permits the use of the polyelectrolyte filter aid. If polyelectrolytes are used on straight sand or mono-media beds they will plug very quickly and terminate after a very short filter run.

One of the methods used to control the automatic backwashing of dual and mixed media filter beds is a combination of head loss across the bed and the turbidity of the filter effluent. It is not unusual to have a continuously recording turbidity meter on the filter effluent or one meter monitoring several filters, each for a prescribed time interval. The head loss across the media can be recorded on the same strip or circular chart as the turbidity.

Limits are set on both turbidity and head loss so that the filter will automatically backwash when either exceeds the preset limit. Both these parameters are interrelated to the polyelectrolyte filter aid dosage. For example, if the head loss across the filter media increases rapidly during the filter run and terminates the cycle while the turbidity of the filter effluent remains unchanged, or at a consistently low level, then indications are that the polyelectrolyte dosage is too high. If the head loss across the media remains low while the turbidity of the filter effluent increases rapidly, the polyelectrolyte dosage is too small and should be slightly increased or another type of polyelectrolyte should be substituted. Ideally, both the filter head loss and the effluent turbidity should reach the upper limits at the same time.

BACKWASHING

The backwash rate and the uniformity of the rising backwash water across the entire filter bed is of the utmost importance to all filters whether they be mono, dual, or mixed media. Air scour has many advantages for mono-media filters but has not been recommended for mixed media since the filter is supported on graded gravel and too much filter bed disruption occurs if air pockets in the backwash water blast upwards through the bed. It is essential for all filter beds that the backwash water be introduced gradually so as not to disturb the supporting gravel. Since backwashing is a function of the viscosity characteristics of the backwash water, which is in turn dependent on water temperature, it is necessary to adjust that backwash rate in accordance with the temperature of the water. Table 16-1 gives the normal recommended backwash rates for most multi and dual media filters.

TABLE 16-1 Backwash Rate for Mixed Media and Dual Media (15–20% Expansion)

Temperature		U.S. gpm/ft²	Rise in inches/min
°F	°C		
32°	0°	10.0	16.0
40	4.4	11.5	18.4
50	10.0	13.0	20.8
60	15.6	15.0	24.0
70	21.1	17.0	27.2
80	26.7	19.0	30.4
90	32.2	20.5	32.8
100	38.0	22.5	36.0

A mixed media filter, and to a lesser extent a dual media filter, backwashes much more efficiently than a mono-media filter. It is possible to backwash a mixed media filter in 8 to 10 minutes and sometimes in as little as 6 minutes depending on the nature of the suspended solids removed. Watching a conventional mono-media sand filter go through its backwash cycle, it will be noticed that there is a gradual darkening and discoloration of the backwash water as it rises through the filter bed. It approaches its maximum murkiness and then goes through a prolonged twilight period before it becomes relatively clean, indicating that the backwash cycle is complete.

Backwash water from a mixed media filter quickly reaches a maximum murkiness, remaining at this level for several minutes, and then clears almost as quickly as it becomes dirty. Increasing the backwash flow rate only expands the bed so that the sand grains are further apart and less able to scrub themselves clean of dirt deposits. With the mixed media beds, and to a lesser degree with dual media, the heavy garnet and/or sand in the bottom layers jet upward into the coarser coal at the top. Since they are of different materials, density, and surface characteristics, they are able to scrub each other more effectively to restore the filter media to its original clean condition. Adequate backwashing is important for all types of filters if a full length of filter run between backwashes is to be achieved. Incomplete backwashing facilities will quickly deteriorate a filter, resulting in short filter runs and poor quality effluent.

Backwashing phenomena are dealt with in detail in *Water Quality and Treatment*.[15]

AIR SCOUR

Mono- and dual media filters which do not rest on layers of graded supporting gravel can be designed to use an air scour prior to water backwash. A filter underdrain system designed to use air scour may consist of a concrete plenum in which specially designed nozzles are fitted to protrude through the bottom of the filter slab into the space below. As many as 50 nozzles per square meter (42 per square yard) of filter area are required.[16] Nozzles come in many different designs depending on the manufacturer. They all have numerous small slot openings ranging from 0.3 to 0.7 mm in width to allow the air and backwash water to rise evenly and uniformly through the filter media, and yet are small enough to prevent the grains of filter media from getting back into the plenum underdrain. The nozzle must have low hydraulic head loss in both directions. Since a considerable number are required for each filter, the nozzles are usually made of plastic so as not to corrode, and to minimize the capital costs.

The use of air scour has demonstrated its popularity in France and countries following French engineering practice and is now finding favor in North America. The recommended procedure is to air scour with some backwash water, using both air and water at the same time with a sufficiently low backwash velocity to avoid expansion of the sand. The sand thus remains stable, and the surface crust is completely broken up by the air. In this way, mud balls do not have a chance to form and are actually reported to be unknown with this type of washing process.[16] During air scour, the wash water rate can be varied over a wide range but must not fall below a figure of 5 m^3/m^2 per hour (2.045 U.S.

gpm/ft²). This is equivalent to an upflow velocity of 5 meters per hour. Air on its own followed by water wash should *not* be used since downward currents develop during the air scour phase and hold the impurities in the filter mass. The air scour and wash water must therefore be applied simultaneously. Once the air scour has been stopped, the backwash water rate should be increased to not less than 12 m³/m²/hour (4.9 U.S. gpm/ft²) or in accordance with the filter manufacturer's requirements. One of the advantages claimed for air scour as opposed to water only is that it will backwash to the same degree of cleanliness with less water and with the elimination of mud balls. One of the disadvantages is that it has not been available for mixed media since mixed media is currently supported on graded gravel. The use of filter nozzles has not been quite as successful with mixed media as the older, conventional, underdrain systems using graded gravel. On the other hand, backwashing mixed media with water also results in less backwash water than the conventional media since the garnet has the effect of another rubbing agent to scrub away the dirt. It has been suggested that a separate air wash grid be placed in the media above the gravel, but this has not yet been extensively developed. Surface wash sweeps with mixed media filters have been fairly successful in keeping the mud problem within reasonable limits.

CONSTANT RATE AND DECLINING RATE— FILTRATION

Traditionally most filter systems in North America operate on the constant rate principle where a rate control valve in the filter underdrain discharge maintains a constant rate of throughput for each filter or, alternatively, a constant level of water in the top of the filter. When the filter comes onstream after backwashing, it will have minimum head loss across the filter and therefore maximum throughput capacity. The rate control valve will therefore close in order to create a head loss and to control the flow of water through the filter. As the filter run proceeds, the head loss across the filter media increases and the rate control valve opens a little more to maintain constant flow conditions until eventually it is wide open and all the head loss is across the filter media. One of the alleged disadvantages of this system is the continual increase in streaming velocities as the filter starts to plug with deposited particulate matter. Eventually the increase in shear force imposed on the particles of floc adhering to the media causes the particles to fragment and fall deeper into the filter or pass through to the underdrains. A turbidity meter on the filter effluent would record the increase in suspended matter and terminate the filter run due to breakthrough.

With declining-rate filtration there is constant available head loss across the filter whether it be clean or dirty. The flow through the filter becomes the variable. When the filter is clean, after backwashing, the flow through the media is at a maximum. However, as the filter becomes plugged and the head loss increases, the flow diminishes, hence the name declining-rate. It is said[17] that declining-rate filtration is a more logical approach than constant rate filtration. It is illogical to be forcing the filter bed to work at the end of the run when it is clogged at the same velocity as it did at the beginning of the run when the media is clean. It is also claimed that the elimination of controllers may lead to simpler and less expensive designs.

Certainly declining-rate filtration has merit, and better filter performance as opposed to constant rate has been demonstrated at the Chicago South District water treatment plant[15] where it is claimed that declining-rate units produced 40% more water per day than the conventional units when operated so that the filter run durations were equal on both types. It is difficult to be specifically in favor of one type over the other. Both have their merits in particular circumstances and it would appear that one of the problems in forming an opinion is the lack of truly comparative data. It would appear that with larger gravity filters declining-rate will become more popular as new designs are developed. However, with smaller plants, and particularly package units, constant rate may have some advantages especially where the demand is relatively constant and only one or two filters are installed.

HIGH RATE FILTRATION

The term high rate seems synonymous with many television commercials which leave the viewer with the impression that because a product is given this designation it is reasonable to suppose that this particular product has better dollar value than standard or regular rated products. Several manufacturers use the term simply because they have designed the hydraulics of the filter, usually pressure filters, that enable excessive quantities of water to be pushed through the filter media, having no bearing on the quality of the filter effluent. Achieving high-rate filtration performance requires (a) accurate control of chemical feed, (b) specially designed high-rate filters, and (c) instrumentation designed to monitor continuously all significant plant variables.[18,19]

SLOW SAND FILTRATION

Much has been written on the virtues of slow sand filtration over the last century. A new slow sand filter was constructed as late as 1958 by the London Metropolitan Water Board at Ashford Common.[20] Tests on slow sand filters are reported by Fair and Geyer[21] and E. W. Steel.[22]

Occasionally one hears a recommendation, by members of the medical profession, to return to the use of slow sand filtration on the ground that slow sand filters will remove viruses. It is understood that crustacea, small aquatic organisms that inhabit the water on top of the slow sand filter, will ingest viruses and other bacteria. Since crustacea cannot pass through the filter bed, they are therefore filtered out of the water. Bacteria are usually small enough to pass through a normal granular filter bed and will do so unless they are attached to a particle of floc. Viruses are even smaller and will certainly pass through any normal water filter unless they are hosted to particulate matter. If, therefore, it is necessary to use crustacea to eliminate the bacteria and virus populations, a process could be designed in order that this can take place. It might consist of a biological tank followed by a microstrainer to remove the larger organisms, assuming that any organisms smaller than this would have been ingested by the crustacea and others. This process would be located immediately before either a mixed media or dual media filter.

FILTER CONTAMINATION

Any system of filter media whether it be sand, anthracite, garnet, or any other combination should never be subjected to an invasion of animate or inanimate particulate matter which cannot be removed by normal backwashing. A plant in northern British Columbia takes water from a lake which always has a large population of free-swimming aquatic insects and crustacea. During the summer months there are also large algae populations, some northern pike, and other fish. The treatment process consists of a microstrainer followed by rapid sand gravity anthracite coal (anthrafilt) filters. The microstrainer was partially bypassed in order to increase the throughput of the plant. As a result many thousands and thousands of organisms accumulated on the filter beds. Unfortunately, the filter rate control valve leaked rather badly and when the plant stopped, because the clear-water well was full, the water in the filter bays leaked through the rate control valve leaving the media high and dry. The various organisms entrapped in the filter media died and decomposed. The result was a foul mass of dark brown oily deposit coating the whole of the filter media that could not be washed out. The filter was completely ruined and had to be replaced.

DIATOMACEOUS EARTH FILTERS

Prior to the introduction of dual and mixed media, the use of diatomaceous earth (DE) filters were becoming popular in the smaller municipal water filtration systems. DE filters produce excellent filtrates very low in turbidity. In certain food processing industries DE filters are the only filters used; they are also popular in the filtration of oil field brines prior to reinjection. They are used in many swimming pools and aquarium installations and are particularly useful where some contamination of the filter media must be anticipated. Since the used diatomaceous earth is discarded to waste at the end of the filter run, this type of equipment is ideal under these circumstances; however, in modern municipal water treatment plants, the use of DE filters is declining in favor of dual or mixed media.

In the design of filtration systems, it is always necessary to consider the quality of the water entering the filters as well as the quality of the finished effluent. Additional pretreatment is justified if less expensive and better filter operations can result.

REFERENCES

1. BAKER, M. N. "The Quest for Pure Water." American Water Works Association, N. Y., 1949.
2. STANLEY, D. R. "Sand Filtration Studies with Radio-Tracers." Proc. Amer. Soc. Civ. Engineers, Vol. 81, p. 592, 1955.
3. IVES, K. J. and J. GREGORY, "Basic Concepts of Filtration." Proceedings of the Society for Water Treatment and Examination, Vol. 16 (3), page 147, 1967.
4. MILLER, D. G. "Rapid Filtration Following Coagulation, Including the Use of Multilayer Beds." Proceedings of the Society for Water Treatment and Examination, Vol. 16 (3), p. 192, 1967.
5. O'MELIA, C. R. and W. STUMM, "Theory of Water Filtration." *J.A.W.W.A.*, Vol. 59, p. 1393, 1967.
6. EDWARDS, D. M. and E. J. MONKE, "Electrokinetic Studies of Slow Sand Filtration Process." *J.A.W.W.A.*, Vol. 59, p. 1310, 1967.
7. HUDSON, H. E. "Physical Aspects of Filtration." *J.A.W.W.A.*, Vol. 61, p. 3, 1969.
8. FOESS, G. W. and J. A. BORCHARDT, "Electrokinetic Phenomena in the Filtration of Algae Suspensions." *J.A.W.W.A.*, Vol. 61, p. 333, 1969.
9. CLEASBY, J. L. "Approaches to Filtrability Index for Granular Filters." *J.A.W.W.A.*, Vol. 61, p. 372, 1969.
10. MOHANKA, S. S. "Multilayer Filtration." *J.A.W.W.A.*, Vol. 61, p. 504, 1969.
11. AMIRTHARAJAH, A. and J. L. CLEASBY, "Predicting Expansion of Filters During Backwash." *J.A.W.W.A.*, Vol. 64, p. 52, 1972.
12. SAKTHIVADIVEL, R.; V. THANIKACHALAM, and S. SEETHARAMAN, "Head Loss Theories in Filtration." *J.A.W.W.A.*, Vol. 64, p. 233, 1972.
13. AWWA Committee Report. "State of the Art of Water Filtration." *J.A.W.W.A.*, Vol. 64, p. 662, 1972.
14. MINTZ, D. M. "Modern Theory of Filtration, Special Subject No. 10", International Water Supply Congress and Exhibition, Barcelona, Spain. Oct. 3–7, 1966.
15. *Water Quality and Treatment*, 3rd Ed. Prepared by the American Water Works Association. New York: McGraw-Hill Book Company, 1971.
16. *Water Treatment Handbook*, 4th Ed. France: Degremont, S. A. Distributed by Taylor and Carlisle, New York. (1973)
17. ARBOLEDA, J. "Hydraulic Control Systems of Constant and Declining Rate Filtration." Paper presented at the 93rd Annual Conference of the AWWA. May 1973, Las Vegas, Nevada.

18. CONLEY, W. R. "High-Rate Filtration." *J.A.W.W.A.*, Vol. 64 (3), p. 203, March, 1972.
19. WESTERHOFF, G. P. "Experience with Higher Filtration Rates", *J.A.W.W.A.*, Vol. 63 (6), p. 376, June, 1971.
20. TWORT, A. C. *A Textbook of Water Supply*. London: Edward Arnold Ltd., 1963.
21. FAIR, G. M.; J. C. GEYER, and D. A. OKUN, *Water and Wastewater Engineering*, Vol. 2. New York: John Wiley & Sons Inc., 1968.
22. STEEL, E. W. *Water Supply and Sewerage*, 2nd Ed. New York: McGraw-Hill Book Company, 1947.

17

IRON AND MANGANESE REMOVAL

IRON AND MANGANESE REMOVAL The large volume of literature on the subject of iron and manganese removal is indicative of the magnitude of the world-wide problems they have caused, but as far as is known, humans suffer no harmful effects from drinking water containing normal quantities of iron and manganese.[1]

When clear, sparkling well waters which contain iron and manganese are exposed to air, the iron and sometimes the manganese is oxidized and they become turbid and highly unacceptable from the esthetic viewpoint. The oxidation of iron and manganese to the Fe^{3+} and Mn^{4+} status, form colloidal precipitates. The rates of oxidation may not be rapid, and the reduced forms which are usually soluble can persist in aerated water for some time before precipitating in the insoluble oxidized form. This is essentially true when the pH is below 6 with iron and below 9 with manganese. The oxygenation rates may be increased by the presence of certain inorganic catalysts or through the action of microorganisms. Both iron and manganese interfere with laundering operations, impart objectionable stains to plumbing fixtures, and cause difficulties in distribution systems by supporting growths of iron bacteria. Iron also imparts a taste to water which is detectable at very low concentrations. Iron-bearing waters are often called *ferruginous waters*. They have a bitter taste and, in combination with the tannin, impart an inky color to tea infusions. They also impart a brown-colored deposit on vegetables during washing and cooking. Linens washed in ferruginous waters are stained with iron molds. For these reasons the U.S. Public Health Service Standards of 1962 recommended that public water supplies should not contain more than 0.3 mg/ℓ of iron or 0.05 mg/ℓ of manganese. However, the AWWA Recommended Potable Quality Water Goals for 1968 suggest upper limits of 0.05 mg/ℓ for iron and 0.01 mg/ℓ for manganese.[2]

It must be realized that there are two different types of iron found in water supplies and the methods used to remove them are entirely different. For convenience one is known as *inorganic iron* and refers to the clear and sparkling well waters that turn turbid on exposure to air; the other may be called *organic iron* which is colored with humic acids that chelate the iron. Organic iron may be present in colored well waters as well as colored surface waters. Many journals refuse to use the term *organic iron* even unscientifically, but it is a popular term. The first thing to do when examining a water that contains iron is to find out if it is ordinary iron (inorganic) or the organic variety; that is to say, the molecules of iron are held in solution by the organic humic substances present in the water. This organic iron is also known as chelated iron.

Chapter 13, in "Water Treatment Plant Design,"[3] points out that "Chemical analysis of water provides no clue to the removal difficulties that may be encountered. For this reason, laboratory and pilot plant tests are advisable in the design of a removal plant for a new or unfamiliar water supply source." Whenever color is present together with iron it is reasonable to assume that the iron may be chelated. A visit to the site and a close visual examination of the water may be all that is required. However, in some cases water may be relatively colorless and yet capable of chelating iron and manganese compounds. For these reasons, it is essential that the waters are analyzed in accordance with *Standard Methods*[4] in order to insure that all iron present is detected. Some field kits do not always agree with the *Standard Methods* tests and, if in doubt, acidifying the sample to break down the chelating agent before performing the test is advisable.

Removal of Inorganic Iron and Manganese

The removal of inorganic iron and manganese is an oxidation process for precipitation followed by settling and filtration. Alternatively, many water treatment plants oxidize and filter out the iron and manganese all on one filter. Since the inorganic iron and manganese are relatively easily oxidized, the only natural source of water containing these elements in the ferrous and manganese states are ground waters from springs or wells.

Iron in Ground Waters

Iron in ground waters containing bicarbonate alkalinity is in the form of ferrous bicarbonate ($Fe(HCO_3)_2$). Iron is very soluble in cold water, particularly waters of low pH free from oxygen and saturated with carbon dioxide, even in excess of 150 mg/ℓ expressed as (Fe).

Usually well waters containing iron are saturated with carbon dioxide (CO_2). The quantity of carbon dioxide present can be determined from the pH and alkalinity (see *Standard Methods*).[4] If there is a large amount of dissolved carbon dioxide present in the well water it would be wise to aerate it first. In this way the free carbon dioxide will be liberated and the pH slightly elevated.

Aeration

Simple aeration may be all that is required to precipitate the ferrous bicarbonate to ferric hydroxide in accordance with the following equations.

$$\underset{\text{(ferrous bicarbonate)}}{Fe(HCO_3)_2} \xrightarrow{\text{aeration}} \underset{\text{(ferrous hydroxide)}}{Fe(OH)_2} + 2CO_2$$

further aeration:

$$\underset{\text{(ferrous hydroxide)}}{4Fe(OH)_2} + O_2 + 2H_2O \xrightarrow{\text{aeration}} \underset{\text{(ferric hydroxide)}}{4Fe(OH)_3}$$

In order that the reaction will go to completion and precipitate the ferric hydroxide, it is necessary that the pH be approximately 7 or higher. If possible the pH should be raised to 7.5 to 8.0, but even so the reaction may take 15 minutes retention before it is complete[5] and in some cases as much as 1 hour retention has been necessary. The length of retention time depends upon the degree of aeration and the dissolved oxygen content of the aerated water as it enters the retention zone. Theoretically 1 mg of oxygen will oxidize 7 mg of Fe^{++} but the pH must be 7.5 or higher. Depending on the temperature, the water should be easily aerated to saturation providing there are no organics, ammonia, or other oxygen demands.

Chlorine and Chlorine Dioxide[6]

Both these gases are powerful oxidizing agents and can be used to oxidize iron. Normally only chlorine is found to be necessary since chlorine dioxide is expensive. The reactions with chlorine in the presence of calcium bicarbonate alkalinity are as follows.

$$2Fe(HCO_3)_2 + Cl_2 + Ca(HCO_3)_2 \longrightarrow 2Fe(OH)_3\downarrow + CaCl_2 + 6CO_2$$

Theoretically, 1 mg of chlorine will oxidize 1.6 mg of Fe^{++}. This reaction will take place over a wide range of pH from 4 to 10 but the optimum pH is 7. The colder the water, the slower are the reactions which may take as long as 60 minutes retention time to complete. Pre-aeration is always worth investigating, particularly to remove any carbon dioxide that may be present. Oxidation by chlorination alone, without aeration, can also work effectively with colored waters and chelated iron.

Potassium Permanganate (KMnO$_4$)

Potassium permanganate is also a powerful oxidizing agent and in addition precipitates manganese dioxide (MnO$_2$) on to the granular surfaces of the filter media which acts as a catalyst to accelerate the reaction to completion. The reactions are as follows:

$$3Fe(HCO_3)_2 + KMnO_4 + 7H_2O \longrightarrow MnO_2 + \underline{Fe(OH)_3} + KHCO_3 + 5H_2CO_3$$
(ferrous bicarbonate)   (ferric hydroxide)

    The products of the reaction are manganese dioxide which precipitates on to the surface of the filter media, ferric hydroxide which is also insoluble, potassium bicarbonate, and carbonic acid, which are soluble and remain in the effluent. The carbonic acid can be broken down into carbon dioxide and water. The process usually consists of a pressure filter containing filter media. Certain types of media are claimed to be more effective in the removal of iron and manganese than others. While these claims may or may not always be correct, one of the major problems is a filter medium that has all the fines on the top of the bed and inevitably creates a high head loss early in the filter run followed by problems of leakage and break-through. It is believed that dual or multi media can be effective providing they are initially treated in place with an aqueous solution of potassium permanganate and a continous feed of potassium permanganate during the whole of the filter run. However, the treatment of the media with potassium permanganate may take some considerable time before it becomes fully effective.

    The amount of potassium permanganate theoretically required in this process is 1 mg KMnO$_4$ for each 1.06 mg of iron as Fe^{++}; but because of the catalytic effect of the precipitated manganese dioxide, it is found that less than the theoretical amount of potassium permanganate is required and 0.6 mg/ℓ of KMnO$_4$ per 1.0 mg/ℓ of Fe^{++} is usually sufficient.

Lime Softening

Lime-softening reactions take place in excess of pH 9 and the process usually operates at pH 10 to 11 depending on how much magnesium has to be removed. At these high pH values the iron and also the manganese are precipitated and removed with the lime sludge. Most of these processes take place in open tanks where atmospheric oxidation obviously plays an important role.

"Organic" or Chelated Iron

The iron content of highly colored waters containing humic substances from muskeg, rotting vegetation, etc. can often exceed 10 mg/ℓ. The process of removal is almost diametrically opposed to the processes used for inorganic iron. Simple aeration is not effective and since color is usually pH-sensitive, raising the pH only intensifies the color and "fixes" the iron.

Usually these waters are low in pH from 4.0 to 6.5 and are also low in alkalinity. In order to remove the iron, it is necessary to remove the color. In one experiment it was found that reducing the color by flocculation to about 50% of its original value reduced the iron content by the same amount. One of the methods of treating highly colored waters high in iron is to add chlorine from a gas chlorinator (it often requires 5 or more mg/ℓ).

The use of sodium hypochlorite is not always as effective as gaseous chlorine since the hypochlorite is made from caustic soda and does not lower the pH. The chlorine seems to break down the organic chelating agent and destabilizes the colloidal color with the result that the iron together with the organic color forms a floc which can be removed by tube settlers and filtration. The floc is very light and will not readily settle. Using the 5° from horizontal tube settlers it is found that there is nearly as much floc adhering to the top of the tubes as there is to the bottom. Alum, or a polyelectrolyte, is usually needed together with some retention time. In cold waters longer retention is needed than with warmer waters. This is one of the reasons why on-site laboratory, bench-scale, and pilot plant studies are an essential prerequisite to the optimum design of iron and manganese removal plants.

MANGANESE REMOVAL

Manganese is often present in both surface and ground water supplies but usually in smaller amounts than iron. It is harder to remove since it only becomes insoluble at pH values in excess of pH 9 and is not easily oxydized even in the inorganic form.

Most of the processes used for iron removal (ie., aeration, chlorine and chlorine dioxide, potassium permanganate, lime softening) are equally applicable to manganese removal, but in each case some on-site laboratory, bench-scale, or pilot test is desirable. For aeration to be effective the pH should be in excess of 9.0 or 10.0. Theoretically 1 mg of oxygen will oxidize 3.5 mg of manganese (Mn^{++} to Mn^{++++}). Both chlorine and chlorine dioxide[6] can be used and theoretically 1 mg of chlorine will oxidize 1.3 mg of manganese.

Potassium Permanganate

This process is by far the most popular method of removing manganese. Large-scale development work was done by R. B. Adams[7] at the Wilkinsburg water treatment plant on the Allegheny River near Pittsburg, Pa. This river has a normal manganese content of 0.5 to 8.0 mg/ℓ. Adams found that manganese removal was the result of the oxidation to manganese dioxide (MnO_2) at pH 7.2 to 8.3. Chlorination before permanganate treatment appreciably reduces the amount of permanganate required because it will oxidize most of the iron and some organics. The reaction is rapid enough to allow coagulation of the fine, precipitated manganese during the normal coagulation process. Polyelectrolytes have aided the alum coagulation in producing in their clarifiers a high-quality filterable water. In this plant consisting of the usual flash mix, flocculation, and clarification, the floc had the appearance of discarded tea leaves. It settled easily, resulting in a low turbidity clarifier effluent which gave good filter runs.

One of the advantages of potassium permanganate is the versatility of this oxidizing agent in water.[7] It will oxidize a great variety of substances that are acid, neutral, and alkaline, including sulfide, sulfite, thiosulfate, cyanide, ferrous iron, manganous manganese, chromic chromium, and numerous organic compounds such as organic acids, alcohols, aldehydes, ketones, phenols, and various nitrogen compounds. Generally speaking, the oxidation of an organic compound with permanganate proceeds more rapidly in an alkaline solution than in an acid solution.

Some of these reactions are presented as follows:

$$3Mn(HCO_3)_2 + 2KMnO_4 + 2H_2O \longrightarrow 5MnO_2 + 2KHCO_3 + 4H_2CO_3$$

$$3MnSO_4 + 2KMnO_4 + 2H_2O \longrightarrow 5MnO_2 + K_2SO_4 + 2H_2SO_4$$

The oxidation of soluble manganous ions is similar to the oxidation of iron. Theoretically 1 part of $KMnO_4$ will oxidize 0.52 parts of divalent manganese to insoluble tetravalent manganese. Again, in practice, the theoretical relationship between the amount of soluble manganese present and the amount of $KMnO_4$ required does not hold true. The quantity of permanganate necessary to effect total oxidation of the soluble manganese is always less than theoretical, assuming other contaminants having a permanganate demand are absent. The theoretical quantity is only approached when all, or a portion of, the manganese is organically bound.[8]

Shull[9] found in a number of Philadelphia treatment plants that water treated with 2.0 mg/ℓ of potassium permanganate showed a manganese reduction of 92.5 percent, together with a significant odor reduction. Optimum removal of manganese occurred at a permanganate-to-manganese ratio of 2.5 to 1, and this ratio proved to be constant regardless of the concentration of manganese in the untreated water. If, however, prior to the potassium permanganate dosage the water is pretreated with chlorine, the same end result can usually be achieved with considerably less permanganate.

Iron and Manganese Removal by Ion Exchange

Iron and manganese removal on a filter bed by potassium permanganate oxidation is not an ion exchange process. However, both sodium-cation-zeolite exchangers and hydrogen-cation exchanger units will remove ferrous iron (Fe^{++}) and manganous manganese (Mn^{++}). For the process to operate effectively, it is essential that there be no dissolved oxygen in the water before it enters the zeolite beds. Even if a very small amount of oxygen is sucked into the pump stuffing box, it will oxidize the ferrous bicarbonate to ferric hydroxide thus fouling the zeolite and stopping the ion exchange reactions. Provided that all the iron and manganese is in the nonoxidized ionic form, the process is said to be effective particularly for traces, but to be economic, other cations, i.e., calcium, magnesium, etc., will be removed at the same time. Ion exchange for the removal of iron and manganese from municipal water supplies without the need to soften is not a particularly attractive proposition compared to the simpler oxidation processes.

Sequestering or Chelating Process

As an alternative to removing the iron and manganese from municipal water supplies, hexameta-phosphates, tri-sodium phosphates, sodium silicates, and other compounds are sometimes used to hold them in solution. This is known as chelating, sequestering, or threshold treatment. For this process to be effective, it is essential that the iron and manganese are in the ionic state. This condition usually exists in a well in which there is sufficient free carbon dioxide to insure that the iron is as ferrous bicarbonate and similarly with the manganese. To apply this technique it is necessary to pump the sequestering agent down into the well in proportion to water flow to allow contact with the iron and manganese ions before oxidation.

The same treatment can be used to sequester calcium and magnesium ions. For the process to be completely effective with ferrous ions, some calcium must be present to form the calcium-phosphate complex. Usually this process is used where the iron content is relatively small, ie., less than 1 mg/ℓ. It is still high enough to cause consumer complaints but this is marginal compared to the cost of removal. The use of sodium silicate has been successful in some of the Ontario Water Resources Commission (OWRC) installations in Canada, and a paper describing its use at the Long Island Water Corporation, New York,[10] proclaims the advantages of the process.

Apart from sequestering iron and manganese, the same process is also effective in depositing a microscopic film of phosphates or silicates on the inside surfaces of unlined steel or cast iron water mains. When a water supply is slightly corrosive, particularly after alum clarification, it may pick up iron from older unlined water mains. The use of these chemicals will inhibit the attack and reduce consumer complaints of red water.

It should be realized however, that the sequestering agents, while effective in cold water systems, can break down and lose some of their sequestering capacity when heated or boiled. The addition of phosphates and silicates in the proportions recommended is not considered to constitute a health hazard. Neither is this process considered to be a substitute for iron and manganese removal; nevertheless, it does serve a useful purpose in the treatment of municipal water supplies.

REFERENCES

1. Taylor, E. W. *The Examination of Waters and Water Supplies*, 6th Ed. Philadelphia: The Blakeston Company, 1949.
2. *Water Quality and Treatment*, 3rd Ed. Prepared by the AWWA Inc. New York: McGraw-Hill Book Company. 1971.
3. "Water Treatment Plant Design." Prepared by the AWWA Inc., published by the AWWA, Inc. 1969.
4. "Standard Methods for the Examination of Water and Wastewater," 13th Ed. Published jointly by AWWA, American Public Health Association and Water Pollution Control Federation, 1971. Available from the Publications Sales Department, AWWA, 2, Park Avenue, New York.
5. Connelly, E. J. "Removal of Iron and Manganese." *AWWA*, Vol. 50, p. 697, May, 1958.
6. White, G. C. *Handbook of Chlorination*. New York: Van Nostrand Reinhold Company, 1972.
7. Adams, R. B. "Manganese Removal by Oxidation with Potassium Permanganate." *AWWA*, Vol. 52 (2), p. 219, February, 1960.
8. Humphrey, S. B. and M. A. Eikleberry, "Iron and Manganese Removal Using $KMnO_4$." Water and Sewage Works Reference Number, page R176, 1962.
9. Shull, K. E. "Operating Experience at Philadelphia Suburban Treatment Plants." *J.A.W.W.A.*, Vol. 54, p. 1232, 1962.
10. Mirando, L. and R. C. Marini, "Improving Water Quality at the Long Island Water Corporation, New York." Presented at the 93rd annual conference of the American Water Works Association, May, 1973, Las Vegas, Nevada.

18

CHEMICALS AND CHEMICAL FEEDING

CHEMICALS AND CHEMICAL FEEDING

Chemicals do not always react with each other simply because the equation for the reaction can be written on paper. Temperature, solubility, pH, ionization, reactivity of the particular ion or radicals, "buffering" actions, physical and chemical inhibitors, all influence the rate of reaction as well as the degree of completion. In water chemistry the variables most commonly encountered are: pH, time of reaction, temperature, concentration of the chemicals in solution, ionic mobility, and degree of ionization. When the temperature is close to 0°C (32°F) chemical reactions in aquatic environments approach a standstill.

Many chemical reactions are associated with the surface properties of the ions and their electrostatic charges, and they can be greatly retarded by the presence of organic matter and microorganisms.

The chemical equations must therefore be considered as a general concept of what appears to take place. However, it must be appreciated that many of these reactions are taking place simultaneously in water softeners or in clarifiers; some are reversible and may be moving in the forward direction in one part of the reactor and in the reverse direction in another. Chemicals must be in close contact with each other before they can react and this is probably one of the biggest problems in water treatment. For example a dosage of 25 mg/ℓ of alum solution may be adequate to coagulate a turbid water but it must be thoroughly mixed with the water. This means that 25 pounds of dry alum must be dissolved in solution and uniformly dispersed in 1,000,000 pounds of water (500 tons). The alum may be fed into the water flow as a 5% solution, and 25 pounds of dry alum would be dissolved in approximately 500 pounds of water. The 525 pounds of alum solution must therefore be evenly mixed and dispersed into 1,000,000 pounds of water in order to fulfill the prescription of 25 mg/ℓ, or 25 ppm.

Failure to uniformly mix the dosing chemicals into the water flow is a major cause of poor plant performance. Nearly all water treatment plants suffer from this problem to some degree, resulting in wastage of chemicals, incomplete reactions, and wide diversity between what is supposed to happen and what actually does happen.

PURITY OF COMMERCIAL CHEMICALS

Although the atomic, molecular, and combining weights are usually given to four significant figures in reference tables, as far as water treatment is concerned it is usually adequate to use only the first two or, at the most, three.

Equivalent weights and those used in the equations describing the chemical reactions are based on the assumption that the chemicals are 100% pure; but in actual fact all chemicals have some impurities, and commercial chemicals have more impurities than analytical laboratory reagents. If, for example, 50% caustic soda is used to neutralize 98% sulfuric acid, then the quantity of commercial caustic soda (NaOH) required would be as follows:

$$H_2SO_4 + 2NaOH = Na_2SO_4 + 2H_2O$$

Molecular weights	(98)	(80)	(142)	(36)
Ratio	1	0.815	1.45	0.365

The equation assumes that the H_2SO_4 and the NaOH are 100% pure. The quantity of 50% NaOH required to neutralize 1 pound of 98% H_2SO_4 would be:

$$1 \times 0.98 \times 0.815 \times \frac{100}{50} = 1.63 \text{ NaOH}$$

EQUIPMENT FOR FEEDING CHEMICALS

A number of equipment manufacturers build chemical feeding equipment specifically designed to suit each application. Fig. 18-1 shows a typical metering, mixing, and feeding system.

FIGURE 18-1 Self-contained Chemical Feeder Equiped for Metering, Mixing and Feeding (Courtesy, Penwatt Corp, Wallace and Tiernan Division).

CHEMICAL FEED TABLES

From laboratory jar test results and pilot plant data, the approximate chemical dosage to treat a specific sample of raw water is determined. It does not always follow, however, that chemical dosages determined in the laboratory will produce the desired result in the full-scale plant since there are many significant differences between the two. Nevertheless, the laboratory results establish a good starting point and plant dosages can be adjusted later when equilibrium conditions have been achieved. Coagulation Control Centers (CCC) are installed in some plants; this usually consists of a small pilot plant model of the full-scale plant, complete with miniature chemical solution feeders and complete instrumentation.

In computerized installations the incoming water flow is metered together with the rate of chemical feed, the two signals are evaluated, and a direct milligrams per liter (mg/ℓ) or parts per million (ppm) read-out, either on a typed sheet or cathode ray tube, is provided.

In all cases it is necessary to check the feeder calibration curves using the commercial-grade chemicals used in the plant. Volumetric feeders often give widely varying results with different batches of chemicals due to changes in particle size, uniformity of grain size, etc. Some chemicals flow very much more freely than others and they are all affected by the moisture content of the chemicals. One advantage of the gravimetric belt is the numerical read-out that can be obtained from the feeder to check its calibration on an hour-to-hour basis.

It is also necessary to be able to quickly and conveniently compile a table or graph giving the pounds or kilograms per hour of chemical necessary to give the prescribed dosage against the flow of raw water into the plant. The calculation is very simple but can be tedious. For example: the flow into the plant is 1300 U.S. gpm; it is required to feed 25 mg/ℓ of soda ash. How many pounds of soda ash are needed?

$$1300 \times 8.33 \times 60 \times \frac{25}{1{,}000{,}000} = 16.25 \text{ pounds/hour}$$

If the flow is likely to vary from 1100 U.S. gpm to 2800 U.S. gpm, it is more convenient to plot a graph of water flow against pounds per hour between the limits of 1000 and 3000 U.S. gpm.

This would be equivalent to

$$1000 \times 8.33 \times 60 \times \frac{25}{1{,}000{,}000} = 12.5 \text{ pounds/hour}$$

and

$$3000 \times 8.33 \times 60 \times \frac{25}{1{,}000{,}000} = 37.5 \text{ pounds/hour}$$

If the dosage of soda ash is likely to change from 25 mg/ℓ to 35 mg/ℓ, it would be convenient to plot these curves also on the same graph (see Figure 18-2).

In order to facilitate these calculations a series of chemical feed tables have been included in the appendices as follows:

A-7   pounds per hour/U.S. gpm
A-8   pounds per hour/U.K. gpm
A-9   grams per hour/liters per second
A-10 grams per hour/cubic meters per day
A-11 kilograms per hour/cubic meters per day

Since the graphs are always straight-line linear relationships, it is only necessary to compute the two end values, for example, at flows of 1000 U.S. gpm and 3000 U.S. gpm. The intermediate values can be read directly from the graphs. Using the same example as before to illustrate the use of the tables, turn to Appendices Table A-7. What are the amounts of chemicals per hour for flows of 1000 to 3000 U.S. gpm with dosage rates of 25, 30, and 35 mg/ℓ from table A-7?

Take the 1000 U.S. gpm flows first. For 25 mg/ℓ and 1000 U.S. gpm:

$$20 \text{ mg/}\ell = 0.000{,}999{,}24 \times 1000 \times 10 = 9.9924$$
$$5 \text{ mg/}\ell = 0.002{,}498{,}10 \times 1000 \times 1 = \underline{2.4981}$$
$$12.4905$$

Say 12.5 pounds/hour

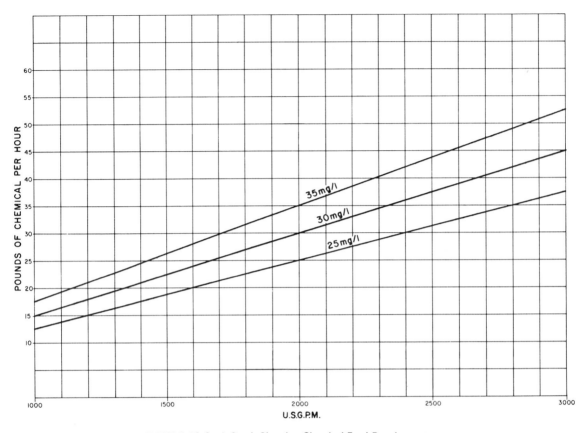

FIGURE 18-2 A Graph Showing Chemical Feed Requirements.

For 30 mg/ℓ and 1000 U.S. gpm:

$$30 \text{ mg/}\ell = 0.001{,}498{,}860 \times 1000 \times 10 = 14.988$$

Say 15.0 pounds/hour

For 35 mg/ℓ and 1000 U.S. gpm:

$$30 \text{ mg/}\ell = 0.001{,}498{,}860 \times 1000 \times 10 = 14.988$$
$$5 \text{ mg/}\ell = 0.002{,}498{,}100 \times 1000 = 2.498$$
$$\overline{17.486}$$

Say 17.5 pounds/hour

Now consider the other end of the graph with a flow of 3000 U.S. gpm. For 25 mg/ℓ and 3000 U.S. gpm:

$$20 \text{ mg}/\ell = 0.002{,}997{,}72 \times 1000 \times 10 = 29.977$$
$$5.\text{mg}/\ell = 0.007{,}494{,}30 \times 1000 \times 1 = \underline{7.494}$$
$$37.471$$

Say 37.5 pounds/hour

For 30 mg/ℓ and 3000 U.S. gpm:

$$30 \text{ mg}/\ell = 0.004{,}496{,}58 \times 1000 \times 10 = 44.9658$$

Say 45 pounds/hour

For 35 mg/ℓ and 3000 U.S. gpm:

$$30 \text{ mg}/\ell = 0.004{,}496{,}58 \times 1000 \times 10 = 44.9658$$
$$5 \text{ mg}/\ell = 0.007{,}494{,}30 \times 1000 \times 1 = \underline{7.4943}$$
$$52.4601$$

Say 52.5 pounds/hour

The system is really quite simple providing the right ordinates are selected in order to multiply only by factors of 10.

Table A-8 is somewhat less laborious since a U.K. gallon weighs 10 pounds instead of the 8.33 pounds of the U.S. gallon. Metric systems of measurement are also simple and convenient.

RECTANGULAR METHOD FOR MAKING DILUTIONS

Whenever it is necessary to dilute a liquid chemical or chemical solution to a lower concentration, the rectangular method of calculation is both quick and accurate. Two examples will serve to illustrate this method.

EXAMPLE 1

How many pounds of 36° Baume liquid alum and how many pounds of water are required to make 100 pounds of 10% aluminum sulfate solution?

Liquid alum of 36° Baume contains the equivalent of 48.18% dry aluminum sulfate. Draw a rectangle with intersecting diagonals and insert figures expressing the strengths of the two solutions in the left-hand corners of the triangle as shown. The concentration of water is, of course, zero. In the center insert the desired strength of the solution. Now subtract the figures on the diagonals, the smaller from the larger, and write the results at the other end of the respective diagonals. These figures then indicate how many parts by weight of the solution shown on the other end of the horizontal line must be taken to obtain a solution of the desired strength.

In this example, 10 pounds of 36° Baume liquid alum (48.18%) of dry alum and 38.18 pounds of water will be required to make 48.18 pounds of

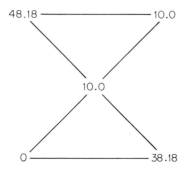

10% aluminum sulfate. To convert this to 100 pounds of 10% aluminum sulfate:

$$\frac{10}{48.18} \times 100 = 20.75 \text{ pounds of } 36° \text{ Baume}$$

$$\frac{38.18}{48.18} \times 100 = 79.25 \text{ pounds of water}$$

100.0 pounds of 10% aluminum sulfate

EXAMPLE 2

How much 90% sulfuric acid and how much water is needed to make a 60% sulfuric acid solution?

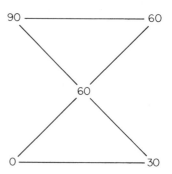

From the rectangle it will be seen that 60 pounds of 90% sulfuric acid are to be mixed with 30 pounds of water in order to produce 90 pounds of 60% sulfuric acid. If however 100 pounds of sulfuric acid are needed then the proportions by weight will be:

$$\frac{60}{90} \times 100 = 66.7 \text{ pounds of } 90\% \text{ H}_2\text{SO}_4$$

$$\frac{30}{90} \times 100 = 33.3 \text{ pounds of water}$$

$$100.0 \text{ pounds of } 60\% \text{ H}_2\text{SO}_4$$

If the stronger acid is to be diluted by a weaker acid, i.e., if 10% acid is used to dilute 90% acid, then proceed as follows:

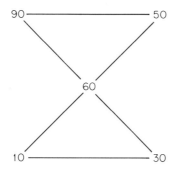

Then 50 pounds of 90% acid are mixed with 30 pounds of 10% acid to produce 80 pounds of 60% acid. For 100 pounds of 60% acid:

$$\frac{50}{80} \times 100 = 62.5 \text{ pounds of } 90\% \text{ acid}$$

$$\frac{30}{80} \times 100 = 37.5 \text{ pounds of } 10\% \text{ acid}$$

$$100.0 \text{ pounds of } 60\% \text{ acid}$$

Alternatively, the dilution equations can be calculated. Where

$x =$ quantity of weak solution or water used in the mixture

$y =$ quantity of strong solution or water to be used in the mixture

$D =$ desired quantity of the final solution

$A =$ strength of the strong solution in percent by weight

$B =$ strength of the desired solution in percent by weight

$C =$ strength of the weak solution or water in percent by weight (if water is used $C = 0$)

FORMULA METHOD OF MAKING DILUTIONS

1. To prepare a definite amount of dilute solution by mixing a strong solution with water or a weak solution, use the following formulae.

$$x = \frac{D(A-B)}{A-C}$$

$$y = D-X$$

2. To prepare a dilute solution of a desired concentration from a definite amount of strong solution and an unknown quantity of water or a weak solution, use the following formulae.

$$x = \frac{y(A-B)}{B-C}$$

$$D = x + y$$

SOLUTION STRENGTH

Recommended strengths of solutions are often given in pounds of chemical per gallon of water. In U.S. literature the gallons refer to approximately 8.33 pounds of water. Imperial or U.K. gallons contain approximately 10 pounds of water, and, since 1.0 U.K. gallon is approximately equal to 1.2 U.S. gallon, the error could be 20% if U.K. gallons are mistaken for U.S. gallons.

To avoid confusion it is not unusual to give solution strengths in terms of percentages, but this can also be misleading if incorrectly calculated. For example 1 pound of chemical per U.S. gallon is equivalent to a solution strength of:

$$\frac{1.0}{1.0 + 8.33} \times 100 = 10.7\%$$

whereas 1 pound of chemical per U.K. gallon is equivalent to a solution strength of:

$$\frac{1.0}{1.0 + 10.0} \times 100 = 9.1\%$$

HANDLING OF DANGEROUS CHEMICALS

The use of highly corrosive and dangerous chemicals such as caustic soda, sulfuric acid, sodium chlorite, and fluosilicic acid should only be used in water treatment plants after all personnel have been adequately trained in the necessary safety precautions.

All personnel working or visiting dangerous chemical areas must wear adequate protective clothing. Access to these areas by untrained personnel should be prohibited, with notices to this effect. In most cities in North America the local worker's compensation boards are anxious to provide literature and wall posters to constantly remind workers of the possible dangers of the chemicals they are handling.

19

DISINFECTION AND FLUORIDATION

DISINFECTION Municipal water supplies should be disinfected but they are never sterilized. Sterilization would be desirable but impractical for such large quantities of water. Chlorine is almost the universal disinfectant for water supplies. Other chemicals of importance are chlorine dioxide, ozone, potassium permanganate, silver, and others. Other disinfection processes are heat and irradiation with ultraviolet light.

The ideal water supply system uses several disinfection barriers together in series. Ground water, for example, passes through a biological barrier as it percolates through the topsoil on its way into the aquifer. The life styles of the organisms in the water will have changed and some of these will be eliminated, others will remain alive, but not necessarily reproducing. Facultative adjustment will occur as the environment changes from contact with the atmosphere and dissolved oxygen to where only chemical oxygen is available. Some organisms will be added to the water as it passes through the topsoil into the aquifer. A properly constructed and maintained well should produce good water free from pathogenic organisms, nevertheless, it should be chlorinated before distribution.

Chlorine is a powerful oxidizing disinfectant which will inactivate most bacteria providing it can get into contact, but normal dosages are not effective against most viruses.[1] However, it is reported[2] that ultraviolet light (UV) is effective against viruses and it is possible that UV disinfection together with chlorination may be one answer to the problem. Minimum water turbidity is essential if UV light is to be effective.

Chlorination

Factors which contribute to good chlorine disinfection are:

(a) Concentration

(b) Low pH

(c) Adequate retention time

(d) Low turbidity

(e) Free chlorine residuals

Most of these subjects have been discussed elsewhere in this book but are correlated here for convenience.

Concentration

In terms of concentration, it is not the dosage of chlorine that is important, but the quantity of free chlorine residual remaining in the water after the chlorine demand has been satisfied. It has often been stated that all public water supplies should carry a measurable residual of free chlorine which is detectable at any point in the system. It is desirable that the concentration of free residual chlorine in the treated water be at a level not less than 0.5 mg/ℓ, after a minimum contact time of 20 minutes.[3]

If the distribution system is such that it is impossible to have a minimum of 20 to 30 minutes retention time between the point of chlorine application and the first consumer, then the system should be redesigned to insure that the prescribed retention time is available. This can be done by installing a holding reservoir, or enlarging a section of the water main to provide the retention and contact time needed. Often a treatment plant clear-well serves this purpose.

Where tastes and odors in water supply systems are aggravated by chlorination, the following corrective actions should be taken:

(a) If the taste is due to phenols or certain algae, then either go to breakpoint chlorination or use chlorine dioxide, ozone, potassium permanganate, or activated carbon.

(b) If the water is tasteless at the water treatment plant after the addition of chlorine, but picks up a taste from the storage reservoirs or distribution systems, then they may be contaminated with nuisance bacteria and require cleaning and disinfecting. New reservoirs may cause taste and odor problems when they are first put into service but this situation only lasts for a short period of time and gradually disappears.

(c) If the water is normally tasteless and suddenly develops a taste in the distribution system, there may be a cross-connection or back siphonage. Failure to take the appropriate action may result in serious consequences.

Most textbooks show the relationship illustrated in Figure 19-1 between percentages of $HOCl^-$, OCl^-, and pH. When chlorine is added to chemically pure water, a dilute mixture of hypochlorous (HOCl) and hydrochloric acids is formed, and very little chlorine exists in solution under normal conditions. Hypochlorous acid ionizes or dissociates practically instantaneously into hydrogen ions (H^+) and hypochlorite ions (OCl^-) depending on pH and to a small degree on temperature. Free available hypochlorous (HOCl) and hypochlorite (OCl^-) that might exist in water are not equally germicidal. The markedly superior effectiveness of HOCl is attributable not only to its stronger oxidizing power but also to its small molecular size and electrical neutrality, which allows it to readily penetrate bacteria cells. Hydrochlorite ion has little if any germicidal effect, and its negative charge is presumed to impede cell penetration.

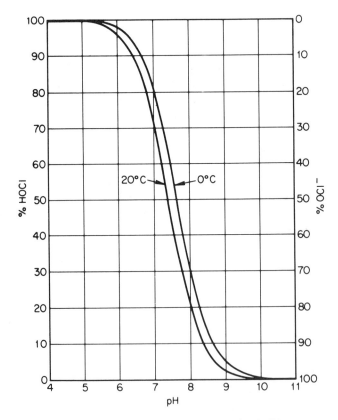

FIGURE 19-1 Distribution of HOCL and OCL⁻ in Water.

pH

The germicidal properties of HOCl are much greater than OCl$^-$. This subject is dealt with in *Water Quality and Treatment*[5] and it is stated that "with free available residual chlorine at a pH of 6.0 to 8.0, a safe residual for complete destruction of bacteria after 10 minutes contact would be not less than 0.2 mg/ℓ, and at pH of 8.0 to 9.0 at least 0.4 mg/ℓ, and at a pH of 9.0 to 10.0 at least 0.8 mg/ℓ, and above a pH of 10.0, more than 1.0 mg/ℓ would be required." The important issue here is to realize that if x mg/ℓ of free chlorine residual is necessary to disinfect a water supply at a pH of 6.0 to 8.0, then the following dosages would be required to do the same job at higher pH values:

pH 6.0 to 8.0 requires x mg/ℓ of chlorine.

pH 8.0 to 9.0 requires $2x$ mg/ℓ of chlorine.

pH 9.0 to 10.0 requires $4x$ mg/ℓ of chlorine.

pH 10.0 and above requires $5x$ mg/ℓ of chlorine.

Retention Time

The recommended values for retention time as given in the literature vary extensively from a few minutes to over an hour. With the less effective chloramines, which make up most of a combined residual, time becomes increasingly important. One of the reasons for longer retention is to insure that the chlorine solution has an opportunity to make contact with the organisms. If the water was completely free of turbidity and all the bacteria were free-swimming, then they would be inactivated very quickly by efficient mixing of aqueous chlorine solution into the water. Organisms which have a protective covering require more time for the chlorine to be effective. This is why the free chlorine residual should exist up to the end of the contact period.

Turbidity

Organisms in turbid water can protect themselves from contact with the chlorine, particularly for short periods. The chemical nature of the suspended particulate matter may also substantially change the immediate environment of the organism by raising the pH value which will change the HOCl to OCl$^-$ and seriously reduce the germicidal effect, or by chemically absorbing the free chlorine.

Free and Combined Residuals

Ammonia may be added in order to avoid chlorine substitution products which often result in taste problems. The chlorine reacts with the ammonia forming chloramines generally as follows:

$$HOCl + NH_3 \longrightarrow H_2O + NH_2Cl$$
$$\text{(monochloramine)}$$

$$HOCl + NH_2Cl \longrightarrow H_2O + NHCl_2$$
$$\text{(dichloramine)}$$

The process, however, is not trouble free. It is essential to add the ammonia first so that the chlorine will react with it before it can form other substitution products. This eliminates any possibility of a free chlorine residual. Chloramines are less effective in the inactivation of microorganisms, (especially virus) compared to free chlorine, so that larger dosages with up to a 2 hour contact period may be necessary for adequate disinfection. One advantage of combined residuals is their persistence in the distribution system. Not that the combined chlorine residual is a particularly effective germicide to protect the water from any pollution entering into the distribution system through a cross-connection or back siphonage, but perhaps as indications of the chlorine/chloramine demand of the system. If a monitoring program indicates that the chloramine residual has drastically changed in a short period of time, there may be a cross-connection or back siphonage.

Flow Measurement

One of the problems in chlorination and also with fluoridation, is accurate water flow measurement and use of this data either electrically or pneumatically to pace and control the rate of chlorine and fluoride feeds. If the water treatment plant is operated at a constant rate, on a start-stop system, where it starts when the clear water well or reservoir is low, and stops when it is full, there is little difficulty since the rate of flow is relatively constant. If however, the flow rate is variable, the problem of accurately dosing becomes involved. Most flow meters are accurate to 1 or 2% of the full scale reading. A variable flow system may have a meter that can measure with 2% accuracy at 1000 gpm under full flow conditions. If this occurs at a velocity of 10 feet per second, then at 100 gpm the velocity will be 1 foot per second and the meter accuracy of 2% of full scale will be ± 20 gpm on a flow of 100 gpm which is really an accuracy of $\pm 20\%$ at the lower end of the register.

Breakpoint Chlorination

Breakpoint chlorination consists of adding sufficient chlorine to satisfy the chlorine demand of the water and to establish a free chlorine residual at the end of a suitable contact time.

This system has several advantages where the raw water supply is taken from a river which may be highly polluted with industrial and domestic waters. The chlorine demand can vary considerably from hour to hour during the day when the industrial plants upstream of the water treatment plant intake are discharging their various effluents. The rate of river flow is also an important dilution factor.

In these circumstances, chlorine is added in sufficient quantities to maintain a free chlorine residual at the end of the contact period. Since the demand is variable, and in order to conserve chemicals, a chlorine residual analyzer should continuously monitor the water at the end of the chlorine contact period. The output signal from the chlorine residual analyzer should correlate with the output signal from the flow meter. The two signals computed together, free chlorine residual and flow, give one signal to the chlorinator. This system has its own problems largely due to sampling difficulties and time lag, equivalent to the retention time between the introduction of the chlorine and its measurement at the end of the contact period. The mechanism of the reaction is traditionally portrayed as illustrated in Figure 19-2.

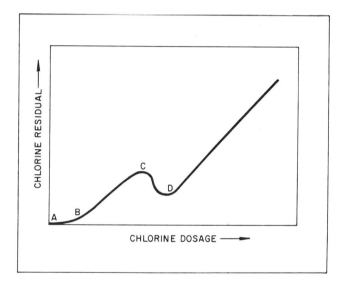

FIGURE 19-2 Breakpoint Chlorination.

The chlorine dosage is indicated along the bottom abscissa and the measured residual on the left ordinate, to equal scale in mg/ℓ.

From A to B the chlorine is absorbed by the organic compounds. The amount of chlorine at dosage point B represents the amount required to meet the demand by the reducing agents, nitrites, ferrous ions, hydrogen sulphite, and others. From B to C chloramines are formed from the reaction of chlorine with the ammonia compounds:

$$HOCl + NH_3 \longrightarrow H_2O + NH_2Cl$$
$$\text{(monochloramine)}$$

$$HOCl + NH_2Cl \longrightarrow H_2O + NHCl_2$$
$$\text{(dichloramine)}$$

$$HOCl + NHCl_2 \longrightarrow H_2O + NCl_3$$
$$\text{(trichloramine)}$$

The quantities of mono-, di- and trichloramines are functions of pH. The chloramines show a chlorine residual available for disinfection but only about 3 to 4% as effective as free chlorine. When all the ammonia has reacted at point C, the oxidation of the nitrogen compounds reduces the chlorine residual and the curve drops to point D. When all the chloramines, initially formed, have been completely oxidized to nitrogen oxides, a "free" or uncombined chlorine residual develops and from point D upwards the curve is at a 45° angle with the abscissa, that is to say, for every mg/ℓ of chlorine dosage beyond point D there will be one more mg/ℓ of chlorine residual shown on the ordinate.

Point D is therefore known as the "breakpoint" and the process of oxidizing the organic and the ammonia compounds with chlorine is known as "breakpoint chlorination."

Breakpoint, or super chlorination, is a very useful method of removing taste problems, oxidizing other impurities, and generally giving good disinfection.

Self-Contained Hypochlorinators

A relatively new development in the last few years is the self-contained electrolytic hypochlorinator which decomposes sea water, natural brine or brackish water, or sodium chloride solution, from dissolved salt to produce sodium hypochlorite. Available equipment using sea water can produce over 100 pounds of chlorine per hour. This type of equipment has considerable possibilities since it only requires power and salt. It can be used on board ship or in remote areas where freight costs would make liquid chlorine prohibitive, since the containers have to be transported both full and empty and the freight costs can be significant.

Chlorine Dioxide (ClO_2)[1]

If the taste problems are not consistant all the year round, the use of chlorine dioxide should be considered. The chemical is somewhat more expensive and troublesome to handle compared to straight chlorine, but it can be very effective in the right circumstances.

Chlorination Equipment—
Installation and Safety

The Ontario Water Resources Commission (OWRC) of Toronto, Ontario, have compiled a "Basic Gas Chlorination Manual" which has been published in four parts in the Journal of the American Water Works Association.[6] Separate rooms should be provided for chlorination equipment which can only be entered from outside the building. An exhaust fan with the intake close to the floor, capable of 20 to 30 air changes per hour, is necessary in all chlorine rooms. Where possible, ton containers or 150 pound cylinders should be stored in a separate room and kept at 5° to 10°F colder than the chlorinator room. Safety equipment including both canister and self-contained air-supplied gas masks should be readily available, together with suitable wrenches and ammonia squeeze bottles to locate the source of leakage. It is also advisable, and mandatory in many areas, to have a complete repair kit immediately available to cap off leaky valves.

Portable Chlorination Equipment

A typical portable packaged chlorination system, completely piped and wired, suitable for use in emergencies or temporary installations is shown in Fig. 19-3.

FIGURE 19-3 Portable Packaged Chlorination System Completely Assembled Ready for Use (Courtesy, Penwatt Corp., Wallace and Tiernan Division).

Ozone (O_3)

Ozone, also known as ozonized air, has proved to be an efficient disinfectant and is used fairly extensively in France. It requires electric power and fairly sophisticated equipment but has not been economically competitive with chlorine in North America. However, it can be produced on site, it is independent of transportation, and for new installations remote from chlorine producers, may have its advantages.

Ozone is produced at the treatment plant site from dry clean air, or from pure oxygen, passed between two electrically charged surfaces. A potential of 15,000 to 20,000 volts causes a rearrangement of the oxygen atoms from the normal O_2 molecule to O_3 molecules. The amount of ozone (O_3) produced can be as high as 30 gms per cubic meter of air which is approximately 2.4% by weight, equivalent to 1.4% by volume.

Although ozone can be liquified, it is presently used in the gaseous form in the water and wastewater industry. It is an extremely powerful oxidizing agent, with a potential nearly twice that of chlorine. Not only a powerful disinfectant, it is very effective in dealing with tastes, odors, and color.

However, ozone residuals only last up to 30 minutes at the most, and are of little value as added protection to the water in the distribution system, but post-chlorination can always be added to the system if necessary.

Ozone is said to be more effective than chlorine for controlling very resistant bacteria in water. Tests have shown that 10 mg/ℓ of chlorine was required for a contact period of 3 minutes to kill an aqueous suspension of bacillus spores. With ozone, the same results were achieved in 14 minutes using only 0.35 mg/ℓ of ozone.[7] It is reported that under certain specific conditions ozone has been more effective than chlorine in the inactivation of viruses.[8]

Dosages Equipment capacity must be selected to provide dosages in excess of the anticipated maximum values, sometimes as high as 4 ppm. An excess will produce a free ozone residual which can be detected with the orthotolidine reagent used for measuring free chlorine residuals. It is reported that 0.1 ppm ozone produces an intensity of color with orthotolidine equal to that given by 0.15 ppm residual chlorine. In the orthotolidine arsenite test however, the intensity of color is the same as that from free residual chlorine.[9]

Care must be taken when discussing dosage rates as to whether or not the units are in terms of parts per million (ppm) by volume in air, or ppm by weight, since there is a significant difference between the two. It is preferred not to use mg/ℓ since this is weight/volume, but rather use parts per million (ppm) by weight.

One of the problems with ozone is its low solubility in water.[11] The process involves the mixing of a gas into a liquid, and unless the mixing of the ozone is at optimum efficiency, higher dosages will be required. Furthermore, with ozonized air there are large volumes of nitrogen and other gases to be bubbled into the water and vented to atmosphere.

Typical dosages are between 8 to 50 lb of ozone per million U.S. gallons of water. In France, the primary purpose of ozonation is disinfection with the removal of tastes, with odors and color as secondary effects, whereas in North America, where chlorine is regarded as the universal disinfectant, ozone would only be considered for difficult waters where taste, odor, and color were problems.

Physical and Chemical Properties Ozone ranges from colorless to a bluish tinge, and has a density of approximately 1.6 times that of air. When produced from air the concentration of ozone in the ozonized air is said to be between 0.5 and 1% by volume. If pure oxygen is used instead of air, the concentration can be as high as 3 to 4%.

Since ozone is an extremely active oxidizing agent and relatively unstable, it is important to handle it in non-corrosive materials such as 316 stainless steel, glass, ceramics, aluminum, Teflon, etc.

Ozone is toxic, is a powerful irritant, and exposure to concentrated atmospheres of ozonized air can be fatal. Fortunately, however, small concentrations, 0.01 to 0.02 ppm by volume, can be detected by smell long before a critical concentration is reached. While 0.1 ppm by volume can be tolerated indefinitely, 1.0 ppm can only be tolerated for 8 minutes and up to 4 ppm can be tolerated up to 1 minute without producing symptoms of coughing, eye watering, and irritation of the nasal passages. Exposure to 10,000 ppm which is equivalent to 1% by volume, the anticipated concentration of ozonated air immediately downstream of the ozonizer, is likely to be fatal in less than 1 minute.

Ultraviolet Light (UV)

The development of equipment for producing and utilizing invisible ultraviolet light for the disinfection of liquids, gases, and the surface areas of solid objects has made considerable advances during the last decade. The use of UV light in the water industry will increase as time goes on.

Ultraviolet light is invisible to the eye but can vividly illuminate certain white objects and other colors. The ultraviolet spectrum includes wave lengths from 2000 to 3900 Angstrom units (Å) (where one Angstrom unit is equal to a millimeter $\times 10^{-6}$). The entire UV spectrum can be divided into three sequences:

(a) Long Wave UV is 3250 to 3900 Å; these rays naturally occur in sunlight but they have little germicidal value.

(b) Middle Wave UV is 2950 to 3250 Å, also found in sunlight. Middle wave UV is best known for its sun-tanning effect, and it provides some germicidal action with sufficient exposure.

(c) Short Wave UV is 2000 to 2950 Å, and has the greatest germicidal effectiveness of all ultraviolet wave lengths. It is employed extensively to destroy bacteria, viruses, molds, spores, etc., both air and water-borne.

Ultraviolet lamps of low pressure mercury vapor provide the most efficient source of short-wave ultraviolet energy at a wave length of 2537 Angstrom units. See Figure 19-4 which shows the most effective zone of germicidal range. The low pressure vapor lamp is almost at the peak of the curve. This particular graph is from the trade literature provided by Ultradynamics Corp., manufacturers of UV equipment. Other suppliers also use the same 2537 Å light source.

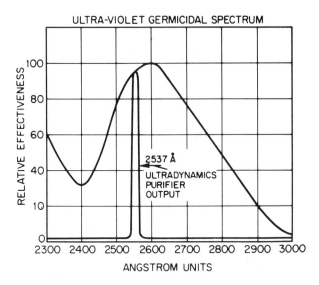

FIGURE 19-4 Ultra-violet Germicidal Spectrum[12].

Ordinary glass has poor UV light transmission and it is important that the lamps and other glass components of a UV cell be made of special high silica ultraviolet transmitting glass. (See Figure 19-5.)

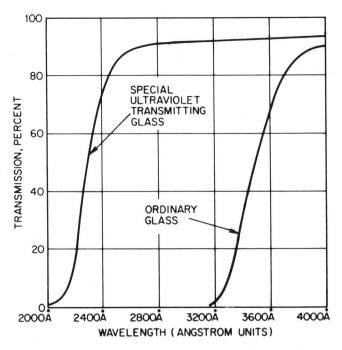

FIGURE 19-5 Transmission Curve for High Silica Glass[13].

Exposure for the inactivation of common bacteria is reported to be 20,000 microwatts per sq cm/second, but to allow for less than 100% transmission a purification plant should be designed to deliver 2537 Angstrom energy in excess of 30,000 microwatts per sq cm/second.[14]

The tubular lamps range in length from 12 to 48 inches, are rated for 7500 hours effective life and consume power in general accordance with Table 19-1.

TABLE 19-1 Power Usage[13]

Lamp Length	Power Consumption	2537 A Output
12″	8 W	1.3 W
18″	18 W	5.8 W
36″	39 W	14.6 W
48″	110 W	51.5 W

A single purifier can be designed to handle 300 U.S. gpm and by placing multiple units in parallel, high capacities can be obtained. A typical one lamp unit is shown in Figure 19-6.

The effectiveness of a germicidal ultraviolet light to inactivate water-borne pathogens is shown in Table 19-2 as reported by Ultradynamics Corporation.

TABLE 19-2 Ultraviolet Water-borne Pathogens[19]

Microorganism	Disease	Ultrads	(a)
Salmonella typhosa	Typhoid fever	4100	
Salmonella paratyphi	Enteric fever	6100	
Shigella disenteriae	Dysentery	4200	
Shigella flexneri	Dysentery	3400	
Vibrio comma	Cholera	6500	(b)
Leptospira spp.	Infectious jaundice	6000	(b)
Poliovirus	Poliomyelitis	6000	(c)
Unidentified	Infectious hepatitis	8000	(d)

Note (a) An ultrad is a unit of ultraviolet radiation with a wavelength equal to 2537 Angstroms with an intensity of one microwatt per second per square centimeter.

Note (b) These values are estimated.

Note (c) Values based on the "American Journal of Hygiene," Vol. 53, p. 131, 1951.

Note (d) Since viruses in general are believed to be more susceptible to UV than bacteria.

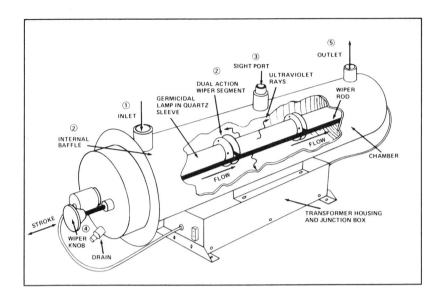

PRINCIPLE OF OPERATION

① The water enters the purifier and flows into the annular space between the quartz sleeve and the outside chamber wall.

② The internal baffle and wiper segments induce turbulence in the flowing liquid to insure uniform exposure of suspended micro-organisms to the lethal ultraviolet rays.

③ The sight port enables visual observation of lamp operation.

④ The wiper assembly facilitates periodic cleaning of the quartz sleeve without any disassembly or interruption of purifier operation.

⑤ Water leaving the purifier is instantly ready for use.

FIGURE 19-6 Typical Water Purifier[14].

Criteria for Acceptability of an Ultraviolet Disinfection Unit[16]

1. Ultraviolet radiation at a level of 2537 Angstrom units must be applied at a minimum dosage of 16,000 microwatt-seconds per square centimeter at all points throughout the water disinfection chamber.

2. Maximum water depth in the chamber, measured from the tube surface to the chamber wall, shall not exceed 3 inches.

3. The ultraviolet tubes shall be:
   (a) Jacketed so that a proper operating tube temperature of about 105°F is maintained, and
   (b) The jacket shall be of quartz or high silica glass with similar optical characteristics.

4. A flow or time delay mechanism shall be provided to permit a 2 minute tube warm-up period before water flows from the unit.

5. The unit shall be designed to permit frequent mechanical cleaning of the water-contact surface of the jacket without disassembly of the unit.

6. An automatic flow control valve, accurate within the expected pressure range, shall be installed to restrict flow to the maximum design flow of the treatment unit.

7. An accurately calibrated ultraviolet intensity meter, properly filtered to restrict its sensitivity to the disinfection spectrum, shall be installed in the wall of the disinfection chamber at the point of greatest water depth from the tube or tubes.

8. A flow diversion value or automatic shut-off valve shall be installed which will permit flow into the potable water system only when at least the minimum ultraviolet dosage is applied. When power is not being supplied to the unit, the valve should be in a closed (fail-safe) position which prevents the flow of water into the potable water system.

9. An automatic, audible alarm shall be installed to warn of malfunction or impending shutdown if considered necessary by the control or regulatory agency.

10. The materials of construction shall not impart toxic materials into the water either as a result of the presence of toxic constituents in materials of construction, or as a result of physical or chemical changes resulting from exposure to ultraviolet energy.

11. The unit shall be designed to protect the operator against electrical shock or excessive radiation. As with any potable water treatment process, due consideration must be given to the reliability, economics, and competent operation of the disinfection process and related equipment, including:
    (a) Installation of the unit in a protected enclosure not subject to extremes of temperature which could cause malfunction.
    (b) Provision of a spare UV tube and other necessary equipment to effect prompt repair by qualified personnel properly instructed in the operation and maintenance of the equipment.
    (c) Frequent inspection of the unit and a record kept of all operations, including maintenance problems.

Ultraviolet light has a future in water disinfection, but since it cannot kill an organism unless it can see it, it obviously is not much use for turbid and highly-colored waters, but neither is chlorine. Slimes and silts can deposit on the tubes and reduce the transmission of UV light. One of the early objections to this process was the inability to detect when the equipment was malfunctioning due to a faulty tube, or a coating of deposit which would tend to mask the transmission of UV, but equipment is now available which will indicate the level of radiation.

Ultraviolet has the advantages that it does not cause taste problems, but on the other hand it does not leave a residual. The retention time necessary to inactivate the organisms are a function of the intensity of the light. These aspects must be carefully considered in the design of a UV irradiation installation.

Potassium Permanganate ($KMnO_4$)

Potassium permanganate is a powerful oxidizing agent and can also be used as a disinfectant. It is fairly expensive and its use as an alternative to chlorine or hypochlorite would only be justified in an emergency.

It is useful for disinfecting new water mains. A stock solution of $\frac{1}{4}$ lb of $KMnO_4$ per U.S. gallon is added to newly installed water mains in the ratio of 1 U.S. gallon of stock solution to 1000 feet of 16-inch diameter main. The equivalent to a dosage of 3 mg/ℓ is reported to be more convenient and equally as effective as hypochlorites.[18]

Silver

For centuries it has been known that silver will inactivate bacteria and a number of processes to implement its use have been developed, i.e., Katadyn-Sand filter, and the Electro-Katadyn process. A fairly detailed account appears in Taylor (7th Edition)[19] and is updated by Holden.[20]

Although the process may have some significance with waters of zero color and low turbidity in the way that King Cyrus, 2500 years ago, used silver flagons to carry his drinking water, there is no indication that the use of silver is likely to be of commercial importance for disinfecting municipal water supplies. There is no apparent evidence to indicate that silver is harmful to humans; it is nevertheless accumulative in the body and large dosages can cause skin discoloration.[21]

Its effectiveness as a bactericide depends upon whether or not the bacteria are free-swimming and in contact with dissolved silver ions. If the bacteria are hosted on particles of suspended matter, so that they are protected from contact with the silver, they will not be inactivated. On the other hand, it is reported that once the water has been in contact with silver and silver ions are in solution, the water will remain bacteria-free for a month or more in spite of daily re-infection. It would appear that there is a "free residual" similar to a free chlorine residual.

FLUORIDATION The artificial fluoridation of public water supplies has significantly reduced the DMF index of growing children and adults who have been fortunate enough to have grown up in areas where the drinking water is naturally or artificially fluoridated. The DMF index refers to the number of decayed, missing, and filled teeth.

Much research has been directed to means of administering fluorides other than into the public water supplies. Unfortunately, none of these schemes have materialized. One suggestion was to add fluorides to children's milk, but since the fluoride precipitates calcium fluoride with the milk instead of in the teeth, the scheme was abandoned. Summaries of the results of fluoridation are given in *Water Quality and Treatment*,[22] and the *Manual of Water Fluoridation Practice*.[23]

The correct fluoride dosage to reduce DMF is very small and this presents a problem if left to individuals using tablets in their drinking water who may have the impression that if one tablet a day does a good job, two will do a better job and they may find they have mottled teeth resulting from an overdose of fluoride.

It seems the only safe and sure method of administering fluorides is to add them to the municipal water supply system.

Optimum Fluoride Levels

The optimum fluoride level which produces the greatest protection against dental cavities, with the least hazard of fluorosis, is dependent on how much drinking water is consumed. Obviously, children in hot climates will drink more water than children in colder climates. The following Table 19-3 is based on the year round maximum local air temperatures of the municipality or the water supply district.

TABLE 19-3 Recommended Fluoride Levels[22]

Average Daily Air Temperatures		Recommended Fluoride Content
°F	°C	
50.0–53.7°	10.0–12.0°	1.2 mg/ℓ
53.8–58.3°	12.1–14.6°	1.1 mg/ℓ
58.4–63.8°	14.7–17.6°	1.0 mg/ℓ
63.9–70.6°	17.7–21.4°	0.9 mg/ℓ
70.7–79.2°	21.5–26.2°	0.8 mg/ℓ
79.3–90.5°	26.3–32.5°	0.7 mg/ℓ

It is important that the fluoride level for each temperature range be held within close tolerances if optimum prophylactic results are to be obtained. In accordance with the EPA report—National Interim Primary Drinking Water Regulations—the maximum levels should never exceed the Recommended Fluoride Levels by more than a factor of two.[24]

It is essential to determine the natural fluoride levels of the raw water before setting the dosages for artificial fluoridation. Waters sometimes have surprisingly high background levels of natural fluorides and in some cases treatment plants must be installed to reduce the level and thus avoid mottled teeth.[23] It is also essential to perform the fluoride analysis strictly in accordance with *Standard*

Methods[25] or "Methods for Chemical Analysis of Water and Wastes,"[26] using a pre-distillation unit to avoid interference from other elements.

Chemicals used in Fluoridation Processes

1. Sodium fluoride (NaF)
2. Sodium silicofluoride (Na_2SiF_6)
3. Fluosilicic acid (H_2SiF_6)
4. Magnesium silicofluoride ($MgSiF_6 \cdot 6H_2O$)
5. Ammonium silicofluoride ($(NH_4)_2SiF_6$)
6. Calcium fluoride (CaF_2)
7. Potassium fluoride ($KF \cdot 2H_2O$)

The first three are the most commonly used and the only ones that will be discussed in the book. For further information see F. J. Maier, *Manual of Water Fluoridation Practice.*[23]

Sodium Fluoride (NaF)   Molecular composition

Na	22.99
F	18.99
	41.98

Fluoride ion (100%) = $\frac{18.99}{41.98} \times 100 = 45.2\%$

Commericial sodium fluoride is between 95 and 98% NaF.

The fluoride ion content therefore varies between 43.0 and 44.3% F ion. Assuming an average fluoride content of 44%, it will require, for a dosage of 1 mg/ℓ, the following feed rates:

19.0 lb of NaF per 1,000,000 U.S. gallons

22.7 lb of NaF per 1,000,000 U.K. gallons

Sodium Silicofluoride (Na_2SiF_6)   Molecular composition

Na_2	45.98
Si	28.09
F_6	113.94
	188.01

Fluoride ion (100%) = $\frac{113.94}{188.01} \times 100 = 60.6\%$

Commerical sodium silicofluoride is about 98.5% Na_2SiF_6.

Assuming an average fluoride content of 59.7%, it will require, for a dosage of 1 mg/ℓ, the following feed rates:

13.95 lb of Na_2SiF_6 per 1,000,000 U.S. gallons

16.76 lb of Na_2SiF_6 per 1,000,000 U.K. gallons

Fluosilicic Acid (H_2SiF_6)   Molecular composition

H_2	2.016
Si	28.909
F_6	113.940
	144.865

Fluoride ion $= \dfrac{113.94}{144.86} \times 100 = 79\%$

Commerical fluosilicic acid varies in strength from 20 to 35% H_2SiF_6 in aqueous solution depending upon the manufacturer. Twenty-five percent strength is fairly common and it will require, for a dosage of 1 mg/ℓ, the following rates:

42.2 lb of H_2SiF_6 per 1,000,000 U.S. gallons

50.55 lb of H_2SiF_6 per 1,000,000 U.K. gallons

Physical and Chemical Properties

Sodium Fluoride (NaF)   Sodium fluoride is a white, odorless, free-flowing material available either as a powder or in finely divided crystals. Its specific gravity is 2.79 and the bulk density of powdered sodium fluoride varies from a light grade (about 50 lb per cubic foot) to a heavier grade (90 lb per cubic foot) but normally about 65 lb per cubic foot.

NaF is non-hydroscopic and when dry is free-flowing, free from lumps. The important characteristic of sodium fluoride is its practically constant solubility of 4.05 gram per 100 mℓ of water at normal room temperatures. This property of sodium fluoride has made it possible to use sodium fluoride saturators for small plants. This equipment consists of a plastic tank with a loose plastic funnel lying upside down inside the tank, resting on the bottom. Through the spout of the plastic funnel is the suction hose of a diaphragm-proportioning pump which is paced in proportion to the treated water flow. Sodium fluoride crystals are deposited into the tank. Water is added, preferably base-exchange softened water or equivalent, since calcium and fluoride ions react together to form insoluble calcium fluoride, and when this occurs, through using hard water, the sodium fluoride is prevented from dissolving.

With soft water, free from calcium and magnesium ions, the solution strength is consistently 4% by weight. Sodium fluoride used for water supplies should be in accordance with the requirements of American Water Works Association, Standard (AWWA B701-71).

Sodium Silicofluoride (Na_2SiF_6) Sodium silicofluoride is a white, free-flowing, odorless non-hydroscopic crystalline powder. Its specific gravity is 2.679 and the bulk density is between 72–97.5 lb per cubic foot, the regular grade is about 72 lb per cubic foot. Unless it is delivered in plastic bags or waterproof inner containers it will cause problems with the feeders.

The solubility of sodium silicofluoride is relatively low:

60°F (15.6°C) = 0.62 gms per 100 mℓ of water

77°F (25°C)   = 0.762 gms per 100 mℓ of water

90°F (32.2°C) = 0.86 gms per 100 mℓ of water

Because of this low solubility a dry feeder and an agitated dissolving tank are needed to prepare the solution for proportional feeding into the water stream. In many cases, because of its poor solubility, it is in partial suspension in a solution.

Sodium silicofluoride used for fluoridating drinking water supplies should be in accordance with the requirements of the American Water Works Association, Standard (AWWA B702-74).

Fluosilicic Acid (H_2SiF_6) Fluosilicic acid is always in aqueous solution between 20 and 35% H_2SiF_6. The anhydrous acid is not known. It is a colorless, transparent, fuming, corrosive liquid having a pungent odor and an irritating action on the skin. It is corrosive to steel and is handled in rubber or lead-lined steel or plastics. The acid is only slightly volatile but the vapors can be poisonous if inhaled. It is highly toxic if ingested. Areas subjected to occasional emission of fumes from weigh tanks are soon coated with a white deposit.

A typical product specification for fluosilicic acid is as follows:

H_2SiF_6	= 25% ±1%
Acid other than H_2SiF_6	= 1.5% max
Heavy metals such as lead (Pb)	= 0.02% max
Iodine as (I)	= 0.025%
Organic material	= Nil
Specific gravity	= 1.24 at 67.5°F
Viscosity (centipoise)	= 4.0 @ 70°F
Crystallizing point @ 25% conc.	= −2°F (−19°C)
(Ice crystals form) @ 26% conc.	= −4°F (−20°C)
Boiling point	= 222.5°F (106°C)

The acid is corrosive to all metals except copper and silver.

Fluosilicic acid used in drinking water supplies should be in accordance with the requirements of the American Water Works Association, Standard (AWWA B703-71).

Hazards As far as it is known there have been no reported cases of illness due to fluoridation among water treatment plant operators handling fluoride chemicals. Dust masks, rubber aprons, and gloves are recommended and should be worn in the fluoride rooms.

Fluosilicic acid however must be handled with great respect and only by operators who have been trained in the handling of dangerous chemicals. It must be treated in the same way as sulphuric acid or caustic soda. Avoid breathing the vapors and use self-contained gas masks with a built-in air supply if it is necessary to enter a room where acid vapors exist. The use of fluosilicic acid in small plants operated by untrained operators and unskilled casual labor should be avoided.

Small Systems

Sodium fluoride (NaF) has many advantages for small towns and villages. It has an almost constant solubility of 4% at all normal room temperatures. A plastic container with granulated sodium fluoride wetted with base-exchange softened water in which the calcium and magnesium have been removed will attain a constant 4% fluoride composition. This solution can then be pumped with a diaphragm chemical pump in proportion to water flow. One problem, however, with this type of system is the tendency of slugs of highly concentrated fluoride ion to pass along the water distribution pipe at each stroke of the diaphragm pump, followed by slugs of water free of fluoride ion. The average daily dosage with this system may be well within the normal accuracy of good practice i.e., ± 0.1 mg/ℓ, but if a sample is taken from the system before these slugs have had an opportunity to become completely homogeneous, the laboratory results may range from 0 or 10 to 12 mg/ℓ. This problem can be overcome by inserting a small cylinder, half filled with air under pressure, between the proportioning pump and the distribution system. This cylinder performs the duty of a hydro-pneumatic tank and allows an almost constant flow of fluoride solution under pressure to flow into the distribution system.

This scheme is ideal for small systems but it is essential that the softened water fed into the sodium fluoride saturator is free of color, turbidity, or iron compounds which would prevent the sodium fluoride from going into solution.

Medium Sized Systems

Sodium silicofluoride would probably be the preferred choice for a 4–5 mgd plant since it is the least expensive of the fluoride chemicals. Providing the sodium silicofluoride is dry and free from lumps, it will feed uniformly through a dry volumetric feeder into an agitated dissolving tank. The solubility of sodium silicofluoride ranges from 0.43% at 32°F (0°C) up to 2.45% at 212°F (100°C).

At normal water treatment plant temperatures of 60°F (15.6°C) it has a solubility of approximately 0.62%. Because of its poor solubility, the dissolving tank should be capable of handling 60 U.S. gallons of water per pound of sodium silicofluoride, with a retention time of at least 5 minutes and it is understood that some state regulations call for 15 minutes of agitated retention.[27] The solution may still contain undissolved silicofluoride which can be abrasive to pumps and valves; if the solution can flow from the dissolving tank to the point of application by gravity, so much the better. If not, a slow speed rubber-lined centrifugal pump with an impeller periphery speed of less than 5000 feet per minute would be satisfactory. Since the flow of dry chemical is metered through the volumetric or gravimeter dry feeder, the quantity of water that is pumped is of no consequence to the accuracy of the dosage, providing it is equal or in excess of the normally accepted 60 U.S. gallons per pound of sodium silicofluoride.

Large Systems

The problem with larger systems is in the handling of the bags and the general inconvenience unless the sodium silicofluoride can be purchased in bulk containers, which is not very often. Fluosilicic acid (25% H_2SiF_6) is convenient to use, particularly if it is available in tank trucks. It is usually more expensive than sodium silicofluoride pound for pound of fluoride ion, depending on location and transportation costs, but the ease of handling reduces the operating costs considerably. The acid is considerably more dangerous to handle than either the sodium fluoride or the sodium silicofluoride. It is usual to provide underground storage tanks in a concrete vault to hold at least 3 or more tank truck loads. The tank must be rubber lined together with all pumps, pipes, and fittings. A small transfer pump is used from the storage tanks to a 30 hour capacity tank installed on a weigh scale in the fluoride equipment room. Accurate records are kept of each daily transfer and usage. The acid is pumped at 25% concentration directly to the point of injection by means of a chemical-proportioning diaphragm pump in accordance with the rate of water flow. As an added refinement, a second pump is used in parallel known as the trimmer. With this arrangement, the first proportioning pump supplies 80 to 90% of the fluosilicic acid demand in accordance with the rate of water flow, while the second pump handles the remainder in response to a signal from the flow meter plus the signal from the fluoride residual analyzer.

REFERENCES

1. WHITE, G. C. *Handbook of Chlorination.* New York: Van Nostrand, Reinhold Company, 1972.
2. VAJDIC, ANN H. "The Inactivation of Viruses in Water Supplies by Ultra-Violet Irradiation." Report R. P. No. 2015 dated June, 1969. Prepared by the Division of Research, The Ontario Water Resources Commission, Toronto, Canada.

3. "Canadian Drinking Water Standards and Objectives, 1968." Published by the Department of National Health and Welfare, Ottawa, Canada, October, 1969.

4. "Water Treatment Plant Design." Prepared by the American Water Works Association, 1969, New York, N.Y.

5. *Water Quality and Treatment.* Prepared by the American Water Works Association, New York: McGraw-Hill Book Company, 1971.

6. "Basic Gas Chlorination Manual." Published in four parts in *J.A.W.W.A.* (64) p. 319, May 1972; p. 395 July 1972; p. 523 August 1972 and p. 683 October 1972.

7. KARLSON, V. P. "Ozone Friend or Foe?" *Pollution Engineering*, Vol. 4, page 32, May/June 1972.

8. THIRUMURTHI, D. "Ozone in Water Treatment and Wastewater Renovation." Water and Sewage Works. Reference Number, page R-106, 1968.

9. COX, C. R. *Operation and Control of Water Treatment Processes.* Published by World Health Organization, Geneva 1969.

10. EVANS, F. L. *Ozone in Water and Wastewater Treatment.* Ann Arbor, Michigan: Ann Arbor Science Publishers, Inc., 1973.

11. NEBEL, C; P. C. UNANGST; R. D. GOTTSCHLING, "An Evaluation of Various Mixing Devices for Dispersing Ozone in Water." Water and Sewage Work. Reference Number, page R-6, 1973.

12. From trade literature published by Key Environmental Control Ltd., North Vancouver, B. C. agents for Ultra Dynamics Corporation, U.S.

13. From trade literature published by Atlantic Ultra-Violet Corporation, Long Island, New York, U.S.

14. MONE, J. G. "Ultraviolet Water Purification." *Pollution Engineering*, Vol. 5 (12), p. 33, December 1973.

15. HUFF, C. B. *et al.* "Study of Ultraviolet Disinfection of Water and Factors in Treatment Efficiency." Public Health Reports, Public Health Services. U.S. Department of Health, Education and Welfare, Vol. 80 (8), p. 695, August 1965.

16. Anon. "Policy Statement on Use of the Ultraviolet Process for Disinfection of Water." Published by—Division of Environmental Engineering and Food Protection—Department of Health, Education and Welfare, Public Health Service, Washington, D.C., April 1, 1966.

17. ODA, A. "Ultraviolet Disinfection of Potable Water Supplies." Division of Research Paper #2012 dated March 1969. Published by the Ontario Water Resources Commission, Toronto, Canada.

18. HAMILTON, J. J. "Potassium Permanganate as a Main Disinfectant." Paper presented at the 93rd Annual Conference of the AWWA, Las Vegas, 1973.

19. TAYLOR, E. W. *The Examination of Waters and Water Supplies*, 7th Ed. London: J & A Churchill Ltd., 1968.

20. HOLDEN, W. S. *Water Treatment and Examination.* London: J & A Churchill Ltd., 1970.

21. McKee, J. E.; H. W. Wolf, *Water Quality Criteria*, 2nd Ed. Published by the Resources Agency of California, State Water Quality Control Board, Sacramento, California. Publication No. 3-A, 1963.

22. *Water Quality and Treatment*, 3rd Ed. Prepared by the American Water Works Association. New York: McGraw-Hill Book Company, 1971.

23. Maier, F. J. *Manual of Water Fluoridation Practice*. New York: McGraw-Hill Book Company Inc., 1963.

24. EPA Report—"National Interim Primary Drinking Water Regulations." *J.A.W.W.A.*, Vol. 68 (2), p. 57, February 1976.

25. "Standard Methods for the Examination of Water and Wastewater," 13th Ed. Published jointly by AWWA, American Public Health Association, and Water Pollution Control Federation (1971). Available from the Publication Sales Department, American Water Works Association, Denver.

26. U.S. Environmental Protection Agency—"Methods for Chemical Analysis of Water and Wastes." 1974.

27. Trade literature cat. File 85.400 dated Nov/70, entitled "Fluoridation Systems," from Wallace and Tiernan, Belleville, New Jersey.

20

WASTE DISPOSAL

In recent years there has been an increasing awareness that the old habits of dumping the water treatment plant wastes back into the stream, river, or lake from where they originated must be more carefully examined. When towns were few, small, and far between, the dumping of filter backwash water and clarifier sludges into creeks and rivers posed few specific problems. But when there are a number of large water treatment plants on one river removing the color, silt, hardness, iron, and manganese and then returning all the unwanted wastes back into the river again, together with the city sewage treated or untreated, it is obvious that the river will seriously deteriorate unless this practice is curtailed.

Each case should be judged on its merits and to enforce legislation demanding that all plant wastes be treated is an unnecessary hardship, and in many cases, almost impossible to implement.

One of the problems in areas where water treatment is required is the large number of small water systems, each with their own treatment plant. One state has more than 89 small municipal water supply systems in a relatively small area. Each system handles its own problems and administration. The nature of the water supply and sewage collection industry necessitates that each area must inevitably be unique and cannot operate in quite the same way as the power, gas, or telephone companies. Imagine the same small area with 89 separate power, gas, and telephone systems!

But many small water supply systems have grown to the limits of their boundaries and the lands adjacent to them, once open fields, are now heavily populated. The water treatment plants were never designed to treat their wastes and now, in many cases, they have no longer the opportunity. Mary Fulmer, in her article on the disposal of waste sludge from water softening plants,[1] has several good suggestions.

(a) Each water treatment plant should have a sludge holding tank.

(b) The sludge holding tank should be emptied periodically by a tank truck, rail tanker or barge.

(c) A sludge processing plant should be conveniently located in the area to process the sludge either into a marketable product if possible, or to reduce its bulk in order to dispose of it in old mine workings, worked out gravel pits, or sanitary landfills.

Much of the lime sludge can be processed for building materials. A regional system such as this could have other ramifications including laboratory, technical services, accouting systems, and an interchange of maintenance crews if needed.

FILTER BACKWASH WATER

The largest single effluent in terms of volume is the filter backwash water. Depending on the amount of pretreatment prior to the filters this can range from 2 to 20% of the plant throughput. Treatment requirements vary with the seasons, particularly on lakes or rivers in which algae blooms prevail.

There is little economic advantage in recycling backwash water providing it is less than 5% of the plant capacity, unless the cost of raw water is abnormally high, and the problems of disposal are unusually difficult, or for reasons of environmental impact. One of the major problems with backwash water is the high flow rate at which it occurs. Most sand or dual media filters backwash at 10 to 15 U.S. gpm per square foot. Mixed media may be 18 U.S. gpm per square foot and in warmer climates with lower water viscosities the backwash water rates increase for the same percentage of filter bed expansion; the backwash rate may be as high as 24 U.S. gpm per square foot. A filter 45 feet long by 15 feet wide may have a backwash water discharge rate of 10,000 U.S. gpm, which is too much to pour down a sanitary drain. However, it could be disposed of in this fashion if first of all it was piped by gravity into a holding tank and then pumped at a lower uniform rate. If the plant has preclarification facilities, the supernatant can be decanted off by means of the floating suction and pumped at a uniform rate into the clarifier inlet. The settled sludge in the bottom of the holding tank could be pumped to the clarifier sludge waste tank. Mechanical collectors can be used to facilitate sludge removal.

COAGULATED CLARIFIER SLUDGES

Alum sludge from color and silt removal clarifiers is difficult to handle. The sludge rarely exceeds 1% by volume unless polyelectrolytes are used, when solids concentration may reach 4% by volume. Small and medium sized plants may for various reasons use lagoons to contain water clarifier sludge.

LAGOONS

Lagoons are large open earth-work or concrete reservoirs 8 to 12 feet deep into which the clarifier sludge is pumped. In most cases they are unsatisfactory, and can only provide a temporary solution. Since they may have less than 5% solids, the alum sludge is a greasy, gelatinous material which does not readily consolidate. The sludge usually contains a large proportion of organic matter which does not help in the consolidation process. Frequently the supernatant is allowed to overflow into surface water runoff ditches, or it may be pumped back into the plant clarifier influent at a slow uniform rate similar to backwash water. The big problem arises when it becomes necessary to clean out the sludge. Drag lines stretched across from berm to berm are normally not very effective. The bucket will make one good pass through the sludge and after that will tend to follow along the same groove. It is usually too thin for a bulldozer and about the only method of emptying it is with a clam-shell bucket or backhoe. The wet sludge is so bulky and low in solids that the transportation costs are excessive.

DRYING BEDS— FREEZING

Drying beds are distinguished from lagoons in that they incorporate an underdrain system to facilitate dewatering of the sludge. A lagoon can be converted into a drying bed by first cleaning it out and then installing a number of perforated laterals across the bottom leading into a gravel-packed sump. A well screen can be installed in the sump and a submersible pump used to keep it dewatered. The laterals are covered with graded gravel and sand similar to a filter. The drying bed should then be filled with water preferably through the dewatering sump, through the laterals, up through the gravel and sand, for a depth of several feet. The clarifier sludge should then be pumped onto the top of the water and allowed to spread as evenly as possible over the whole area. As the level of water and sludge rises, the pump should be operated to dewater it, but only sufficiently to keep several feet of water above the top of the sludge. In cold climates ice will form over the top and if necessary the water level should be allowed to rise sufficiently to keep several feet of water between the top of the sludge and the underneath side of the ice. When the drying beds are full they should be taken out of service and slowly dewatered in the autumn before freeze-up. The semi-dewatered sludge should be allowed to freeze. Palin[2] reports that after freezing and thawing a 2% alum sludge settled by gravity to 20.2% and could be filtered to a concentration of 33.9% by weight.

Other investigators[3] have reported that freezing is an economically attractive system of handling alum sludges in northern climates but it is not economical if mechanical refrigeration has to be used. Natural freezing has been used successfully in Copenhagen to dewater alum in sludge lagoons.[3]

A U.K. research panel reported in 1973[4] that drying bed underdrain systems clogged, but on the other hand "the freezing and thawing of gelatinous sludges results in a granular product with excellent drainage properties, so that the final product can be disposed of to a land-fill project with subsequent top-soiling and grassing. This process is therefore very attractive from a disposal point of view." One of the problems of implementing the "drying bed-freezing" process in the United Kingdom is the variable weather and many areas of the country do not have regular frosts.

In conclusion, it is generally agreed by most investigators that when regular natural freezing is normal the use of freeze-drying schemes for small and medium sized water treatment plants are worth investigating. They could be troublesome in very cold climates and unsuitable for warm and mild climates with heavy rainfalls and unreliable freezing.

VACUUM FILTRATION

Vacuum filters have not been as successful on alum-coagulated sludges as they have on lime-softened sludges. For a vacuum filter to be effective on alum sludge, it is usually necessary to use a precoat or to add lime or another filling agent in order to get good results.[4]

CENTRIFUGES

Centrifuges have also had their share of grief with coagulated alum sludges because of the low solids to liquid weight ratio.[5] But recent developments have improved their performance and they are now becoming widely accepted.

FILTER PRESSES

Filter presses, also known as plate-and-frame filter presses or leaf filters, have been successful in handling alum sludges.

Filter presses have been used in the chemical and processing industries for a century or more and in many cases are the only filters that will do specific jobs. Their introduction into the sanitary engineering industry was met with resistance. After visiting a modern unit in operation, the most skeptical people have often become filter press promoters. The early objections to the use of filter presses were:

(a) Batch process.

(b) Messy.

(c) Manual opening and disposal.

(d) Cost of filter cloths and general high maintenance.

However, most, if not all these problems have been overcome. It is still a batch process rather than continuous, but if two or more presses are installed this should be no more of a problem than sand filters which are also batch processes. With modern designs they can be automated, and much of the objections to "messy" operating conditions and high input of manual labor is no longer as serious an objection as it was. With new synthetic fabrics, the replacement of cloth has been extended to 12 to 18 months service life. The use of polyelectrolytes before introducing the gelatinous sludge into the presses, has greatly improved their filtering characteristics.

The Santa Clara County Flood Control Water District examined various sludge disposal schemes for their Rinconada water treatment plant and concluded that filter presses were economical and technically feasible.

A recent panel of investigators in England[4] commented that "The amount of polyelectrolytes for conditioning sludges, new materials for filter cloths and a certain amount of automation in the opening and closing of filter presses has changed the situation in recent years and the process is now finding much more favor. A great deal of literature is available on the theory of pressing and producing cakes." The investigation concluded that after conditioning with a suitable polyelectrolyte any water works sludge could be economically dewatered in a filter press and is a recommended process for coagulant sludges.

QUALITY OF SLUDGE

Sludge quantities are difficult to calculate with precise accuracy but the following formula has been suggested.[4]

Using aluminum sulfate ($Al_2(SO_4)_3 21 H_2O$):

$$W = S + 0.07H + (0.015 \times 1 \times D).$$

and using ferrous sulfate ($FeSO_4 7H_2O$):

$$W = S + 0.07H + (0.013 \times 28 \times D).$$

Where:

W = Weight of dry solids in mg/ℓ.

S = Suspended solids in mg/ℓ.

H = Color of water in Hazen units.

D = Dosage of coagulant used in mg/ℓ.

Factors are not included for softening sludges and reference should be made to the original paper.

ALUM RECOVERY

Alum recovery processes have been used in England, and in Japan[5] and elsewhere using sulfuric acid:

$$2Al(OH)_3 + 3H_2SO_4 \longrightarrow Al_2(SO_4)_3 + 6H_2O$$

It takes 1.9 pounds of sulfuric acid for each pound of sludge treated in the recovery process. It is reported that it is a costly and difficult process to operate. The reclaimed alum is less effective in water treatment than fresh alum which means that the dosages must be increased and therefore supplemented by adding commercial alum. It is unlikely that this process will find favor.

SOFTENING SLUDGES

Cold lime-soda-softening sludges are usually easier to handle than coagulant sludges, i.e., color and silt removal only. Softening sludges contain calcium carbonate, magnesium hydroxide, and others. In multi-clarifier plants the first clarifier may be producing coagulated sludges containing organics, aluminum hydroxide, and silt, but the second, and possibly the third, clarifiers will contain the calcium and magnesium compounds. It is usual to pump the sludge from the softening clarifiers into a thickener which is a small sludge clarifier where polyelectrolytes may be added to thicken the sludge as much as possible prior to centrifuging or filter pressing. Larger plants may economically justify a lime recovery or recalcining plant. The effluent carbon dioxide gas together with the nitrogen and excess air can be used in the recarbonation process for the acidification of the water following lime treatment. Sludge may be handled by the following processes:

DRYING BEDS

Drying beds work well with lime-softening sludge and can be an economical method of disposal since the granular sludge drains well through the sand and the underdrain system. Freezing is not necessary with softening sludges as it is with coagulated sludges since the precipitate is already granular and not gelatinous.

THICKENING

If a thin sludge of 4% solids can be improved to 8% by thickening, then the total weight of the sludge to be handled by the succeeding processes of centrifuging or filtration is roughly halved. In many cases it is cheaper to thicken to 15 or 20% solids and truck the thickened sludge to a disposal site rather than incur the expense of further treatment.

The advanced wastewater treatment process[6] such as the South Tahoe Public Utility District plant in California where excess lime (dosage of about 400 mg/ℓ) is added to a secondary sewage treatment plant effluent produces a lime-softening sludge.

A practical example based on papers by Vesilind and Larsen[7,8] will serve to demonstrate the procedures to design a thickener suitable for handling 170,000 pounds per day of lime-softened sludge. Fortunately, this project is an addition to a lime-softening treatment plant, and actual sludge settling characteristics were determined. The thin sludge off the water clarifiers is between 1 and 2% solids as it flows into the mud clarifier or thickener where it can be thickened up to as high as 20% total solids. However, it is almost impossible to pump the sludge to the centrifuges at this concentration and they normally only thicken to 8 to 15% prior to centrifuging.

SIZE OF THICKENER

To determine the settling characteristics of the sludge, a small pilot plant was constructed consisting of a number of old 45 gallon oil drums with their tops cut out. These were filled with fresh samples of the thin clarifier sludge. A number of vertical glass settling tubes, about 48 inches high, complete with measuring scales attached to the side of the tubes, were located adjacent to the 45 gallon drums. The thin sludge was allowed to settle to some degree in the 45 gallon barrels, occasionally stirred with a wooden paddle, and the clear supernatant transferred from one drum to another as necessary to provide different concentrations. A bucket of sludge was taken from one barrel and rapidly poured into a settling tube as quickly as possible and immediately the time and the height of the liquid in the tube was noted. While this was being done, a 100 mℓ beaker of sludge was grab-sampled out of the bucket for drying and weighing to determine the total solids content.

With varying degrees of stirring and mixing from one container to another, many samples can be taken with varying concentrations of total solids. Their concentration can be determined by taking a weighed empty 100 mℓ beaker filled with sludge and then evaporated to dryness, expressing the total solids which include not only suspended but also dissolved solids. A correction can be made to the results if considered necessary, but the percentage of dissolved solids is so small compared to the weight of suspended sludge that this refinement is not justified.

The settling results are shown in Figure 20-1, Graph 1 entitled Settling Curves Sludge Thickening. The results are then tabulated in Table 20-1.

Of the 14 settling tests it will be observed from both Table 20-1 and Figure 20-1 that the thinner sludges (i.e., Tests No. 1, 2, 3, and 4) have much faster settling rates than the heavier sludges, (i.e., Test No. 9) where the concentration of incoming sludge was equivalent to 7.65% solids and had an almost negligible settling velocity.

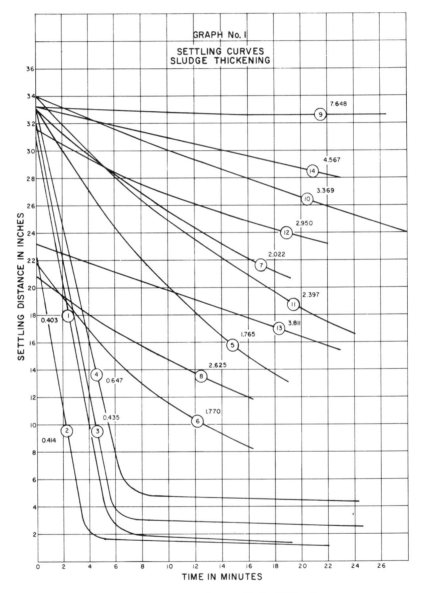

FIGURE 20-1 Settling Curves—Sludge Thickening.

TABLE 20-1 Initial (Uncorrected) Flux Calculations

From Settling Tube: Initial (Uncorrected) Flux Calculations

Test No.	Total Solids			% Solids	Settling Velocity ($V =$ ft/min)	$G = VC$ (lb/ft²/ min)	G (lb/ft²/ day)
	gm/ 100 ml	gm/ℓ	$C =$ lb/cu ft				
1	0.403	4.03	0.252	0.40	0.438	0.110	158
2	0.414	4.14	0.259	0.41	0.435	0.113	162
3	0.435	4.35	0.272	0.44	0.417	0.114	164
4	0.647	6.47	0.405	0.65	0.350	0.142	205
5	1.765	17.65	1.052	1.76	0.132	0.138	198
6	1.767	17.67	1.054	1.77	0.1015	0.107	154
7	2.022	20.22	1.260	2.02	0.0626	0.0788	114
8	2.625	26.25	1.640	2.63	0.0518	0.0847	122
9	7.648	76.48	4.780	7.65	0.00375	0.0180	26
10	3.369	33.69	2.110	3.37	0.0316	0.0670	97
11	2.397	23.97	1.500	2.40	0.0817	0.1230	177
12	2.950	29.50	1.840	2.95	0.0333	0.0610	88
13	3.811	38.11	2.390	3.80	0.0288	0.0690	99
14	4.567	45.67	2.860	4.60	0.0197	0.0562	81

gm/ℓ (0.0624) = lb/cu ft

Referring to Figure 20-1, only the steep portion of the curve is considered to be significant and the settling rates are calculated on the following time intervals:

TABLE 20-2 Settling Times

Test No.	Inches Settled	Times In Minutes	Velocity Feet/Minute
1	21.00	4	0.438
2	15.75	3	0.435
3	20.00	4	0.417
4	21.00	5	0.350
5	4.75	3	0.132
6	3.65	3	0.1015
7	3.00	4	0.0626
8	3.10	5	0.0518
9	0.45	10	0.00375
10	1.90	5	0.0316
11	4.90	5	0.0817
12	2.80	7	0.0333
13	4.85	14	0.0288
14	2.60	11	0.0197

In order to form the experimental data into a smooth curve, it is desirable to replot the data on semi-log paper. From the data calculated in Table 20-2, plot the settling velocity in feet/minute against the concentration of the sludge expressed in terms of pounds of solids per cubic foot as per Figure 20-2.

When the points from Table 20-2 are plotted on semi-log paper (which

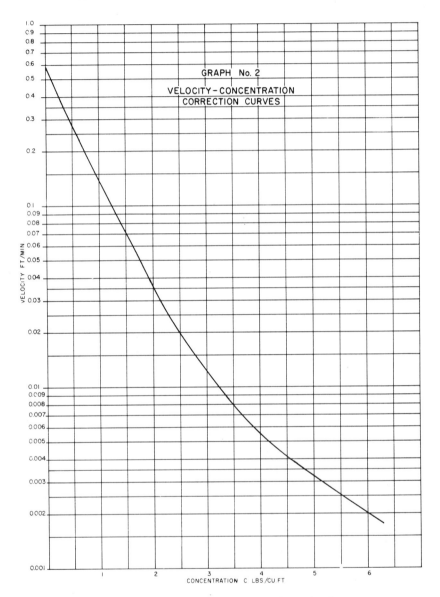

FIGURE 20-2 Velocity—Concentration Correction Curve.

usually straightens out the most curvaceous lines) a smooth curve is drawn through the points to represent the settling characteristics of the sludge.

From Graph No. 2 (Fig. 20-2) Table 20-3 is drawn. In this case, 21 points have been taken from the semi-log curve in Graph No. 2. From the velocity of settling (V) and the concentration (C) the solid flux $G = VC$ is calculated in terms of lb/ft² per minute. For example at point No. 1, a physical cubic foot of sludge measuring $12'' \times 12'' \times 12''$ would contain 0.1 pounds of solids and the solids would settle in the sludge at the rate of 0.5 feet per minute. This is known as the solids flux and bears a direct relationship to the required surface area of the thickener.

TABLE 20-3 Corrected Velocity and Concentration Values from Graph No. 2 (Fig. 20-2) Semi-Log

Point No.	Settling Velocity $V = ft/min$	Concentration $C =$ lb/cu ft	Solids Flux $G = VC$ $(lb/ft^2/min)$	$G =$ $lb/ft^2/day$	mg/ℓ	% Solids
1	0.5	0.1	0.050	72	1.6	0.16
2	0.4	0.27	0.108	156	4.32	0.43
3	0.3	0.48	0.144	207	7.68	0.77
4	0.2	0.77	0.154	222	12.35	1.24
5	0.15	0.96	0.144	207	15.40	1.54
6	0.10	1.25	0.125	180	20.0	2.00
7	0.08	1.41	0.113	163	22.5	2.30
8	0.06	1.61	0.0966	139	25.7	2.60
9	0.05	1.75	0.0875	126	28.0	2.80
10	0.04	1.91	0.0764	110	30.5	3.05
11	0.03	2.14	0.0642	92	34.1	3.41
12	0.02	2.50	0.0500	72	40.0	4.00
13	0.015	2.78	0.0418	60	44.5	4.50
14	0.01	3.20	0.0320	46	51.1	5.10
15	0.008	3.47	0.0279	40	55.6	5.60
16	0.006	3.89	0.0234	33.6	62.0	6.20
17	0.005	4.18	0.0210	30.3	67.0	6.70
18	0.004	4.58	0.0184	26.5	73.1	7.30
19	0.003	5.13	0.0154	22.2	82.0	8.20
20	0.0025	5.50	0.0138	19.9	88.0	8.80
21	0.002	6.00	0.0120	17.3	96.0	9.60

From Table 20-3, showing the corrected values, it is possible to plot Graph No. 3, Figure 20-3, of Solids Flux G in lb/ft² per day, against the concentration C in lb/cu ft.

From this data the size of a thickener can be estimated for various overflow concentrations.

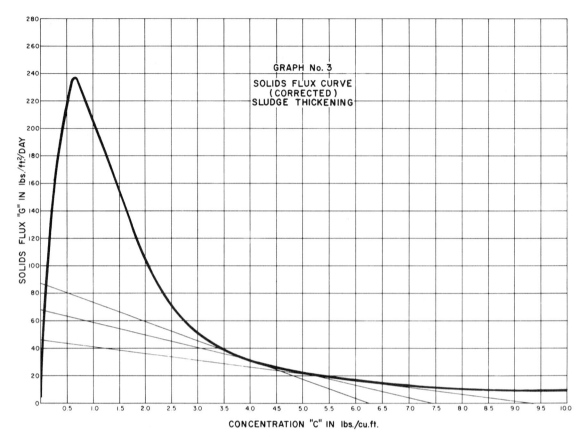

FIGURE 20-3 Solids Flux Curve (Corrected).

DIAMETER OF THICKENER

Suitable for handling 170,000 lb per day = Q_oC_o.

Required area of the thickener is given by $\dfrac{Q_oC_o}{G\,\min}$

Where:

Q_o = Quantity of sludge.

C_o = Concentration.

$G\,\min$ = Solids flux taken from Graph No. 3 (Fig. 20-3)

(a) Assume 10% solids in the underflow. Underflow solids 10% = 100 gms/ℓ = 6.24 lb/cu ft

G(min) from graph #3 = 87 lb/ft² per day Area required = $\dfrac{170,000}{87}$ = 1950 ft² = 50 ft diameter

(b) Assume 12% solids in the underflow. Underflow solids 12% = 120 gm/ℓ = 7.5 lb/cu ft G(min) = 68.2 lb/ft² per day Area Required = $\frac{170,000}{68.2}$ = 2490 ft² = 56 ft diamter

(c) Assume 15% solids in the underflow. Underflow solids 15% = 150 gm/ℓ = 9.375 lb/cu ft G(min) = 46.0 lb/sq ft per day Area Required = $\frac{170,000}{46.0}$ = 3690 ft² = 68.5 ft diameter

From these calculations, plus factors of experience, the size of the thickener can be determined. Some authorities suggest that these calculations would result in too large a diameter of thickener and should therefore be reduced. Larsen and Vesilind[9] suggest an overdesign correction factor of 1.15 on diameter, so that for a 15% underflow concentration for 170,000 pounds of sludge per day, the thickener diameter should be $\frac{68.5}{1.15}$ = 59.7 ft (say 60 ft diameter).

DEPTH OF THICKENER

At 15% underflow solids the limiting concentration C_L = 6.25 lb/cu ft is equivalent to 10% solids C_L (See Graph No. 3).

Underflow volume = $\frac{170,000 \ (100)}{10 \ (24) \ (62.4)}$ = 1135 cu ft/hour

Storage of sludge for 24 hours = 1135 (24) = 27,200 cu ft

Height for sludge = $\frac{27,200}{*2825}$ = 9.6 ft

The side wall depth (SWD) of the clarifier should be not less than:

Sludge depth	9.6 feet
Clear water zone	5.0 feet
Freeboard	2.0 feet
	16.6 feet

Say 17.0 feet.

At the first exposure to this calculation the use of the 10% limiting C_L for a 15% underflow is confusing but when it is realized that only the bottom portion of the sludge is concentrated to 15%, and the superseding layers are gradually less concentrated toward the top of the sludge blanket, it becomes obvious that a lower concentration must be considered; otherwise there will not be enough depth in the thickener to hold all the sludge required. Settling sludge is also very springy; adequate side wall depth must be provided for good control of the

*Area of 60 ft diameter clarifier = 2825 ft².

underflow concentration and clear, low turbidity supernatant off the top launders. From the thickener, with a sludge concentration of 8 to 15% solids, the sludge can then be handled in a variety of ways depending on economic and ecological factors. The clear supernatant water will overflow from the top of the thickener back into the water treatment plant either before or after the clarifiers, depending on the thickener supernatant.

The sludge can be disposed of by trucking to disposal sites. This is often the simplest and cheapest approach for small plants. Depending on the economics, the thickener overflow can be further concentrated by centrifuging, vacuum filtration, or filter pressed, depending on the end use. If the sludge is to be dumped at a suitable disposal site there is no point in further concentration unless there is an advantage in trucking and disposal costs. If, on the other hand, there is a considerable quantity of lime sludge or, as Mary Fulmer[1] suggests, a number of plants sufficiently close together that they could build a sludge plant to handle the residue from several smaller plants, then recalcination for recovery of quicklime may be a suitable solution to the disposal problem.

It is understood that lime can only be recalcined about four times before it has too many impurities, which reduce its effectiveness in water treatment to the stage where it is no longer economical to use.

The point at which it is economical to switch from sludge disposal to recalcination depends on many factors, but it is generally believed, particularly with advanced wastewater treatment where fairly large dosages of lime are used, that the switchover point is approximately 10 mgd and larger.[6] Further information on suitable equipment for lime handling is given in the recent review by the Institution of Water Engineers.[4]

REFERENCES

1. FULMER, M. "Disposal of Waste Sludge from Water Softening Plants." Water and Wastes Engineering, Vol. 7 (8), p. 33, 1970.

2. PALIN, A. T. "Treatment and Disposal of Alum Sludge." Journal of the Society for Water Treatment and Examination, Vol. 3, p. 131, 1954.

3. "Disposal of Wastes from Water Treatment Plants." AWWA—Research Foundation Report, Part I, *J.A.W.W.A.*, Vol. 61, p. 541, articles from the AWWA Research Foundation Report published in the Journal of the AWWA as follows:

    Vol. 61, (10), October 1969

    Vol. 61, (11), November 1969

    Vol. 61, (12), December 1969

    Vol. 62, (1), January 1970

    are available from the AWWA in a bound reprint.

4. "Disposal of Water Works Sludge; Final Report of Research Panel No. 14." Journal of the Institiution of Water Engineers, Vol. 27 (8), p. 399, November, 1973.

5. ALBRECHT, A. E. "Disposal of Alum Sludges." *J.A.W.W.A.*, Vol. 64 (1), p. 46, January, 1972.

6. CULP, R. L. and G. L. CULP, *Advanced Wastewater Treatment.* New York: Van Nostrand Reinhold Company, 1971.

7. DICK, R. I. and B. B. EWING, "Evaluation of Activated Sludge Thickening Theories." Journal of the Sanitary Engineering Division, p. 9, August, 1967.

8. VESILIND, P. A. "Design of Prototype Thickeners from Batch Settling Tests." Water and Sewage Works, Vol. 115, p. 302, July, 1968.

9. LARSEN, I. and P. A. VESILIND, "Evaluation of Activated Sludge Thickener Theories." Journal of Sanitary Engineering. p. 190, February, 1968.

21

STABILIZATION

STABILIZATION Stabilization has many meanings in the water and wastewater industry but in this chapter, it refers to water in the distribution system. Water which is unstable will either (a) deposit scale, (b) corrode the pipes, (c) or do both at the same time.

It is said that bottled beers are at their peak of perfection the day they leave the brewery. Providing they are kept chilled, and in the dark, they will keep their good quality for several months but their biological reactions are not always completed at the time they leave the plant and they will continue to change. So it is with water; almost invariably water is at its best as it leaves the water treatment plant and deteriorates en route to the customer.

WELL WATERS Well waters may contain free carbon dioxide which is released as it is pumped to the surface and reduced in pressure. If a pH reading is taken directly at the pump discharge and again after it has been standing, exposed to the atmosphere, it will often be noticed that the pH has increased. This is due to the liberation of free carbon dioxide, and once this has occurred calcuim carbonate ($CaCO_3$), may start to precipitate.

The amount of free carbon dioxide present can be calculated from the nomograph for evaluation of free carbon dioxide content in *Standard Methods*.[1] It is necessary to know the water temperature, total filtrable residue, pH, and the bicarbonate alkalinity.

MARBLE CHIP TEST

A simple and practical test to determine whether or not a water is corrosive or scale forming with respect to calcium carbonate can be done by the marble chip test. This is described in detail in "Simplified Procedures for Water Examination"[2] under the heading of Calcium Carbonate Stability Test.

It consists of collecting two samples of water. Be careful to avoid aeration and loss of carbon dioxide (CO_2). Add to one sample 0.3–0.5 grams of calcium carbonate and then stopper both samples. The bottle containing the calcium carbonate is shaken at periodic intervals to insure that it has thoroughly contacted the water in the bottle. It is then left overnight and the next day each bottle is carefully opened without disturbing and 100 mℓ of each sample is carefully pipetted from the top and filtered through Whatman No. 50 filter paper to insure that no calcium carbonate passes into the filtrate. The first 25 mℓ of sample through each filter paper is discarded. It may be questioned as to why it is necessary to filter the blank sample since there is no calcium carbonate in this water, but it is essential to treat both samples exactly alike since the results of the total alkalinity test of the filtered waters will be compared. For convenience, call the bottle with no calcium carbonate the blank sample and the other the carbonate sample.

CASE I

If the total alkalinity of the carbonate sample is higher than the total alkalinity of the blank sample after they have been sitting overnight, then it is inferred that the water known as the blank sample is unsaturated with respect to calcium carbonate and may also be corrosive. If the pH of the two waters is taken, it will probably be found that the pH of the carbonate sample is higher than the pH of the blank sample. The pH of the carbonate sample is referred to as the pH of saturation (pH_s) with respect to calcium carbonate. The pH of the blank sample is the actual pH. These two pH values are related to each other in respect to the Langelier Stability Index.

CASE II

If, in the case of lime-softened waters, it is found that there is a decrease in the total alkalinity and pH of the carbonate sample, then it is inferred that the water is over-saturated with respect to calcium carbonate and likely to be scale forming.

CASE III

If, after sitting overnight both the blank and carbonate samples are still identical in total alkalinity and pH, it can be said that the water is stable in equilibrium with calcium carbonate.

In one instance a well water had an actual pH of 6.8 at the well head. When exposed to the atmosphere, left overnight and assumed room temperature, the pH increased to 7.2 by the release of carbon dioxide to the atmosphere. The same water when subjected to the marble chip test increased from 6.8 to 7.7 by the absorption of $CaCO_3$ from the marble. The alkalinity increased from 45 to 104 mg/ℓ as $CaCO_3$ and the total hardness from 56 to 122 mg/ℓ as $CaCO_3$. This same water which had a calcium hardness of 42 mg/ℓ and a pH of 6.8 at the well head was pumped through several miles of 8 in. diameter asbestos cement pipe and when it reached the other end it had a calcium hardness of 62 mg/ℓ and a pH of 7.3 which was still below the marble chip test pH of saturation of 7.7. The calcium hardness had increased 20 mg/ℓ in transit, expressed as $CaCO_3$. This indicates that the water had gained 20 mg/ℓ of $CaCO_3$ in transit, equivalent to 167 pounds of pipe material in terms of $CaCO_3$ per million U.S. gallons (mgd U.S.) passing through the pipe. This material could only come from the pipe walls. It is possible to liberate the carbon dioxide by aeration and increase the alkalinity and pH with sodium carbonate (Na_2CO_3). Slaked lime ($Ca(OH)_2$) is also a possible alternative but is likely to be less popular than the sodium compounds since it increases the calcium hardness.

SATURATION INDICES

Several indices have been compiled to indicate whether a water is likely to be scale forming or corrosive. The best known index is probably the Langelier's Saturation Index followed by Ryznar Stability Index, Larson and Buswell Saturation Index, and Stiff and Davis Stability Index and others. Langelier's index is the simplest and the most fundamental method.

Much has been published on how to calculate the various indices from analytical data. Reference is made to *Water Quality and Treatment*[3], *Water Supply and Pollution Control*[4], *Water and Wastewater Engineering*[5], and there are many trade publications including "Water Supply and Treatment".[6,7]

Langelier Index

Professor Langelier's original paper was published in 1936 in the Journal of the AWWA.[8] Equipment and chemical manufacturers have produced slide rule calculators which will calculate the pH_s (pH of saturation) from the chemical analysis of dissolved solids, calcium hardness, temperature, and total alkalinity. When the pH_s is compared with the actual pH the respective index can be calculated.

Using Langelier's Index, a negative number in the presence of oxygen tends to be corrosive to ferrous metals. A positive number is less likely to be corrosive and may deposit scale.

Ryznar Stability Index

The Ryznar Stability Index is as follows:

$$\text{Stability Index} = 2pH_s - pH$$

For values greater than 6, the water tends to be corrosive and with values less than 6 it has a scale forming tendency.

SPECIFIC CONDUCTIVITY

For calculating stability indices the total dissolved solids (TDS) is one parameter necessary to obtain the pH_s. Sometimes TDS is difficult to obtain and is not always available but Specific Conductance expressed in micro-mhos/cm can be used to determine a close approximation if the following relationship is used.

$$\text{TDS} \simeq \text{Specific Conductance} \times 0.64$$

The Langelier and the Ryznar Stability Index are good general indications of the water characteristics but they can be misleading. A very soft water of low alkalinity and pH from a mountain stream had all the indications of being severely corrosive and yet was perfectly satisfactory with very few problems, due to a "built-in" corrosion inhibitor that the chemical analysis had not identified. The introduction of chlorine destroyed the equilibrium and severe corrosion occurred near the point of chlorine injection.

The presence of oxygen has a significant influence on whether or not a water is corrosive. Aerated water of low pH has been handled in aluminum pipes without difficulty until copper heating coils were installed some distance upstream resulting in severe pitting of the aluminum.

A hard well water of a positive Langelier Index was without problems in cast iron but perforated the copper service lines with tiny holes neatly drilled through the wall of the copper.

A well water high in ferrous bicarbonate, chlorides, and sulfates is both scale forming and corrosive to cast iron. It causes graphitization of cast iron,[9] which softens the metal to the extent that a hole can be scratched through the base of a cast iron water meter with a penknife.

The presence of oxygen, chlorides, and sulfates all tend to aggravate corrosion. If, however, it is necessary to prevent corrosion under specific circumstances the oxygen must be eliminated entirely but this is not practiced for municipal waters. (See Chapter 16.) All these problems are beyond the simplification of assuming that because a water has a positive Langelier Index or Ryznar Stability Index of less than 6.0, it is not corrosive. Stray electric currents can also play havoc with the external surfaces of steel and cast iron pipelines, particularly in high conductivity soils which are frequently the cause of much external corrosion.

ORGANIC DEPOSITS Surface runoff water impounded in reservoirs is often of such good quality in both color and turbidity that it is piped directly to the distribution system with only the addition of chlorine. Many of these waters contain microscopic particles or organic matter, dead algae, and other deposits from the multitude of dead flora and fauna of the aquatic environment. The evidence can be seen in most toilet tanks on distribution systems of this type. The particulate matter is finely divided when it first enters the distribution system and would be difficult to remove except by filtration with polyelectrolytes through mixed and dual media filters. As the water passes through the distribution system the particles coagulate and settle in the water mains. As velocities increase during the peak hour periods these deposits tend to roll along the bottom of the pipe, eventually they build up to the stage where a fast draw-off from the system results in heavy turbidity, causing many customer problems—particularly if the customer happens to be processing photographic film. See the case of Munshaw Color Service Ltd. vs City of Vancouver.[10]

REFERENCES

1. *Standard Methods for the Examination of Water and Wastewater*, 13th Ed. Published jointly by AWWA, American Public Health Association and Water Pollution Control Federation 1971. Available from the Publication Sales Department, AWWA, Inc., Denver, Colorado.

2. "Simplified Procedures for Water Examination." *AWWA*, M12. Simplified Laboratory Manual. Published by AWWA, Inc., Denver, Colorado.

3. *Water Quality and Treatment*, 3rd Ed. Published by the American Waterworks Association Inc., (1971).

4. CLARK, J. W., W. VIESSMAN, and M. J. HAMMER, *Water Supply and Pollution Control*, 2nd Ed. International Textbook Company, 1971.

5. FAIR, G. M., J. C. GEYER, and D. A. OKUN, *Water and Wastewater Engineering*, Vol. 2. New York: John Wiley & Sons Inc., 1967.

6. RIEHL, M. L. "Water Supply and Treatment." Bulletin 211, National Lime Association, Washington, D.C.

7. "Alkalies and Chlorine in the Treatment of Municipal and Industrial Water." Bulletin No. 8, 3rd Ed. (1957), published by Solvay Process Division. Allied Chemical and Dye Corp. 61 Broadway, New York 6, N. Y.

8. LANGELIER, W. F. "The Analytical Control of Anti-Corrosion Water Treatment." *J.A.W.W.A.*, Vol. 28 (1500), 1936.

9. TAYLOR, E. W. *The Examination of Waters and Water Supplies*, 7th Ed. London, J & A Churchill Ltd., 1958.

10. *Damages Action Against Vancouver, B. C.* Canadian Section of the American Waterworks Association, Vol. 10 pp. 1–11 Jan. 1960, Vol. 10 pp. 22–24 June 1961, Vol. 10 pp. 25–32 May, 1962.

22

DEMINERALIZATION

DEMINERALI-ZATION Waters which are high in total dissolved solids (TDS) particularly with the more soluble ions of sodium and potassium which will not precipitate out of solution by lime-soda softening may be treated by demineralization processes. Waters in this category are sea water with a TDS of 30,000–35,000 mg/ℓ or brackish waters from 1000 to 10,000 mg/ℓ of TDS. In the majority of cases the predominent cation is sodium (Na) and the anions are chlorides (Cl), sulfates (SO$_4$), or bicarbonates (HCO$_3$).

For municipal supplies, brackish waters should obviously be avoided if possible in view of the cost of treatment, but sometimes there are no other alternatives.

LOW TOTAL DISSOLVED SOLIDS (TDS) For municipal supplies where the raw water has a total dissolved solids (TDS) content of up to 6000 mg/ℓ, containing principally sodium compounds, there are several available processes which should be evaluated:

Reverse Osmosis (RO)

The Reverse Osmosis process consists of pumping water under pressure (350 to 1500 psi) through very thin membranes made from cellulose acetate or nylon. The membranes are commercially available with capacities from several thousands of gallons per day upwards.

The process of osmosis is as old as life itself and is the basic means by which plants, animals, and humans derive their substance by selective transport of water and certain chemicals through cell walls, which function as semi-permeable membranes. Osmosis is the natural flow of water or solvent relatively low in TDS through the pores of the membranes into the more concentrated solutions of higher TDS on the other side. If the process is reversed, that is to say, if the solution containing the higher TDS on the one side of the membrane is subjected to higher pressures than the water on the other side of the membrane, the molecules of water will pass back through the pores of the membrane into the solution of lower TDS on the other side. See Figure 22-1 (a, b and c) showing the principles of osmosis and reverse osmosis. A number of municipal plants have been built and operated successfully. The Greenfield, Iowa, plant is described in detail.[1]

Reverse osmosis can only reduce the TDS and not eliminate them. It is possible that a raw water with a TDS of 2000 mg/ℓ can be reduced to less than 200 mg/ℓ TDS with 75% of the influent going forward as treated water and the remaining 25% containing the 8000 mg/ℓ TDS discarded to waste. The problem of disposal of concentrated waste must also be considered.

The RO process could be followed by ion or base exchange units using the salt-rich waste from the RO process to regenerate the base exchange resins, thereby reducing the hardness still further.

The water used in the RO process must be relatively free from suspended solids, precipitated iron, oil, and other contaminants which could plug the membranes. The Langelier Saturation Index must be negative and therefore slightly corrosive and the Ryznar Stability Index must be greater than 6, to insure that the semi-permeable membranes do not become encrusted with calcium carbonate.

Although RO is relatively new as water treatment technology, it could develop extensively since the process is ideal for many desalinization projects. The RO process can be used to reduce the TDS content by 80 to 90% and an ion exchanger can polish the water to any degree of quality required.

Cost data for water treated by the RO process is given by Faber et al,[2] but it must be appreciated that the cost is dependent on power supplies, trained maintenance staff, and above all, the water characteristics. It may be necessary to precede the RO process with conventional clarification and filtration pretreatment to reduce color and turbidity. Small pilot plants are available to test the process under actual operating conditions before large sums of capital have to be invested.

Electrodialysis (ED)

Electrodialysis (ED) has a longer record of service than reverse osmosis (RO) and was the first practical process to desalt water other than distillation. Its principles were described in the literature prior to World War II but it was not

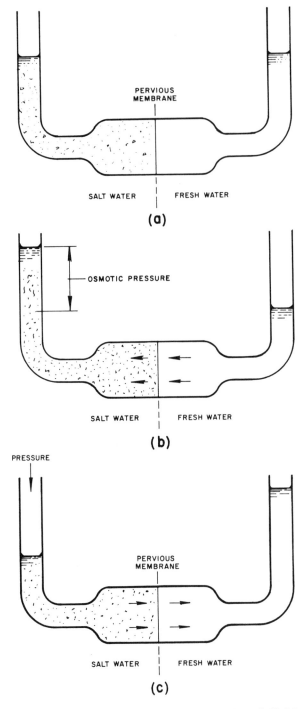

FIGURE 22-1 Principle of Osmosis and Reverse Osmosis (R.O.); (a) Fresh Water Is Separated from Salt Water by a Semi-permeable Membrane. At Zero Time, Both Levels Are the Same; (b) Fresh Water Migrates Through the Membrane in the Salt Water. This is Osmosis; (c) When Pressure is Applied to the Salt Water Side of the Membrane, the Salt Migrates to the Fresh Water Side. This is Reverse Osmosis.

developed commercially until after the war when better semi-permeable membranes became available.³ The basic ED unit consists of an inlet water channel, two semi-permeable membranes, and two direct current (DC) electrodes. One membrane is cathodic and only allows the passage of positively charged anions to pass through, i.e., Ca, Mg, Na, K, etc. and is impermeable to negatively charged and neutral ions.

The other membrane is anodic and will only pass the negatively charged ions, i.e., Cl, SO_4, HCO_3, CO_3, OH, etc. The water passes through the cell with the anode membrane on one side and the cathodic membrane on the other side. As the water reaches the end of the cell, most of the ions will have been removed. It is claimed that the brackish water recovery through an ED cell is limited to about 87%; the unwanted ions are accumulated in the remaining 13% which is discarded to waste.²

Compared to the RO process, ED is a little more complicated and requires AC-DC rectifier equipment. The costs of the water from both RO and ED processes are relatively close to each other. With waters of approximately 1000 mg/ℓ TDS, it would appear that RO is a little cheaper than the ED process but there are many factors which must be evaluated.⁴ Again, it is essential that the water will not contaminate the semi-permeable membranes.

Ion Exchange

Ion exchange processes are designed to produce almost any desired degree of finished water quality. Their prime limitation is the TDS of the incoming water which is limited to an upper value of approximately 1500 mg/ℓ. In some cases the ion exchange process can operate successfully up to 2000 mg/ℓ TDS but the limiting factor is the amount of treated demineralized water that is needed to back-wash the beds free from the regenerative solutions. The more refined the finished water, the greater becomes the quantities of backwash and rinse water, until a state of diminishing returns results when the rinse water just balances with the amount of water the plant produces.

In order for an ion exchange resin to remove the elements from a water, it is essential that it be ionized and that the water be free from oil and other contaminants that would foul the surface of the resins and prevent them from being surface active.

The ion exchange process, originally developed from the observations of two chemists, Thompson and Way, in 1850, when they discovered that a solution of ammonia fertilizer passed through a column containing soil was denuded of ammonia and replaced by calcium.⁵ Later, when synthetic cation exchange materials were manufactured for the base-exchange softening process, whereby calcium and magnesium cations could be removed from the water and replaced with sodium ions, the synthetic cation exchange material was called a "zeolite" a name derived from two Greek words *zein* and *lithos* which together mean boilingstone.

From electrochemical theory, the ions with a positive electrical charge are called cations since in an electrolytic cell they migrate toward the cathode or negative electrode. Similarly, the ions and radicals with a negative electrical charge are called anions since they migrate toward the anode or positive electrode. The most commonly encountered cations and anions in water supplies are listed as follows in Table 22-1:

TABLE 22-1 Most Common Cations and Anions Removed by Ion Exchange[6]

Cations		Anions	
Calcium	(Ca^{++})	Bicarbonate	(HCO_3^-)
Magnesium	(Mg^{++})	Carbonate	(CO_3^{--})
Sodium	(Na^+)	Sulfate	(SO_4^{--})
Potassium	(K^+)	Chloride	(Cl^-)
Iron	(Fe^{++})	Nitrate	(NO_3)
Manganese	(Mn^{++})	Silicate	($HSiO_3^-$)
Aluminum	(Al^{+++})	Silicate	(SiO_3)

Base-Exchange Softening

The earliest ion exchangers were known as base-exchange softeners, using naturally occurring zeolites. They are essentially cation exchangers whereby the relatively insoluble calcium and magnesium ions can be exchanged for sodium ions which are more soluble.

The zeolite is regenerated with common salt (NaCl) and is known as the sodium cycle. It is customary to refer to the zeolite as a chemical symbol, (Z), or sometimes (R). The reactions which take place on the surface of the zeolite beds may be represented in the ionic form as follows:

$$Na_2^+ Z + \begin{cases} Ca^{2+} \\ Mg^{2+} \end{cases} \longrightarrow \begin{cases} Ca^{2+} \\ mg^{2+} \end{cases} Z + 2Na^+$$

Written in the simplified molecular form:

$$\begin{Bmatrix} Ca \\ Mg \end{Bmatrix} + \begin{cases} (HCO_3)_2 \\ SO_4 \\ Cl_2 \end{cases} + Na_2 Z \longrightarrow \begin{Bmatrix} Ca \\ Mg \end{Bmatrix} Z + \begin{cases} 2NaHCO_3 \\ Na_2 SO_4 \\ 2NaCl \end{cases}$$

Hard water will continue to be softened automatically as it passes downward through the bed of the zeolite, until the bed can no longer continue to exchange sodium ions for calcium and magnesium ions. The zeolite is then considered to be exhausted and will require to be taken out of service and regenerated with salt solution (sodium chloride), using ordinary common salt, or in some cases, sea water. The reactions written in ionic form are as follows:

$$2Na^+ + \left. \begin{array}{c} Ca^{2+} \\ \\ Mg^{2+} \end{array} \right\} Z \longrightarrow \left\{ \begin{array}{c} Ca^{2+} \\ \\ Mg^{2+} \end{array} \right. + Na_2^+ Z$$

When expressed in the molecular form:

$$\left. \begin{array}{c} Ca \\ \\ Mg \end{array} \right\} Z + 2NaCl \longrightarrow Na_2 Z + \left. \begin{array}{c} Ca \\ \\ Mg \end{array} \right\} Cl_2$$

The effluent from the regeneration cycle will consist of a solution of calcium and magnesium chloride. It is possible that since these constituents are soluble, they may be discharged to municipal sewage systems, subject to local regulations.

As the softening cycle proceeds, the zeolite becomes exhausted and some leakage takes place and the hardness starts to increase. A simple base-exchange unit is shown in Figure 22-2. If it is essential to remove calcium and magnesium completely, the system will require a secondary softener of smaller diameter but approximately equal depth in series with the primary softener as shown in Figure 22-3. Always regenerate as soon as there is a calcium or magnesium content in the water collected from the sample point between the two units. When the units are in operation, valves (A) and (D) are open and (B) and (C) are closed. When

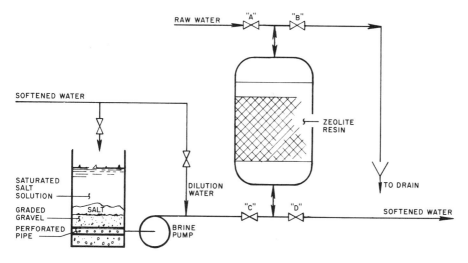

FIGURE 22-2 Simple Base Exchange Softener.

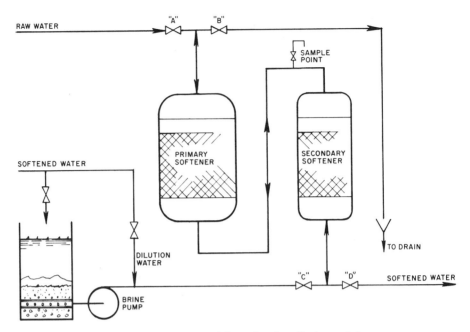

FIGURE 22-3 Primary and Secondary Base Exchange Softener.

the units are regenerating valves, (A) and (D) are closed and (B) and (C) are open.

The brine from the salt saturator is usually 25 to 26% NaCl and is diluted to approximately 10% with softened water before regenerating the zeolite.

Resin Capacity

The capacity of a resin is usually measured in kilograins (kgr) per cubic foot, where one kilograin = 1000 grains and since 7000 grains is equivalent to 1 pound, therefore 1 kilograin is equivalent to 0.143 pounds. Since calcium and magnesium hardness are expressed in terms of mg/ℓ as calcium carbonate ($CaCO_3$), the calculations are relatively simple but would be even simpler if we could get away from the archaic use of grains. A convenient method of calculating the cation or anion content of water for ion exchange purposes:

Milligrams per liter (mg/ℓ) ÷ 17.1 = grains per U.S. gallon

Milligrams per liter (mg/ℓ) ÷ 14.3 = grains per Imperial gallon or U.K. gallon.

Resin capacity is usually quoted in terms of kilograins per cubic foot of resin. It is also necessary to stipulate the number of pounds of salt used to regenerate the resin in order to determine its exchange capacity. For example, if a resin supplier states that a particular resin has an exchange capacity of 29.2 kilograins of $CaCO_3$ per cubic foot when regenerated at the rate of 15 pounds of salt per cubic foot, we can calculate the amount of water that can be treated per cubic foot of resin. Assume that the water has a hardness of 100 mg/ℓ of $CaCO_3$; this is equivalent to $100 \div 17.1 = 5.85$ grains per U.S. gallon. Therefore, 1 cubic foot of resin will theoretically treat

$$\frac{29.2 \times 1000}{5.85} = 4991 \text{ U.S. gallons}$$

This is equivalent to an exchange capacity of:

$$\frac{4991 \times 8.33 \times 100}{1,000,000} = 4.15 \# \text{ of } CaCO_3/15\# \text{ of salt}$$

Therefore, 1 lb of salt will replace $\frac{4.15}{15.0} = 0.275\#$ of $CaCO_3$ or 0.514 lb of NaCl per kilograin of $CaCO_3$ removed.

Lower concentrations of salt solutions are used with better conversion efficiencies. For example, if 10 lb of salt is used to regenerate each cubic foot of resin instead of 15 lb of salt, the capacity of this particular resin will drop from 29.2 to 24.8 kgr of $CaCO_3$ per cubic foot.

$$\frac{24.8 \times 1000}{7000} = 3.54 \text{ lb of } CaCO_3 \text{ per cubic foot of resin}$$

Therefore 1 lb of salt will replace

$$\frac{3.54}{10} = 0.354 \text{ lb of } CaCO_3$$

This represents a conversion efficiency increase of:

$$\frac{(0.354 - 0.275)(100)}{0.275} = 28.73\%$$

It should be noted that the capacity of the base-exchange unit will have dropped from 29.2 to 24.8 Kgr as $CaCO_3/ft^3$ which is a reduction in the length of run of 15.1%.

In a growing community, it is preferable to have larger base-exchange units than initially required so they can initially be operated at a lower salt per cubic foot regeneration level, and hence a higher conversion efficiency. As the water demand increases the regeneration level can be increased to maintain a higher removal rate per cubic foot of resin, but at a lower conversion efficiency.

The regeneration cycle is very important in the interest of good salt economy. Using 8 to 10% NaCl solution the flow rate should be about 1 U.S. gpm per cubic foot. Many operators prefer to give the bed a short initial rinse at the rate of 1.0 to 1.5 U.S. gpm per cubic foot immediately prior to regenerating. This separates the beads of resin from each other and displaces the untreated water from within the bed. After brine regeneration, which should never be hurried, the bed should be rinsed at approximately 1.0 to 1.5 U.S. gpm/ft^3 for 45 to 60 minutes using approximately 60 U.S. gpm per cubic foot. Rinse water samples should be taken at periodic intervals to ascertain if the backwash duration can be reduced.

Base-exchange softening removes the calcium and magnesium by replacing them with sodium ions. Since calcium (atomic wt 40) and magnesium (atomic wt 24.3) are both divalent and are therefore replaced by two atoms of monovalent sodium (atomic wt 23.0) the total dissolved solids (TDS) of base-exchanged softened water is higher than the original water, i.e., each atom of calcium replaced by two atoms of sodium will increase the

$$\text{TDS} = \frac{(2)(23) - 40}{40} \times 100 = 15\%$$

And similarly each atom of magnesium replaced by two atoms of sodium increases the

$$\text{TDS} = \frac{(2)(23) - 24.3}{24.3} \times 100 = 89\%$$

Water for low pressure heating boilers is often softened by base exchange but since this does not change the alkalinity there is a problem due to the thermal breakdown of the bicarbonate in the boiler to carbon dioxide (CO_2) in accordance with the following equation:

$$2NaHCO_3 \longrightarrow Na_2CO_3 + CO_2 + H_2O$$

The carbon dioxide can become a serious problem in that it will boil off with the steam and form carbonic acid (H_2CO_3) in the condensate, corroding the condensate return lines and fittings right back to the boiler where it will again separate into gaseous CO_2. This process will continue as evidenced by a corroded groove in the bottom of the condensate return pipes unless the carbonic acid is neutralized by maintaining caustic alkalinity in the feed water and keeping the pH is excess of 9.0.

The higher the total dissolved solids (TDS) in the make-up feed water the higher must be the rate of blow-down in order to prevent the TDS of the boiler water from becoming too high.

The following Table 22-2 suggests the maximum allowable total dissolved solids in the boiler.

If a water is high in total dissolved solids it may be necessary to demineralize to remove both the anions and cations.

Cation-Hydrogen Ion Exchanger

The same resin can often be used for cation-hydrogen exchangers, but since strong mineral acids are used to regenerate, the equipment must be rubber-

TABLE 22-2 Maximum Allowable Total Dissolved Solids in Boiler Water[7]

Boiler Pressure (psi)	TDS (mg/ℓ)
0–300	3,500
301–450	3,000
451–600	2,500
601–750	2,000

lined. The reactions are as follows:

$$\left.\begin{array}{l}CaSO_4\\Ca(HCO_3)_2\\CaCl_2\end{array}\right\} + HZ \longrightarrow CaZ + \left\{\begin{array}{l}H_2SO_4\\H_2CO_3\\HCl\end{array}\right.$$

Similarly for magnesium compounds.

The H_2CO_3 formed in the ion exchanger will tend to break down to H_2O and free CO_2 and it is usually necessary to install a degasifier immediately after the cation-hydrogen ion exchangers to remove the carbon dioxide.

When the cation-hydrogen resin is exhausted, it is regenerated with sulfuric or hydrochloric acid as follows:

$$CaZ + \left\{\begin{array}{l}H_2SO_4\\\text{or}\\HCl\end{array}\right. \longrightarrow HZ + \left\{\begin{array}{l}CaSO_4\\\\CaCl_2\end{array}\right.$$

producing a waste effluent containing calcium sulfate or calcium chloride.

Strong Basic Anion Exchanger

Following the degasifier is an anion exchanger.

$$\left.\begin{array}{l}H_2SO_4\\H_2CO_3\\HCl\\H_2SiO_2\end{array}\right\} + ZOH \longrightarrow \left\{\begin{array}{l}ZSO_4\\ZCO_3\\\\ZCl\\ZSiO_2\end{array}\right. + H_2O$$

(silaceous acid)

The finished water will be equivalent to distilled water and depending on the equipment used could be considerably better than distilled water. The resin is regenerated with caustic soda (NaOH).

$$\left. \begin{array}{l} Z.SO_4 \\ Z.CO_3 \\ Z.Cl_2 \\ Z.SiO_2 \end{array} \right\} + NaOH \longrightarrow Z.OH + \left\{ \begin{array}{l} Na_2SO_4 \\ Na_2CO_3 \\ NaCl \\ NaSiO_2 \end{array} \right.$$

The waste effluent consists of sodium compounds.

The demineralization of water with ion exchange resins to virtually zero TDS is of interest to readers of industrial water treatment.[5,8,9,10,11]

HIGH TOTAL DISSOLVED SOLIDS (TDS)

When the TDS exceeds 5000 mg/ℓ as TDS, there are two basic processes that should be evaluated: evaporation and freeze desalting.

Evaporation

Distillation equipment is more expensive than RO or ED plants for small capacities but becomes economical as the size increases. One of the advantages with distillation is that it does not matter how high the TDS content of the raw water is, but it does influence the ratio of forward feed to discarded waste. Distillation is a thermal process and if the plant can take advantage of low grade heat from the condensers of a thermal power station or other source of heat, then so much the better.

From the earlier single-effect evaporators, producing less than a pound of fresh water per pound of steam, the art of evaporation has progressed through multiple effects to the two modern concepts of multi-stage-flash (MSF) and vertical tube evaporators (VTE). The vertical tube evaporator is also known as multiple effect falling film evaporator (MEFF). Each type of evaporation has certain advantages over the other and the Office of Saline Water (OSW), and Envirogenics Co. of El Monte, California, have taken the best from both the (MSF) and (VTE) systems and have constructed a large pilot plant using sea water in order to prototype 200 mgd units. This project is described by Krebs et al.[12]

There are a great many problems associated with sea water evaporation in terms of corrosion, fouling of heating and cooling surfaces from marine growths, disposal of heated concentrated brine of 100,000 mg/ℓ TDS and more, cooling the product water, and above all the best use of primary heat. The term performance ratio referring to the pounds of product water per 1000 BTU, approximately equivalent to pounds of water per pound of steam condensed, is often used as a measure of thermal efficiency.

Vapor Recompression

Vapor recompression takes the low temperature, low pressure vapor from the top of an evaporator, compresses it, and uses the hot vapor to boil the water to produce more vapor. The condensed vapor gives up its heat of compression to the heating surfaces of the evaporator and leaves the plant as product.

This process has been successful in relatively small package units of less than 1 mgd.[13]

The process is not new and has been used for many years in areas where electric power is cheap. Diesel and gas engine compressor drives are also used and advantage is taken of the heat from the exhaust gases and jacket cooling. The concept is basically a heat pump where heat is abstracted from a relatively low temperature source and elevated by vapor compression to a higher temperature.

Freeze-Desalting Processes

People in northern lands have relied on cutting blocks of ice from the rivers during the winter and storing them in tanks in the basements of their homes, for year-round use. As the water freezes, most of the dissolved solids in the water are rejected as the ice crystals form. It is therefore relatively lower in TDS.

Commercial applications of freeze-desalting processes extract product water of 50 to 500 mg/ℓ TDS from raw waters varying from 5000 to 40,000 mg/ℓ, TDS (sea water is approximately 35,000 mg/ℓ TDS). The ratio of product water to input varies from 30 to 50% depending on salinity.

There are basically three crystallization processes under investigation by the OSW, direct freezing, secondary refrigerant, and the hydrate processes.[14]

The two processes of vapor freezing and vapor compression have been combined into a process known as vapor freezing, vapor compression (VFVC). This process has many advantages for the production of drinking water from sea water in quantities that would be too small to be economical for the MSF or VTE processes, particularly where a 30 to 50% yield, i.e., 50 to 70% of the total water abstracted is discarded to waste, presents no specific problems.

Solar Evaporation

No appraisal of the various demineralization processes is complete without some mention of solar energy and solar evaporation in particular. Although much has been written on this subject, it has scarcely gotten off the ground, due in large part to the high costs involved.

The amount of solar radiation reaching the earth surface depends on latitude, cloud cover, and other variables. Strock[15] quotes values, for the southern part of the United States, of 1800 to 2000 BTU per square foot per day as a yearly average. But in Washington, D. C., it is only approximately 1200 BTU per square foot per day. In the tropical countries where higher temperatures prevail, the period of daylight hours may be appreciably shorter, particularly compared to our summer months. There are also seasonal variations in sunlight. Average air temperatures are not so critical as hours of sunlight and clear skies, but obviously have a bearing on the overall output of a solar still. Grune, et al[16] in 1960, investigated a number of solar stills on the Georgia Institute of Technology campus, Atlanta, Georgia. Some improvements can be made with forced convection stills but since they need electrical power, this may make the forced convection design impractical in many areas and the fact that they are fixed to the ground, whereas the sun is continually moving, only adds to the reasons for low conversion efficiency.

In 1970, the United Nations published a book entitled *Solar Distillation*[17] to assist the undeveloped countries where fresh water is almost unobtainable.

The present design of solar stills has a very low thermal efficiency. According to *Solar Distillation*,[17] "Productivity is conveniently referred to, in round numbers, as being, typically, 0.1 U.S. gallon per square foot per day, for a good day. It is however highly dependent upon solar radiation and less dependent upon air temperature and other factors. On clear winter days a well-designed still will yield perhaps 0.03 U.S. gallons and on hot clear summer days, perhaps 0.12 U.S. gallons per square foot per day. Summing up these yields over a year, experience shows that an annual solar still yields about 25 U.S. gallons per square foot per year, with some variations due to climate and still design."

The "Journal of Solar Energy Science and Engineering"[18] Vol. 11., Number 3–4, for July and December 1967, featured a photograph of the 30,000 square foot solar still at Coober Pedy, South Australia. This still, according to *Solar Distillation*[17], would only have an annual yield of $30,000 \times 25 = 750,000$ U.S. gallons per year which is only equivalent to an average of 1.43 U.S. gpm, not a very encouraging flow. On the North American continent, with an average per capita consumption of approximately 150 U.S. gpcap day, this would only be sufficient to meet the average annual demand for 13.7 people with no provision for peak day demands. On the other hand, in specific instances, it would be the only logical approach to the water supply problem.

The capital cost of a solar still installation is high since the ground must be carefully leveled and graded, and the bottom troughs are usually in dense concrete to avoid seepage losses and must be water-tight. Plastic covers have been tried but glass is preferred, since it has higher transmissive capabilities, remains cooler, and performs better as a surface condenser.

Prior to 1969 *Solar Distillation*[17] estimated that the capital cost of a solar still was between 60¢ and $2.00 per square foot. Most of their construction cost estimates indicated that the materials and labor for a durable, 20-year life solar still, would be about $1.00 per square foot. There has been considerable inflation over the last 5 to 10 years which would increase this price appreciably.

Solar Distillation[17] goes on to say that "assuming favourable interest rates, such as those granted to a public utility type of venture, and the length of useful service as noted, one obtains water costs of $3—$6 per 1000 U.S. gallons."

SUMMARY OF THE PRESENT STATE OF THE ART

The production of fresh water from the sea or other salty brackish waters falls into two distinct categories.

For low total dissolved solids up to approximately 5000 mg/ℓ TDS, it is more economical to remove the solids from the water, i.e., ion exchange up to 1500 mg/ℓ TDS, reverse osmosis (RO), and electrodialysis (ED). However, when the total dissolved solids exceed approximately 5000 mg/ℓ TDS it is more economical to remove the water from the solids by distillation and freeze separation processes.

COST OF PRODUCING DEMINERALIZED WATER

The cost of producing demineralized water from sea water or brackish water requires careful evaluation. Some recently published cost data[19] for brackish waters (TDS up to 5000 mg/ℓ) has been analysed in Table 22-3. Unfortunately, the reference does not include cost data for sea water plants. (TDS up to 35,000 mg/ℓ.)

One frequent omission in compiling cost data is to neglect the capital and operating costs of the pretreatment plant required ahead of the demineralizer. Waters containing oil, greases, unchelated iron and manganese, scale forming compounds, algae, and other biological organisms must obviously be pretreated before the water enters the demineralizer, otherwise the operating costs of the process will be considerably increased due to the frequent replacement of the diaphragms, exchange media, and the fouling of heat exchangers resulting from surface contamination.

TABLE 22-3 Capital and Operating Costs for MGD (US) Plants

1	2	3	4	5	6	7	8	9	10	11
Process	Water	T.D.S.	Capital Cost Jan. 1972	ENR Adjustment June 1975 × 1.3	Annual Amortization × 0.163	Operating Costs (1972) (1000 US gal)	ENR Adjustment June 1975 × 1.3	Annual Operating Cost ×292,000	Total Annual Cost	Cost per 1000 US gallons ÷292,000
E.D.	CaSO$_4$ & Ca(HCO$_3$)$_2$	5000	$1,100,000	$1,439,000	$233,000	$0.80	$1.04	$304,000	$537,000	$1.84
		3000	900,000	1,170,000	191,000	0.65	0.845	246,000	437,000	1.50
		2000	750,000	975,000	159,000	0.52	0.675	197,000	356,000	1.22
		1000	550,000	715,000	117,000	0.37	0.482	141,000	258,000	0.875
	NaCl	5000	$ 900,000	$1,170,000	$191,000	$0.70	$0.91	$266,000	$457,000	$1.57
		3000	730,000	950,000	151,000	0.55	0.715	209,000	360,000	1.24
		2000	650,000	845,000	137,500	0.47	0.61	178,000	315,500	1.08
		1000	420,000	546,000	89,000	0.33	0.43	126,000	215,000	0.74
Ion Exchange	CaSO$_4$ & Ca(HCO$_3$)$_2$	3000	$1,550,000	$2,020,000	$330,000	$1.25	$1.625	$475,000	$805,000	$2.76
		2000	1,000,000	1,300,000	212,000	0.75	0.975	284,000	496,000	1.70
		1500	800,000	1,040,000	170,000	0.58	0.754	220,000	390,000	1.335
		1000	450,000	585,000	95,200	0.37	0.48	140,900	235,200	0.805
	NaCl	3000	$1,550,000	$2,020,000	$330,000	$1.00	$1.30	$380,000	$710,000	$2.43
		2000	1,100,000	1,300,000	212,000	0.70	0.91	266,000	478,000	1.64
		1500	800,000	1,040,000	170,000	0.54	0.704	205,000	375,000	1.285
		1000	450,000	585,000	95,200	0.34	0.443	129,000	224,200	0.77
R.O.	CaSO$_4$ & Ca(HCO$_3$)$_2$	5000	$ 620,000	$ 805,000	$131,600	$0.60	$0.78	$228,000	$359,600	$1.23
		3000	590,000	767,000	125,000	0.50	0.65	190,000	315,000	1.08
		2000	550,000	715,000	117,000	0.46	0.60	175,000	292,000	1.00
		1000	440,000	572,000	93,200	0.37	0.48	140,000	233,200	0.80
	NaCl	5000	$ 610,000	$ 794,000	$129,000	$0.52	$0.677	$198,000	$327,000	$1.12
		3000	600,000	780,000	127,000	0.48	0.624	182,000	309,000	1.06
		2000	590,000	766,000	125,000	0.45	0.585	171,000	296,000	1.01
		1000	490,000	637,000	104,000	0.38	0.494	144,000	248,000	0.85

FOOTNOTES FOR TABLE 22.3

COLUMN 1 ED Electro-dialysis
Ion Exchange
RO Reverse osmosis

COLUMN 2 Two types of water have been considered, the first having predominently high calcium sulfates and bicarbonates and the second having high sodium chloride contents.

COLUMN 3 Total dissolved solids (TDS) expressed in mg/ℓ.

COLUMN 4 Capital cost for a 1 million U.S. gallon per day plant taken from reference #19.

COLUMN 5 Engineering News Record (ENR). Construction cost index adjustment for January 1972. When the cost data was compiled the ENR was approximately 1700. For June 1975, an ENR index of 2200 has been postulated. The estimated increase in capital cost is taken as the ratio $\frac{2200}{1700} = 1.3$.

COLUMN 6 Annual amortization has been taken over a 10-year period at 10% interest. The annual repayment factor is 0.163.

COLUMN 7 Operating cost per 1000 U.S. gallons of water treated taken from reference #19 for a 1 million U.S. gallon per day plant. This cost consists of chemicals, power, labor, and management. It can therefore vary widely for different installations.

COLUMN 8 Operating cost in column 7 corrected for ENR index adjustment. This is not strictly correct but is a fair approximation.

COLUMN 9 It is assumed that any plant would not be expected to produce more than 80% of its total capacity equal to $365 \times 1 \times 0.80 = 292$ million U.S. gallons per year.

COLUMN 10 Total cost = Column 6 + Column 9.

COLUMN 11 Cost per 1000 U.S. gallons is equal to Column 10 ÷ 292,000.

REFERENCES

1. MOORE, D. H. "Greenfield, Iowa, Reverse Osmosis Desalting Plant." *J.A.W.W.A.*, Vol. 64, p. 781, Nov., 1972.

2. FABER, H. A., S. A. BRESLER, and G. WALTON, "Improving Community Water Supplies with Desalting Technology." *J.A.W.W.A.*, Vol. 64, p. 705, Nov., 1972.

3. *Water Quality and Treatment*. 3rd Ed. Prepared by the AWWA Inc., New York: McGraw-Hill Book Company, 1971.

4. LYNCH, JR., M. A. and S. M. MINTZ, "Membrane and Ion-Exchange Processes—A Review." *J.A.W.W.A.*, Vol. 64, p. 711, 1972.

5. APPLEBAUM, S. B. *Demineralization by Ion Exchange*. New York and London: Academic Press, 1968.

6. Trade literature from the Permutit Company of Paramus, New Jersey entitled: "Demineralization including Silica Removal" by Permutit Ion Exchange. Bulletin 3803E.

7. PINCUS, L. I. *Practical Boiler Water Treatment*. New York: McGraw-Hill Book Company, 1962.

8. DOFNER, K. *Ion Exchangers Properties and Application*. Ann Arbor Science Publishers Inc., 1973.

9. NORDELL, E. *Water Treatment for Industrial and Other Uses*, 2nd Ed. New York: Reinhold Publishing Corporation, 1961.

10. POWELL, S. T. *Water Conditioning for Industry*. New York: McGraw-Hill Book Company, 1954.

11. AWWA—*Water Treatment Plant Design*. Published by American Water Works Association, Inc., 1969.

12. KREBS, F. W., J. R. COFER, and E. H. SIEVEKA, "Sea-Water Distillation Module." *J.A.W.W.A.*, Vol. 64, p. 749, 1972.

13. "Water Desalting Techniques—A Summary." (This is a series of articles) Water and Wastes Engineering Industrial, Vol. 6 (11), p. F-1 to F-26, Nov., 1969.

14. ZABBAN, W., T. FITHIAN, and D. R. MANEVAL, "Conversion of Coal-Mine Drainage to Potable Water by Ion Exchange." *J.A.W.W.A.*, Vol. 64, p. 775, Nov., 1972.

15. STROCK, C. *Handbook of Air Conditioning, Heating and Ventilating*. New York: The Industrial Press, 1959.

16. GRUNE, W. N., R. B. HUGHES, and T. L. THOMPSON, *Solar Stills*. Water and Sewage Works—Vol. 108 (10), page 378 (Oct 1961).

17. *Solar Distillation*. Published by United Nations. Sales No. E.70.11.B.1 available from the U.N. Sales Section, New York or Geneva, 1970.

18. "The Journal of Solar Energy Science and Engineering." Published quarterly by the Solar Energy Soceity, Headquarters, Campus, Arizona State University, Tempe, Arizona.

19. "Desalting Techniques for Water Supply Quality Improvement." Report for office of Saline Water. U.S. Department of the Interior by *AWWA*, 1973.

23

DISTRIBUTION SYSTEMS

"Out of sight and out of mind," was never more truly expressed than with reference to a water distribution system and yet it constitutes the largest proportion of the total capital investment. It includes reservoirs, pumps, chlorination, pressure reducing stations, metering, instrumentation, valving and piping, fire protection, and public relations. There are many moral and legal responsibilities involved in design and operation of a distribution system.

DISTRIBUTION SYSTEMS

Wherever a public water supply system is installed in North America, a person expects to be able to drink without fear of being poisoned or diseased. The water in the distribution system must therefore be free from pathogenic bacteria and other harmful pollution. The fact that there are few instances where serious problems have occurred speaks volumes for the water works industry. But the possibility of pathogenic bacteria is not the only hazard confronting the public or the purveyor.[1] There are property damage claims due to accumulated silt in the mains, damage by flood due to burst pipes, and problems due to pressure fluctuations.

Munshaw Color Service Ltd. versus the City of Vancouver in British Columbia, Canada,[2] is an interesting example of the sort of problems that can

LEGAL ASPECTS

arise. Munshaw Color Service Ltd. operates a photographic film processing establishment. They were aware that the city water mains contained deposits of silt and the water entering their plant was often turbid. To safeguard their process they installed cartridge filters to prevent suspended particulate matter from entering their processing tanks. On the morning of August 1, 1957, the City of Vancouver was using a fire hydrant 1250 feet away from Munshaw's to "flush and drag" a sewer pipe. The heavy draw-off from the water main disturbed and roiled the deposited silts. Some of the silt entered Munshaw's process tanks and ruined batches of film in the process of being developed. There was no explanation as to why their own filter system had failed.

The Chief Justice ruled that the Plaintiff's damages of $3,694.87 and their cost against the defendant—City of Vancouver—were justified on the grounds of *res ipsa loquiter*, which means: the thing speaks for itself, a phrase often used in accident cases where the evidence of negligence on behalf of the defendant is obvious.

Fortunately for the Water Works Industry, the City of Vancouver took the case to a court of appeal and on May 19, 1961, the judge reversed the former ruling, allowed the appeal and dismissed the action as follows:[3]

> "... the (previous) learned chief justice held that the City was negligent in not warning the Plaintiff that there might be an excessive amount of sediment in the water resulting from the use of the hydrant. But if I am right in thinking that neither the plaintiff nor the City has any reason to foresee that the use of the hydrant would cause the unprecedented amount of sediment that descended upon the Plaintiff, then there was no occasion either for the City to give the warning or for the Plaintiff, if it received one, to do more than to rely upon its filters to take care of the sediment as it had done in the past. No warning of the proposed operation would have put the Plaintiff on guard against the unexpected quantity of silt, so the damage would still have occurred.
>
> I would allow the appeal and dismiss the action.
>
> Dated May 19, 1961.

The case was then taken to the Supreme Court of Canada and on April 24, 1962, the court confirmed the decision of the Provincial Appeal Court and dismissed the action against the city.[4]

Had the first judgment been sustained there would have been a precedent which could have resulted in serious consequences for the water supply industry. All distribution systems have some deposits in the water mains and particularly with old cast iron pipes. Apart from any after precipitation from incompleted chemical reactions, incompatibility of two waters from different sources of supply, distribution systems always tend to accumulate deposits of one form or another. One of the arguments in favor of the first judgment in the Munshaw vs. City of Vancouver was that the city should have warned the film processing company that it was flushing the mains and that increased turbidity was to be expected. But what about a fire? There is no way that the services of the fire brigade can be withheld until all consumers in the immediate vicinity can be informed that the high rate of water usage is likely to stir up the silt deposited in the mains and cause an increase in turbidity! It is therefore the consumer's responsibility to take whatever safeguards he believes are necessary to protect his own property against excessive turbidity when hydrants are operated and excessive draw-offs occur.

Dissolved solids, liquids, and other forms of contamination are more difficult to guard against than turbidity. If a water main is inadvertently cross-connected to a source of contamination, the water purveyor may be held responsible.

The water purveyor who owns and operates the system is responsible to his public to supply them with "potable" water. If for any reason the water supplied to the public is not potable, then the purveyor may become liable for any injuries that may result. The water may be contaminated from a cross-connection or from other people's neglect and through no negligence on the part of the purveyor; nevertheless he is subject to liability.

A water purveyor responsible for the supply of safe and wholesome water to the public is in an extremely difficult legal position. He is able to control the water quality through the treatment plant into the distribution systems, but from there on, he has very little control. He relies on the plumbing code and the plumbing inspectors who must insure that the code is adhered to but this is easier said than done. A new building requires a normal domestic water supply for its offices, lawns, and fire protection. The plumbing inspectors check the drawings and finally visit the location to insure that the plumbing code requirements have been complied with before the water supply is turned on at the service connection. If a few months later modifications are made, unbeknown to the plumbing inspector, the whole aspect of the water use may have changed. The building may change hands and the next occupier may have a chrome-plating plant in the basement which definitely requires some cross-connection control. By the time the plumbing inspector makes his next visit the damage may be done. Plumbing inspectors must have the right to visit during working hours, without an appointment, any consumer premises connected to the public water system. If permission to inspect is refused, or the internal piping arrangements are not satisfactory, then the plumbing inspector should turn off the water supply to that consumer until such time as the conditions have been adequately corrected. There have been court cases, supporting the right of a municipality, to shut off the water supply to any customer who fails to provide adequate backflow prevention where there is a danger or possibility of contaminating the public water supply system.

Cross-connections between the public water supply system and source of contamination water are difficult to locate, and in many cases, the fact that bacteria and viruses may have entered a water supply system is based on circumstantial evidence. Unfortunately, whenever typhoid or cholera epidemics occur, the water supply is automatically blamed and the real source of contamination may be overlooked. Nevertheless, the water purveyor must be aware of his obligation to the public he serves, be aware of the pitfalls, and have a lawyer who is versed in this particular aspect of common law.

NETWORK STUDIES

The design of a distribution system starts with a developer's plan. Once the boundaries have been defined, a good contour map of the area on as large a scale as possible is essential, preferably from up-to-date aerial photography. Aerial photogrammic surveys can give contour lines to an accuracy of within 0.1 feet. It is far less costly, particularly for large areas, and much quicker than using ground crews.

Having determined where the distribution system will be located, locate hydrants and check fire flow requirements with the local fire underwriter's offices. Fire insurance risks are based on the demerit point system and the following list of deficiency points applies to a 1973 schedule.

Water supply	39%
Fire department	39
Fire service communications	9
Fire safety control	13
	100%

Previous fire underwriter's recommendations favored large reservoirs to fight a prolonged fire. Now there is more emphasis on the rate of flow which can be as much as 8000 U.S. gpm for old and congested downtown areas on the basis that it is better to try and wash the fire out as quickly as possible before it spreads into a prolonged struggle.

The water supply into the area is then considered both in size, pressure, and location,[5] whether or not service or storage reservoirs are required, booster pumps, pressure sustaining valves, and pressure reducing stations. A skeleton layout of trunk mains and the principal arteries should be shown on the topographical plan. If the terrain is hilly it may be necessary to plot a number of profiles and divide into separate pressure zones. Attention must be given to control systems and how the flows, pressures, and levels in reservoirs can be monitored and controlled.

HARDY CROSS METHOD

The old methods of determining pipe sizes, using the Hardy Cross system[6] for balancing flows and pressures around the various loops, is still applicable for small systems.[7,8,9,10] It may be useful to run an approximate analysis on the skeleton of the system using Hardy Cross or the computer to determine the main parameters. Once the concept of the system has been established a computer should be used to determine the size, head losses, pressures at various rates of flow, using one of several available network programs.[11,12,13] A computer is almost essential in this phase of the design, and once the program has been formulated, literally millons of figures of flow, head loss, velocities, and pressures can be produced very quickly.

ELECTRIC ANALOG

Distribution system network analysis programs can be performed on an electric analog which consists of a series of electrical resistances of known value, arranged together to represent the lengths and the diameters of the pipes in the distribution system. A voltage is impressed across the analog and the current flow is measured. From these readings the head losses throughout the various parts of the system can be calculated. The electric analog was originally designed for the analysis of power distribution systems and is useful to a limited degree on water distribution systems providing the land is perfectly level, but it runs into difficulties with systems serving hilly terrains. With the introduction of modern computer technology, the electric analogs are no longer of interest for water distribution system analysis.

COMPUTER PROGRAMS

A number of programs are available for distribution system analysis. Most of the programs are based on the Hardy Cross Method, which is the least sophisticated and requires less computer memory. Computer application of mathematical modelling to water distribution systems opens new horizons to water utility management, and leads to improved system operation. Once the mathematical model has been written from the schematic diagrams of the system, defining the sources of supply, plumbing, storage, distribution, pipeline hydraulic gradients, and other system features, the model can be used over a period of years to analyze the effects of changes and additions to the system. Some of the many uses of the mathematical model are listed as follows:

1. Analyzing pressure problems or failure of pumps and reservoirs to give desired flows.
2. Developing operating procedures for emergencies such as fire, pipeline break, pump failure, or a reservoir out of service.
3. Indicating short-or-long range water system developments.
4. Determining the order of priority for improvements in system capacity.
5. Evaluating capabilities of an existing water system under anticipated water demands.

6. Comparing alternative water supply, transmissions, or distribution options for a system, including costs.

7. Estimating the effects of new large industrial customers on an existing water system.

8. The benefits to be gained if several water systems are consolidated together.

9. Evaluating the effect of a major change or addition to the system.

THAWING FROZEN PIPES

The problem of thawing frozen pipes and service connections must always be considered when designing water supply systems in cold climates. It is usual to bury the pipes 6-12 inches below the normal frost line, and where there is always frozen ground such as perma-frost areas of the far north, then the pipes must be insulated, provided with heating cables, and usually installed above ground in insulated utilidors. With normal buried installations there is always a problem due to the differences in the depth of frost penetration. Some heavy clay soils are fairly resistant to frost penetration particularly if they are covered with a layer of undisturbed snow. In other areas, within the same municipality, the ground may consist of sand or loose gravel where the frost line may penetrate 15-16 feet particularly at locations where there is vehicular traffic with no continuous snow cover.

Smaller diameter pipes particularly in asbestos cement should be avoided in these areas, apart from the fact that they cannot be thawed out electrically, they also have low beam strength when subjected to frost heave. In frozen ground there are changes in temperature from one section of buried pipes to another and the contraction forces can be considerable, apart from the heaving of the frozen ground itself. The cost of going deep enough over the whole system in order to keep clear of the frost line may be exorbitant and not justified. In areas where troubles are likely to occur the following precautions should be considered:

1. Encasing the pipes inside an insulated culvert. Permanent heating cables should be installed with convenient electrical connections on the surface.

2. Use larger diameter pipes so that if they do tend to build up an ice layer it is less likely to plug the pipe.

3. Heat the water a few degrees to prevent freezing and design the system to maintain continual water circulation.

4. Avoid stagnant sections where there is little or no flow.

The use of heating cables or tapes has many advantages but adds to the cost of the installation and can only be justified in obviously troublesome sections.

Frost penetration is not uniform from year to year. In some years it may only be 6 feet while in other years it can go down to 9 to 10 feet. The worst time is in early spring when the frost is still penetrating deeper into the ground even though the surface is beginning to thaw. Electric welders and low voltage transformers are used for applying current through a frozen line in order to restart the water flow.

The use of electric welding machines for pipe thawing may impress too much current on the system. Pipe joints may not always be good electrical connectors. Rubber pipe joint materials may become overheated, perish, or harden to the extent that they leak. If high voltages are used to thaw service and distribution systems, it is possible that other circuits grounded to the water system may be affected. Special low voltage generators or transformers only should be used for thawing water pipes if the voltage and amperage is competently supervised. Knowlton[14] suggests 40–60 volts and the following amperage:

$\frac{3}{4}''$ to 2'' diameter   100 to 200 amps

2'' to 6'' diameter   200 to 300 amps

6'' to 10'' diameter   300 to 500 amps

The time required for thawing depends on type of pipe, size and length, joints, extent of freeze, condition of the soil, and the soundness of electrical connections and lead wires to and from the source of power and the pipe.[15]

METERING

The importance of conserving water by metering each consumer has already been discussed in Chapter 1. Not only are there savings in the water utility department if less water is wasted, but there are equivalent savings in sewage collection and treatment.

The question as to whether or not to meter each consumer is one of economics. If water is supplied from an impounding reservoir located on a mountain, fed by gravity to a hundred or more houses without even chlorination, then the installation of meters would only add cost to the system and nothing to its value. If on the other hand the same village is supplied from a turbid, highly colored river, where chemical treatment and pumping is essential, then metering presents a different aspect.

One of the early objections to the installation of meters for relatively small communities was the problem of reading them. They were invariably installed in inaccessible locations, difficult to read, plus the problem of people being out when the meter reader called. Dogs and other unpleasant circumstances have prevented meters from being installed but today there are two principal systems which overcome these difficulties.

(a) The meter can be installed almost anywhere inside the building with a remote readout station located on the outside wall. This readout station can have a normal dial where the meter reader notes the figures and writes them in his book. Alternatively, a multipronged plug can be inserted into a suitable receptacle attached to the readout station which will record the data on a magnetic tape. This data can then be translated directly to a billing machine which will type the consumer invoices.

(b) An even more sophisticated system consists of a direct linkage with the telephone company.[16,17,18] This system can also include the gas and electric meters, and the foot-slogging meter readers making door to door visits are no longer required. The chances of human error are largely eliminated and the older objections are no longer valid.

Meter Repairs

As meters wear they pass more water than recorded, and it is necessary to have a meter shop to test and repair them.[19,20]

The frequency of repairs depends on water quality and the meter. In some areas once every 5 years, and in others every 10 years.

The water purveyor should be responsible for the service connection from the water main to the meter. In his own interest he should insure that the service connection is suitable for long and continuous service with respect to the water quality and the soil conditions. They should not develop leaks, which would be regarded as unaccounted losses chargeable to the municipality. To further protect the distribution system, the meters could be provided with check valves, as a minimum protection against possible cross-connections.

LEAKAGE SURVEYS

If the installation of water meters is justified, then periodic leakage surveys are also justified. A $\frac{1}{2}$-inch diameter hole at 50 psi can leak 38,000 U.S. gallons per day, sufficient to keep a village of 300 people in water, at an average consumption of 125 U.S. gpd per capita. It would not be surprising to find several leaks of similar magnitude in most small distribution systems. With meters installed on every service connection it is not difficult to calculate the unaccountable losses from month to month. Organized leakage surveys can appreciably reduce these losses to a minimum.

HEATING OF WATER IN DISTRIBUTION SYSTEMS

In northern towns and cities of North America it is not unusual to heat the water a few degrees, often up to 3°–5°C (37°–41°F) before it enters the distribution system. Care must be taken with both the design and operation of these systems. If the temperature is not accurately controlled, there will be a wastage of fuel and also stresses in the pipe system resulting from uneven expansion.

REFERENCES

1. RHYNE, B. W. "Recent Cases of Water Utility Liability." *J.A.W.W.A.*, Vol. 64 (2), p. 82, 1972.
2. "Damages Action Against Vancouver, B.C." Published in the Canadian Section of the American Water Works Association—Water Works Information Exchange, Vol. 10, page 1 (1), January 1960.
3. "Appeal—Vancouver Damage Action." Reference to 2 above, Vol. 10, p. 22 (3), June 1961.
4. "Vancouver Damage Action—Supreme Court of Canada." Reference to 2 above, Vol. 10, p. 25 (3), May 1962.
5. Joint Discussion. "Balancing Needs, Responsibility and Costs in the Distribution System." *J.A.W.W.A.*, Vol. 62 (10), p. 629, October, 1970.
6. CROSS, HARDY. "Analysis of Flow in Networks of Conduits or Conductors." University of Illinois, Engineering Experiment Station, Bulletin 286, November 1936.
7. FAIR, G. M., J. C. GEYER, and D. A. OKUN, *Water and Wastewater Engineering, Volume 1, Water Supply and Wastewater Removal*. New York: John Wiley & Sons Inc., 1966.
8. CLARK, J. W., W. VIESSMAN, and M. J. HAMMER, *Water Supply and Pollution Control*, 2nd Ed. International Textbook Company, 1971.
9. BABBITT, H. E., J. J. DOLLAND, and J. L. CLEASBY, *Water Supply Engineering*, 6th Ed. New York: McGraw-Hill Book Company, 1962.
10. TONG, A. L., T. F. O'CONNOR, D. E. STEARNS, and W. O. LYNCH, "Analysis of Distribution Networks by Balancing Equivalent Pipe Lengths." *J.A.W.W.A.*, Vol. 53, p. 192, February 1961.
11. ALEXANDER, S. M., N. L. GLENN, and D. W. BIRD, "Advanced Techniques in the Mathematical Modeling of Water Distribution Systems." Presented to the April 1973 annual Pacific Northwest Section Meeting of the AWWA by R. W. Beck and Associates of Seattle, Washington.
12. HANN, R. W. "System Simulation for Analysis and Design." *J.A.W.W.A.*, Vol. 62, p. 279, May 1970.
13. KALLY, E. "Computerized Planning of the Least Cost Water Distribution Network." Water and Sewage Works. Reference Number, page R-121, 1972.
14. KNOWLTON, A. E. *Standard Handbook for Electrical Engineers*, 9th Ed. New York: McGraw-Hill Book Company, 1957.
15. STEEL, E. W. *Water Supply and Sewerage*, 4th Ed. New York: McGraw-Hill Book Company, 1960.
16. BLAHA, J. W. "Water Bills and Computers." *J.A.W.W.A.*, Vol. 62 (10), p. 603, October 1970.
17. LEE, J. H. "Remote Water-Meter Reading Tester." *J.A.W.W.A.*, Vol. 63 (6), p. 329, June 1971.
18. O'LEARY, T. V. "Meter Reading Revolution." *J.A.W.W.A.*, Vol. 63 (8), p. 481, August, 1971.

19. Committee Report. "Meter Installation and Maintenance." *J.A.W.W.A.*, Vol. 62 (4), p. 200, April 1970.
20. AWWA. "Water Meter Manual—M6"—available from the American Water Works Association.
21. CAMPBELL, F. C. "Distribution System Leakage Survey." *J.A.W.W.A.*, Vol. 62 (7), p. 400, July, 1970.
22. Joint Discussion. "Locating Leaks in Mains and Services." *J.A.W.W.A.*, Vol. 62 (10), p. 403, July 1970.

24

CROSS-CONNECTION CONTROL

There are three terms in common usage which refer to this subject; cross-connection, backflow, and backsiphonage, and they are defined as follows:

Cross-connection Means:

(a) any physical connection through which a supply of potable water could be contaminated or polluted.

(b) a connection between a supervised potable water supply and an unsupervised supply of unknown or doubtful potability.

Backflow Means:

(a) a flow condition, induced by a differential in pressure, that causes the reverse flow of water or other liquid into the distribution system of a potable water supply.

(b) the backing up of water through a conduit or channel in the direction opposite to normal flow.

Backsiphonage Means:

A form of backflow caused by a negative or subatmospheric pressure within an underground buried water distribution system.

If the water pressure in the distribution system, upstream of the cross-connection, is higher than the downstream pressure after the cross-connection, then there is no immediate danger. Unfortunately, there are many occasions when the distribution system is under reduced pressure, particularly if there is a main break on a lower level, or there is a fire department pumper connected to a hydrant. Under these conditions pollutants of various descriptions enter the water main and contaminate the water. When the distribution system pressure is restored, the contaminated water is delivered to the service connections. Water utilities often receive consumer complaints which indicate that a backsiphonage condition has occurred. A hotel proprietor reported that he lost all the water out of his central heating boiler during the fighting of a downtown fire. The operator of an automatic coin laundry reported that the washing machines went dry at the same time the hotel lost its boiler water. The water distribution system had been reduced to sub-atmospheric pressure by the fire department's pumps and the mains acted in reverse and sucked the water out of the boiler, out of the laundry, and any ground water through leaky pipes, into the distribution system. When the pressure was restored by stopping the fire pump, the boiler water and the laundry water, and the ground water which was in the main, was fed back to the consumers further down line. If the boiler water was chemically treated to prevent scale and corrosion in the central heating system, then these chemicals, (some of them are highly toxic), would be supplied as drinking water to the other consumers. Fortunately, in the case described there were no reported cases of sickness. But there are many cases where people have been less fortunate and one occurred at Holy Cross College in Worcester, Mass., and is known as the "Holy Cross Episode."[1] Other incidents have also been reported elsewhere.[2,7]

SUMMARY OF PROBLEMS

The problems of cross-connections and backsiphonage are as follows:

1. If the distribution system is always under adequate positive pressure, over and above the pressure in any system to which it could be connected, then there is little or no possibility of backflow occurring. In practice this can never be the case. Water distribution systems are frequently under reduced pressure due to

    (a) Failure of source of supply, power failure, mechanical and instrumentation failure, poor road construction causing main breaks, frost damage, earth tremors, excessive consumer demand, the use of water for fire protection, portable pumpers connected to the hydrants, and many other circumstances which can easily result in sections of the water supply and distribution system being under reduced pressure.

(b) Poor maintenance and the failure to correct excessive distribution system leakage is one of the major problems. It must be appreciated that even a good system can never be bottle tight and a 10% loss is considered to be good for most municipal systems; some of them are reported to leak in excess of 100% of the normal average annual flow. In water supply systems in which the services are not individually metered, it is very difficult to estimate the unaccounted losses in the system.

The water supply and sewage collection systems are normally located in separate trenches and should be at least 10 feet apart in accordance with the requirements of "10-State Standards."[8] It is not always possible however, to have 10 feet of separation and in many cases the water and sewage pipes are in the same trench. In any event, they cross over each other at each roadway intersection. It would seem therefore that one of the biggest risks of contamination can be from backsiphonage of the distribution system itself, but not all water supply authorities accept this concept.

2. The only adequate means of preventing backflow from a process tank back into the water supply system is to install a pipe system with an adequate air gap between the water supply and the tank receiving the water. It should also be recognized that the air gap can be easily bypassed and its effectiveness destroyed. The air gap itself must be designed to avoid its becoming ineffective due to wave action or other reasons.

3. Cross-connections to chemical and petroleum tanks must obviously be avoided.

4. Conduct distribution system leakage surveys to eliminate as many leaks as possible.

5. Contamination from poorly designed and incorrectly installed water-using applicances and toilet fixtures. Competent and adequate inspections of both new and old premises are essential to this end.

6. The installation of adequate backflow prevention devices will reduce the chances of backflow but they must be frequently tested and maintained. These devices also rely on air gaps and they must be installed to prevent the air vent from becoming flooded with non-potable water.

AIR VENT VALVES

It is impossible to insure that the system will never be under reduced pressure. Air and vacuum breakers should be installed to allow air to enter the distribution system, whenever reduced pressure conditions prevail. The air vent valves must be located in such a way that only air can enter the distribution system. In many cases the air valves are located in manholes which can be flooded with contaminated water. This adds to the problem when the vacuum breaker opens to fill the line with air, if it is under reduced pressure. In some cases, by oversight, air valves have been placed in air tight manhole chambers completely devoid of any air vents.

HYDRANTS The misuse of hydrants by members of the fire department can be a significant factor in water main failures. Surge pressures are proportional to changes in water velocity. If, for example, water is moving through a pipeline at 5 feet per second to several open hydrants, and some of these hydrants are closed or the pumpers are stopped, the forward velocity is reduced in a few seconds to 3 feet per second. The shock wave caused by this change in velocity is equivalent to almost 100 psi above the normal pipe pressure. It can be appreciated that if several hydrants are closed at the same time the shock wave or water hammer pressures could easily rupture the pipe. The fractures that result from these surges are not always known to exist, they just add a little more to the "unaccounted losses," but in the event of reduced pressure conditions, they constitute a cross-connection with any contaminated ground water that may be present in the vicinity of the fracture.

CROSS-CONNECTIONS Cross-connections of any sort must be avoided. For example, if a new water distribution system is to be installed in an established district where each home has its own well and septic tank, then the new distribution system should not be connected to the homes until the private water systems have been effectively disconnected. This may mean the removal of the well pump and the filling in of the private wells.

Where cross-connections are inevitable in large industrial and hospital complexes, to take advantage of the water pressure in the distribution system suitable backflow preventors must be installed, preferably two in parallel, to enable one unit to be tested while the other is in service so as not to interrupt the flow of water to the consumer.

BACKFLOW PREVENTION DEVICES

Approved Air Gap

An air gap is the only backflow prevention device, but it is only effective if the vertical height of the gap is not less than twice the diameter of the pipe. An approved air gap can only be installed in systems that are not under pressure.

Any other device such as a vacuum breaker or a Reduced Pressure Backflow Preventer (RPBP) is a mechanical adaptation of an air gap.

Vacuum Breakers

Vacuum breakers are diaphragm or suction-operated valves that are installed in pipelines, hose bibs, flush toilets, and other pressure systems that open to atmosphere whenever the pressure in the pipe drops to below atmospheric pressure. When the valve opens, air is sucked into the valve to prevent backsiphonage. Vacuum breakers should be installed on all hose connections, lawn sprinkling systems, and on any water outlet which could result in water sucked into the distribution system from a contaminated source, i.e., water lying on the lawn, or water in a sink.

It is essential that the vacuum breaker be installed in a location to always maintain the air gap. It would be no advantage to install the vacuum breaker in a pit or in a manhole that could be flooded, since the air gap would be destroyed.

Reduced Pressure Backflow Preventers

For closed-in pressurized systems, the only acceptable device is a Reduced Pressure Backflow Preventer (RPBP). Backflow prevention devices like safety valves on boilers must work 100% of the time without failure. But since all mechanical devices fail sooner or later, the RPBP, like safety valves, must fail-safe in the open position calling attention to itself by spilling out water.

By courtesy of Hersey-Sparling the diagram in Figure 24-1 explains the mode of operation of Reduced Pressure Backflow Preventers.

The laboratory and field testing of backflow prevention devices is an essential function if equipment is to be capable of operating for long periods without failure. Much of the work in this field is done by the Foundation for Cross-Connection Control Research, at the University of Southern California. They publish lists of equipment that they have approved. Equipment that was originally approved but has since failed in the field is removed from the approved list until corrections are made.

Fire underwriters' approved RPBP devices are also available which have low head losses.

Similar to the simple vacuum breaker, the RPBP device is only a mechanically operated air gap and must therefore be installed in a location above ground level that cannot be flooded.

Double Check Valves

Double check valve assemblies, consisting of two ordinary spring loaded or balance-weight assisted swing check valves in series, are sometimes used in locations where a backflow would not be injurious to health. They offer a degree

OPERATION AND INSTALLATION INSTRUCTIONS
BEECO REDUCED PRESSURE PRINCIPLE BACKFLOW PREVENTERS

DESCRIPTION

The BEECO Reduced Pressure Backflow Preventer operates on the principle that water will not flow from a zone of lower pressure to one of higher pressure. It provides maximum protection against backflow caused by both backpressure and backsiphonage.

The device consists of two spring-loaded check valves (A and B) and a spring-loaded, diaphragm actuated differential pressure relief valve (C) located in the zone between the check valves.

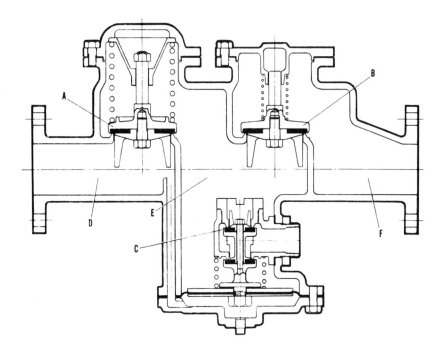

OPERATION

NORMAL

The first check valve (A) causes all water passing through it to be automatically reduced in pressure by approximately 8 psi.

The second check valve (B) is lightly spring-loaded and forms the "double check" feature of the device. It acts to prevent unnecessary drainage of the domestic system in case a backflow condition occurs.

FIGURE 24-1 Backflow Preventer

The relief valve (C) is spring-loaded to remain open, and diaphragm actuated to close by means of differential pressure.

To illustrate the operation, assume water, having a supply pressure of 60 psi, is flowing in a normal direction through the device. If all valves beyond area F are closed, creating a static condition, the water pressure in area D will be 60 psi and water pressure between the check valves (E) will be 52 psi.

The inlet pressure of 60 psi is transmitted through a cored passageway to the underside of the diaphragm of the relief valve (C). This valve is spring-loaded to remain in an open position until the differential pressure amounts to approximately 4 psi.

During normal operation, therefore, the 8 psi differential pressure produced by the first check valve (A) exceeds the spring-loading of the relief valve (C) and causes the relief valve (C) to remain closed.

There are two conditions that tend to produce backflow:

Backsiphonage — where the pressure in the drinking water system becomes less than atmospheric due to a vacuum or partial vacuum in that system.

Backpressure — where the pressure in the non-potable system exceeds that in the drinking water system.

Backsiphonage

As the supply pressure drops in area D, it also drops in the area **below** the diaphragm of the relief valve (C). When this drop amounts to approximately 4 psi, the relief valve (C) will start to open. This happens because the spring above the diaphragm of the relief valve (C), which is trying to force the valve open, is designed to compress with a differential pressure of 8 psi. When that differential is decreased to 4 psi, the spring will extend and cause the relief valve (C) to start to open.

This spring-loaded relief valve is designed to eliminate intermittent discharges and "spitting" with normal minor fluctuations in the line pressure.

As the supply pressure continues to drop, the relief valve (C) automatically continues to drain and, regardless of the pressure on the supply side, approximately 4 psi **less** pressure will be maintained between the check valves (zone E). This will cause continual drainage which will be readily visible at the drain outlet.

Backpressure

Assume that pressure at the discharge side (F) increases to 80 psi, while the supply pressure (D) remains at 60 psi:

1. **If the second check valve (B) does not leak,** water under higher pressure in area F will not enter the area between the check valves (zone E), and the pressure in this zone will remain at 52 psi. Under these conditions, the relief valve (C) will remain closed since the 8 psi differential pressure is still being maintained between the supply pressure (area D) and the area of reduced pressure between the check valves (zone E).

2. **If the second check valve (B) does leak,** water under high pressure (area F) will flow into zone E. If the pressure in this zone increases to approximately 56 psi — still 4 psi **lower** than the supply pressure (area D) — the relief valve will start to open and discharge this reversely flowing water to atmosphere, maintaining the pressure in zone E approximately 4 psi lower than supply pressure. The relief valve will automatically continue to drain as long as this backflow condition exists and as long as the second check valve (B) is leaking.

    If for any reason the first check valve (A) should leak during a shutoff beyond area F, the water under higher pressure in area D will leak into zone E. This will cause the relief valve to open as previously described and, again, provide visual indication at the drain outlet.

    In the unlikely event that the relief valve diaphragm should rupture, an unbalanced condition between area D and zone E will occur, and the relief valve will immediately discharge to atmosphere.

FIGURE 24-1 (Continued)

of protection to the system but the check valves can leak and backflow can occur with a double check valve assembly. Since it does not incorporate an air gap, it cannot really be regarded as a backflow preventer.

REFERENCES

1. TAYLOR, F. B. "The Holy Cross Episode." *J.A.W.W.A.*, Vol. 64 (4), page 230, 1972.
2. "Water Supply and Plumbing Cross-Connections." Published by U.S. Department of Health, Education and Welfare, Washington, 25, D.C. 1963.
3. GOUDEY, R. F. "Elimination of Cross-Connections in Los Angeles." Water Works and Sewage, Vol. 92 (3), March, 1945.
4. RAMEY, J. C. and N. R. AUGVIK, "Cross-Connection Control Program." *J.A.W.W.A.*, Vol. 60 (2), page 213, 1968.
5. ANGELE, SR., G. J. "Cross-Connections and Backflow Prevention." Published by the AWWA, 1970.
6. "Cross-Connection Control Manual." Published by U.S. Environmental Protection Agency Office of Water Programs Water Supply Division, 1973.
7. SOBOLEV, A. "European Plumbing Notes." Published by the Building Research Station, Bucknalls Lane, Gaston, Watford WD2 7Jr, England, 1971.
8. *Recommended Standards for Waterworks.* Published by Health Education Service, P.O. Box 7283, Albany, N.Y. 12224—This standard represents the requirements in Illinois, Indiana, Iowa, Michigan, Minnesota, Missouri, New York, Ohio, Pennsylvania and Wisconsin, hence the term "10 State Standards."
9. Joint Committee Report. "Use of Backflow Preventers for Cross-Connection Control." *J.A.W.W.A.*, Vol. 50, p. 1589, December, 1958.
10. BAIRD, J. N., W. R. SANFORD, and G. A. CRISTY, "Reduced Pressure Principle Backflow Preventer Evaluation and Use at Oak Ridge National Laboratory." Reprinted from *Health Physics*, New York: Pergamon Press, Vol. 11, p. 743, 1965.
11. "Recommended Practice for Backflow Prevention and Cross-Connection Control." AWWA, M14. Backflow Prevention Manual. Published by *AWWA*, 1966.
12. "AWWA Standard for Backflow Prevention Devices—Reduced Pressure Principle and Double Check Valve Types." *AWWA*, C506-69.
13. Trade literature from Hersey Sparling Meter Company, Dedham, Massachusetts.
14. "Accepted Procedure and Practice in Cross-Connection Control Manual," 2nd Ed., July, 1973. Published by Pacific Northwest Section. American Water Works Association, Inc.

25

PIPELINES

Many formulas have been devised for estimating the head losses due to friction when water and other liquids flow through pipes. Resistance to flow is caused by the roughness of the pipe wall, the viscosity of the fluid, and the relative velocity between the fluid and the pipe wall. In water works engineering there are three formulas in popular use, William-Hazen,[1] Manning,[2] and Colebrook-White.[3]

PIPELINE FRICTION

William-Hazen Equation

The William-Hazen formula is:

$$V = (1.318)\,(C)(r^{0.63})(S^{0.54})$$

where:

$V =$ velocity of water in the pipe in feet per second

$C =$ the roughness coefficient and is a factor depending on the inside surface conditions of the pipe

$r =$ the hydraulic radius of the wetted surface in feet and is the ratio of the cross-sectional area of flow divided by the wetted perimeter. For circular pipes $r = \dfrac{D}{4}$

$S =$ hydraulic slope in feet.

297

A number of tables, slide rules, and monographs have been published on the basis of this formula[1] and for this reason the equation is in popular use. The *Hydraulic Tables*[1] were first published in 1905, and have been used extensively in the industry for the last 70 years, but they do have a number of disadvantages. When the metric system becomes universal in North America, it is possible that the new tables will be published using one of the formulas based on the Reynold's number criteria such as the Colebrook-White equation.[4,5] The faults with the Hazen and Williams formula are associated with the use of the C factor for pipe friction. The surface roughness coefficient of the pipe wall is expressed by C values 80 to 160 for very rough tuberculated to very smooth pipe. The head loss due to friction inside a rough 4-inch diameter pipe will be very different compared to an equally rough 48-inch diameter pipe. In order to express the equivalent head loss it is necessary to employ a lower C value for the 4-inch diameter compared to the 48-inch diameter pipe, since the pipe wall effect of the bigger pipe is less, although both pipes have exactly the same roughness on the inside wall surface. With the William-Hazen formula the difficulty is in determining, from visual inspection of the pipe, what is the C value. Whereas with Colebrook-White equations the K value for roughness can be actually measured.

One advantage of the William-Hazen equation is that it can be readily calculated if tables are not available.

Manning Equation

The Manning equation is named after Robert Manning, an Irish engineer (1816–1897), and is used for calculating flows in open channels and corrugated culverts, rather than smooth bore pipe. The manufacturers of culverts usually supply a set of tables from which the flows and the head losses can be calculated.[2]

Colebrook-White Equation

One of the disadvantages of the Colebrook-White equation is its complexity, but it is no problem to a computer and fortunately tables have been published in both English and metric units. The equation is written as follows:

$$\frac{1}{\sqrt{f}} = -2 \log \left(\frac{k}{3.7D} + \frac{2.51}{Re \sqrt{f}} \right)$$

where:

f = friction coefficient $2g Di/V^2$
k = linear measure of effective roughness
D = pipe diameter
Re = Reynolds number $(DV)/v$
V = velocity in feet per second
v = kinematic viscosity of the fluid

g = gravitational constant (32.2)

i = hydraulic gradient

The Institution of Water Engineers in their "Manual of British Water Engineering Practice"[3] has published a nomograph entitled "Universal Pipe Friction Diagram" based on the work of Prandtl, Van Karman, Nikuradse, and Colebrook. This nomograph is sufficiently accurate for most practical purposes and is shown in Figure 25-1, (see the fold-out at the inside of the front cover).

Greater accuracy can be achieved with the use of this nomograph if the velocity (V) is precalculated and plugged into the nomograph together with the internal pipe diameter (D) instead of using (Q) (quantity: thousands of imperial gallons per hour) and the pipe diameter (D), which has too short a "length of sight" for accurate alignment.[6]

The roughness coefficient K expressed in inches is a measure of the actual roughness of the pipe surface. Recommended roughness values of K inches are given in Table 25-1.

TABLE 25-1 Recommended Roughness Values[5]

Pipe Materials	Values of K in inches		
	Good	Normal	Poor
Class 1			
Smooth materials—drawn copper, aluminum, brass, plastic, glass, fiberglass		0.0005	
Class 2			
Asbestos cement		0.001	
Class 3			
Bitumen lined cast iron		0.0015	
Cement mortar lined steel		0.0015	
Uncoated steel	0.001	0.0015	0.003
Galvanized iron	0.003	0.008	0.015
Uncoated cast iron	0.007	0.015	0.030
Class 4			
Old tuberculated water mains, with the following degrees of attack:			
Slight	0.025	0.06	0.15
Moderate	0.06	0.15	0.25
Severe	0.60	1.5	2.5
Class 5			
Woodstave pipe	0.015	0.030	0.060
Class 6			
Smooth surface precast concrete pipe in lengths over 6 feet with spigot and socket joints internally pointed.	0.003	0.006	0.015
Precast pipes with mortar squeeze at the joints		0.15	0.30
Class 7			
Gravity sewer pipes (new)	0.030	0.060	0.15
Gravity sewer pipes (dirty)	0.25	0.50	1.0

LARGE DIAMETER PIPES[14]

Steel and concrete cylinders are the usual choices. The steel pipe can be welded and coated with cement mortar or epoxy enamel. The outside can be wrapped with tapes of various formulations or continuous polyethylene sleeves. Much information is available to make the best possible use of steel pipe.[7,8,9]

Data on concrete cylinder pipe is also available from a number of manufacturers.[10,11] Concrete pipe is usually delivered in 16 foot lengths.

MEDIUM SIZED PIPES

In addition to the steel and concrete cylinder is cast iron, ductile iron, and asbestos cement. Cast iron pipe has the longest history for general waterworks distribution systems. It is reported that King Louis XIV of France in 1664 ordered 15 miles of cast iron pipe to supply water for the town of Versailles. Sections of this pipe are said to be still in use. Ductile iron pipe has become popular for waterworks use and has superior properties to the older cast iron. It is available in 18 foot lengths and can be cut in the field with an abrasive saw.

Asbestos cement pipe is available from 4-inch up to 36-inch diameter, in 13 foot lengths. It is readily jointed with rubber rings in bell and spigot pipe ends. One of the advantages of asbestos cement pipe, apart from its low cost, is its usual inertness to most corrosive waters and soils, but it will be attacked by free carbon dioxide (CO_2). It will not discolor the water, it maintains a very good friction coefficient, and it can be used for water pressures comparable to its alternatives.

SMALLER SIZED PIPES

All the other materials of construction, steel, cast iron, ductile iron, asbestos cement (AC), are all available in the smaller sizes. In addition there is aluminum, polyethylene, polyvinyl-chloride (PVC), and copper.

Aluminum pipe has been used in a number of instances for the transmission of water. One of the advantages of aluminum is its resistance to corrosion, particularly with aerated waters of low pH, but the presence of chlorides and other elements can reduce its resistance. It can be welded using inert gas or it can be joined with mechanical couplings. Care must be taken to exclude all traces of copper from the system. One installation suffered severe pitting within the first few years of operation because a copper heating coil had been installed at the intake pumphouse. Traces of copper from this coil were sufficient to cause serious corrosion. Much data and experience has been obtained by the companies manufacturing aluminum pipe.[12,13]

CATHODIC PROTECTION

Buried steel, cast and ductile iron pipe are subject to external corrosion. In soils where the electrical conductivity is high, the attack can be accentuated. Once certain areas become activated or anodic to other areas then current flow takes place through the soil to the pipe, through the pipe and back into the soil again, forming a completed loop. The transfer of electrons through the soil removes metal from one area and may deposit it in another.

Wrapping the pipe with a polyethylene coating, tape, or sheath, will prevent the passage of electrons and stifle electrolytic corrosion. Unfortunately however, these protective and insulated membranes are never 100% impervious, and if moisture can get to the pipe so can electrons. If there are two or more penetrations in reasonably close proximity to each other, then electrolytic cells will become established and active corrosion will follow. A concentrated attack of this type can be worse than if the whole pipe were exposed.

To overcome this problem a technique known as Cathodic Protection is used, consisting of applying an impressed current to the pipe relative to the ground. Isolated electrolytic cells cannot develop since the pipe becomes cathodic to its environment. This is done either with magnesium electrodes buried in the ground and electrically connected to the pipe or by applying a constant low voltage impressed current to the pipe.

The design of the cathodic protective system is the work of specialists in this field.[15] As soon as the pipeline route has been staked out, the corrosion engineer should be advised since it may be necessary for him to conduct soil resistivity surveys and investigate for possible stray currents that may orient themselves along the line when it is installed. There have been numerous problems of this nature particularly since high tension single wire electrical transmission has become popular. The return to the generator takes place through the ground and will follow lines of least resistance, which may cause havoc with buried steel, cast, and ductile iron pipes. When the resistivity survey is completed the use of certain classes of pipes may be unsuitable for that particular area.

REFERENCES

1. WILLIAMS, G. S. and A. HAZEN, *Hydraulic Tables*, 3rd Ed. (Revised). New York: John Wiley & Sons Inc., 1963.
2. "Handbook of Drainage and Construction Products." Published by Armco Drainage & Metal Products of Canada Ltd., 1958.
3. SKEAT, W. O. *Manual of British Water Engineering Practice*. Volume II, 1969. Published for the Institution of Water Engineers, London, by W. Heffer & Sons Ltd., Cambridge, England.

4. ACKERS, P. "Charts for the Hydraulic Design of Channels and Pipes," 3rd Ed. (Metric Units). Hydraulics Research Paper No. 2. Her Majesty's Stationery Office (HMSO) London, 1969.

5. ACKERS, P. "Tables for the Hydraulic Design of Storm-drains, Sewers and Pipelines." Ministry of Technology, Hydraulics Research Station; Hydraulics Research Paper No. 4. Published by HMSO 1963.

6. WALKER, R. *Pump Selection*. Ann Arbor, Michigan: Ann Arbor Science Publishing Inc., 1972.

7. "Steel Pipe Design and Installations." *AWWA*, Manual No. 11 published by the American Water Works Association, Inc., 1964.

8. "Welded Steel Water Pipe Manual." Published by Steel Plate Fabricators Association Inc., Hinsdale, Illinois, USA.

9. "Piping Design and Engineering." Grinnell Company Inc., Providence Rhode Island.

10. Trade Literature—"Hyprescon Concrete Pressure Pipe." Canada Iron Foundries Limited, Peel Street, Montreal, P.Q., Canada.

11. DODD, D. N. "Concrete Pipe for Water Distribution." Presented at the AWWA Annual Convention Las Vegas—Engineering and construction committee, May, 1973.

12. "Aluminum Pipe for Fluid Flow." Published by Aluminum Company of Canada Limited, Montreal, Canada. 1962.

13. Trade literature from Alcan International Ltd., Montreal, Canada. Entitled "Sales Development Bulletin No. 12—A Series of Reprints on Aluminum Pipelines." 1964.

14. The Journal of the AWWA, Vol. 64, pages 410–466, July 1972, are devoted to the subject of "Pipe for Water Supply."

15. PARKER, M. E. *Pipeline Corrosion and Cathodic Protection*, 2nd Ed., Houston, Texas: Gulf Publishing Company, 1962.

26

WATER HAMMER

Water hammer is a term used to describe transient pressure surges in pipelines. The surges may not always be accompanied by audible noises but can still be of sufficient magnitude to result in pipe failure.

WATER HAMMER

Water hammer occurs when there are rapid changes in the velocity of the liquid flowing in the pipeline. The following concepts are basic.

1. When water is flowing in a pipeline and has its forward velocity suddenly reduced or stopped by the closing of a valve, the pressure in the pipeline upstream of the valve will increase due to the energy in the moving column of water being brought to a stop or a change in velocity. The pressure on the downstream side of the valve will be reduced by an equal amount, but since the lowest pressure in a pipeline cannot be less than full vacuum the reduced pressure will therefore be between a partial and a full vacuum depending on the rate of change in velocity.

2. The pressure increase on the upstream side of the valve caused by the change in the flow is proportional to the change in velocity and the period of time in which this change occurs. Obviously if the change in flow is very gradual, then the increase in pressure will also be very gradual and of low magnitude. But if the change is rapid the increase in pressure will also

be rapid and can be 100 feet of head for every foot per second change in velocity. (Use a rule of thumb of 50 psi (115 ft head) per foot per second change in velocity.) For example, if the pressure in a water main was 100 psi and the water flowing at 5 feet per second was suddenly brought to a halt, the increase in pressure would be approximately 250 psi plus the 100 psi steady-state condition making a total pressure of 350 psi.

3. On the downstream side of this valve the negative pressure would theoretically be a decrease of 250 psi. Since the pressure was originally 100 psi, the theoretical negative pressure would be a negative 150 psi. It is impossible to have such a condition, therefore the pipe would be reduced to full vacuum.

4. On the upstream side of the valve, when the pressure increased to 350 psi, the pipe wall stretched in circumference. If the induced stress is less than the elastic limit or the lower yield point of the pipe material the pipe will expand. There will be an increase in volume in the enlarged stretched section. The energy originally manifested in the moving water column has now been transferred into the increase in pipe wall stress, and the resulting strain will increase the pipe diameter. Once the column of water is brought to rest the water pressure returns to the steady-state condition of 100 psi. But the pipe walls are still extended and under strain, but within the elastic limit. The pipe then contracts and closes in onto the column of water, thus causing a reversed wave back along the pipe, opposite in direction to the original forward flow. This pressure surge moves backwards and forwards along the pipeline until the strain energy and friction losses dissipate the energy change. In large pipelines the surges may continue for some time, each wave diminishing in magnitude until all the energy of the original shock wave has disappeared.

5. What happens on the downstream side of the valve when the pressure is reduced to a full vacuum? The pipe tends to collapse inward due to the external pressure on the outside. If the pipe is small in diameter and has thick rigid walls, then the pipe will probably withstand the collapsing pressure. If the pipe is large in diameter, thin walled, and unable to withstand a full vacuum, the pipe walls will collapse inward.

6. What happens to the water column on the downstream side of the valve under these reduced pressure conditions? The water column tends to continue down the pipeline due to its momentum. If the pipe is able to withstand the reduced pressure conditions without collapsing then the water column will separate, rather like the mercury thread in a broken thermometer. If the water is warm it will tend to boil momentarily at the interface. When the moving segments of the separated column rejoin each other, the pressure at impact may be greatly in excess of the 50 psi per foot per second change in velocity, due to the force of impact of the two portions of separated water column.

The magnitude of the shock wave is directly proportional to the change in liquid velocity according to Allievi's water hammer equation.[1]

ALLIEVI'S EQUATION

$$H_1 - H_0 = \frac{a}{g}(v_0 - v_1)$$

Where:

H_1 = new head in feet

H_0 = steady state head in feet at velocity V_0

a = velocity of the shock wave in feet per second

g = 32.2 fps/sec

v_0 = is the initial steady state velocity in feet per second

v_1 = new velocity in feet per second

The wave velocity a is calculated by the following equation:

$$a = \frac{12}{\sqrt{\frac{w}{g}\left(\frac{1}{k} + \frac{d}{Ee}\right)}}$$

Where:

a = wave velocity in feet per second

w = weight of fluid (lb/cu ft)

g = 32.2 fps/sec

k = bulk modulus of compressibility of liquid (psi)
(for water k = 294,000 psi)

E = Young's Modulus of elasticity for pipe wall material
(for steel) E = 29,400,000 psi
(for cast iron) E = 15,000,000 psi
(for asbestos cement) E = 3,400,000 psi

d = inside diameter of pipe (ins)

e = thickness of pipe (ins)

Wave velocity a for various diameters and wall thicknesses of pipe is graphically given in reference.[1] For steel pipe the value of a ranges from 3500 to 4660 fps. It will be seen from Allievi's water hammer equation that if a value of 3700 fps is taken for the velocity a, then the increase in pressure for each foot per second change in flow velocity is approximately equal to 115 ft head or 50 psi, i.e.,

$$H_1 - H_0 = \frac{3700}{32.2}(v_1 - v_0)$$

$$\Delta H = 115 \text{ ft } (v_1 - v_0)$$

It should be noted that the pressures due to water hammer increase with a decrease in the $\frac{d}{e}$ ratio. That is to say, the pressure increases with the increase in wall thickness. The maximum shock wave pressure, in accordance with Allievi's equation, only occurs if the change in the flow velocity takes place in less than the time required for the shock wave to travel from one end of the pipe to the other and reflect back again. This is known as:

$$\text{Reflection time} = \frac{2L}{a} \text{ seconds}$$

Where:

L = length of pipe in feet

a = velocity of the shock wave in feet per second

If, for example, water is flowing in a 10,000 foot long steel pipe at 5 fps and the wave velocity for the pipe is 3700 fps, then the time required for the shock wave to travel the length of the pipe and reflect back again is:

$$\frac{2L}{a} = \frac{2 \times 10,000}{3700} = 5.4 \text{ seconds}$$

Now if the initial steady state velocity of 5 fps is reduced to 0 in 5.4 seconds or less then the magnitude of the shock wave developed is approximately $50 \times 5 = 250$ psi over and above the static head. Once the positive shock wave passes through the pipe line, a negative shock wave develops in exactly the same way and is theoretically of equal but of opposite magnitude.

This sequence of events is illustrated in Figure 26-1, Water Hammer Conditions in Pipeline. For simplification, the pipe friction under the initial steady-state conditions has been ignored.

The conditions in the penstock in Figure 26-1(a) show the steady-state condition with water flowing from the reservoirs A to the valve B at the velocity of V_0 fps. Pipe friction is ignored and the Hydraulic Grade Line (HGL) is parallel to the pipeline and equal to the level in the reservoir.

In Figure 26-1(b) the valve at B is closed in less than $\frac{2L}{a}$ seconds. The water is still moving along the pipeline at velocity V_0 with the result that the pressure in the line is increased in accordance with Allievi's water hammer equation.

The diameter of the pipe is increased as shown by the dotted lines. The front of the shock wave moves towards the reservoir at the wave velocity a which we will assume is approximately 3700 fps and as the wave confronts the forward flow, the pipewall continues to stretch until the velocity of flow in the line is reduced to 0 as shown in Figure 26-1(d) and the time taken for this condition to occur is $\frac{L}{a}$ seconds. For the 10,000 foot pipeline this would be approximately 2.7 seconds.

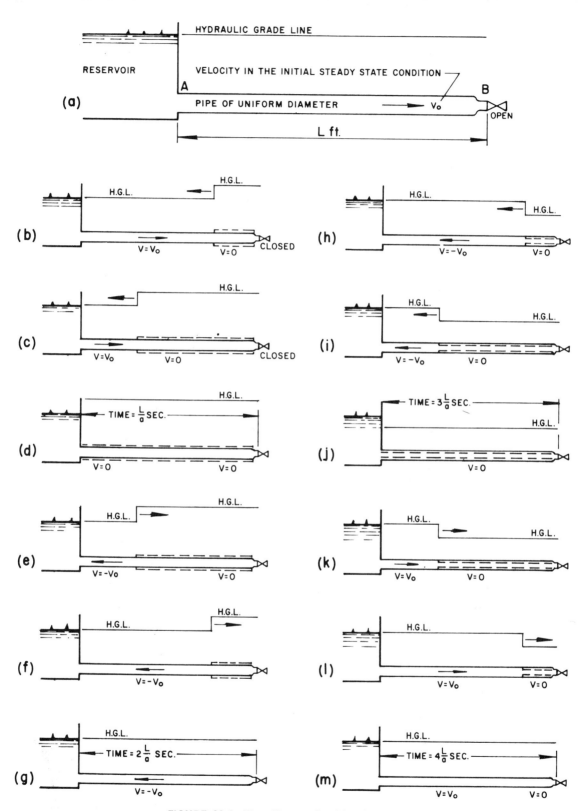

FIGURE 26-1 Water Hammer Conditions in a Pipeline.

All the kinetic energy of the original steady-state of flow in the forward direction is now converted into strain energy in the pipe wall. The pipe then proceeds to contract to its normal diameter. This strain energy within the pipe walls causes the extra volume of water contained in the pipe to be rejected back into the reservoir. Assuming that there are no energy losses in converting from kinetic energy again, the velocity of flow in the pipeline will be restored to (V_0 fps) but in the opposite direction ($-V_0$ fps).

This situation will continue until the original HGL is restored at the valve as illustrated by Figure 26-1(g). The pressure at B is now the same as at A and as it was under the initial steady-state conditions, however, the water continues to flow with ($-V_0$ fps) velocity from the pipeline into the reservoir. The time is now $\frac{2L}{a}$ seconds.

This will cause a negative pressure in the pipeline as illustrated by Figure 26-1(i) and the pipe will contract to less than its original diameter by the atmospheric pressure exerted on the outside of the pipe. If it is a large pipe with relatively thin walls, it may collapse.

Once the negative wave front has reached the reservoir the velocity in the pipeline is again at 0 and the time is $\frac{3L}{a}$ seconds. The head of water in the reservoir exerts a restoring effect on the pipeline which proceeds to return to the diameter it was under steady-state conditions with the water flowing in the forward direction at V_0 fps. This will continue until the forward velocity reaches the closed valve B in $\frac{4L}{a}$ seconds. The forward velocity results in an increase in pressure as per Figure 26-1(g) and the whole cycle is repeated. In our ideal conditions the shock wave would continue in perpetual motion indefinitely; however, in actual practice the friction in the pipe and the heat generated by strain energy dissipate the surge and a new steady-state condition prevails.

GRAVITY SYSTEMS

A typical example of the problems that can occur is illustrated in Figure 26-2. The water from the dam with an overflow spillway at an elevation of 820 feet flows through approximately 13 miles of pipeline down to the reservoir some 440 feet lower in elevation. The line is 30 inches in diameter for most of the route but reduces to 22-inch and 18-inch as the slope increases. With the motorized valve wide open the line will carry approximately 15 mgd (U.K.) (18 mgd U.S.).

The flow velocities at 15 mgd (U.K.) are as follows:

30-inch diameter	6.0 fps
22-inch diameter	11.0 fps
18-inch diameter	15.0 fps

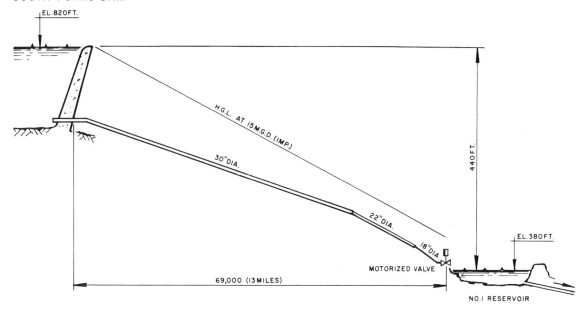

FIGURE 26-2 Gravity Water Supply.

If the motorized valve was closed from fully open to fully closed in less than $\frac{2L}{a}$ seconds which in this case would be:

$$\text{Reflection time} = \frac{2L}{a} = \frac{2 \times 69{,}000}{3700} = 37.3 \text{ seconds}$$

the shock wave in the 18-inch diameter pipe would be approximately $115 \times 15 = 1725$ feet above the elevation of the valve which is equivalent to approximately 750 psi. But this is not all. The normal hydrostatic pressure at the motorized valve is 440 feet or 190 psi when the valve is closed and there is no flow. The shock wave pressure is in addition to this so that the total pressure at the motorized valve is then equal to $1725 + 440 = 2165$ feet or 940 psi. It does not require much imagination to envisage what the end result would be with the pipeline, valve, and fittings designed for 250 psi service.

PUMPING STATIONS

Another type of problem exists where there is a pump at one end of the line and a valve at the other. See Figure 26-3. When the pump is running, the conditions in the pipeline are similar to the previous examples. Under controlled conditions shock waves can be kept to a minimum by the use of slow operating valves,

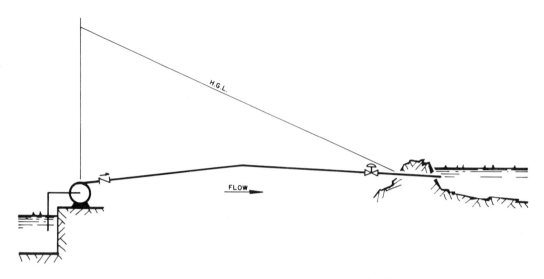

FIGURE 26-3 Pumped Water Supply.

variable speed pumps, etc. However, unscheduled stoppage during power failure at the pumphouse is a different story. The condition in the pipeline, immediately after the pump stops due to power failure, is one of reduced pressure. This reduced pressure causes the pipe to contract due to external pressure and the column of water proceeds to separate, probably at several locations along the line where there are high spots.

The empty spaces due to column separation are approximately at full vacuum. If the forward velocity was initially 5 fps this drops to 0. The pipe is fully contracted along its entire length and the water column has several separations. The external pressure may cause the pipe to collapse inwards particularly in large diameter thin wall sections.

Once the negative shock wave has reached the reservoir in $\frac{L}{a}$ seconds and the forward flow velocity is 0, the water in the reservoir starts to flow back through the valve into the pipeline, in the opposite direction, at its former steady-state flow of 5 fps. The separated water columns close together with considerable impact as the empty pockets are filled. The return shock wave travels back along the pipe again with the velocity of sound and reaches the pump in $\frac{2L}{a}$ seconds after the power failure.

Several methods of water hammer protection are suggested for this type of installation but only two have any practical significance. The various suggestions are as follows:

(a) Flywheel at the pump coupling.

(b) Vacuum and pressure relief valves.

(c) Air inlet valves along the line.

(d) Surge buffer tank with air inlet.

(e) Hydropneumatic tank.

Flywheel at the Pump Coupling

This would be an ideal solution if a reasonably sized flywheel would keep the pump in operation long enough to gradually reduce the flow to 0. It must be remembered however that any change in pipe velocity in less than $\frac{2L}{a}$ seconds is regarded as an instantaneous change.

Supposing the pump operates at 1750 rpm, it is possible that the developed head at 1200 rpm is equal to 0 flow on the system head curve. It follows, therefore, that if the flywheel is to be effective it must maintain the pump rotation between 1750 and 1200 rpm for as many $\frac{2L}{a}$ seconds as necessary to reduce the shock wave to reasonable proportions. It will also be observed that the longer the line the longer the $\frac{2L}{a}$ time period becomes, and therefore the size of the flywheel must be proportional to the length of the line. It soon becomes obvious that a flywheel may be a possible solution for a very short pipeline but it is impractical for a long one.

Vacuum and Pressure Relief Valve

A vacuum and pressure relief valve might be appropriate but the speed at which a shock wave travels and the fact that separation can occur at any point on the line tends to limit the validity of this suggestion.

Air Vent Valves Along the Pipeline

This helps the situation but it is not the complete answer.

Surge Buffer Tank with an Air Inlet and Relief Valve[2]

This suggestion is a practical possibility and consists of a small enclosed pressure tank connected to the pipeline at the pump end and possibly other tanks along the pipeline as required. A vacuum release valve is sometimes fitted to the top of the tank to draw air into the vessel as the water leaves the tank to fill the line. When the positive shock wave returns, the air is vented out of the tank and the tank is refilled with water from the pipe ready for the next negative surge. This system of water hammer control is used to prevent column separation; however, it has limited capacity and is only suitable for relatively small systems.

Hydropneumatic Tanks

These tanks are located at the pump end of the line and serve a similar function to the Surge Buffer Tanks only they have more capacity. A good example of the use of hydropneumatic tanks is the Lake Huron to London Aqueduct.[3] This system, consisting of 30 miles of 48-inch diameter pipe, is protected by two 12′–6″ diameter × 90′–0″ long pressure tanks.

The size of the pressurized surge tanks is usually between 0.5 and 1.0% of the volume of the pipeline. For short high velocity lines 2 to 5% of the pipeline volume may be needed. For very long lines some other device may be more economical. In some cases no surge tanks are required if the pumps have flooded suctions and adequate net positive suction head (NPSH) so that the water can continue to be drawn through the pump and its check valve by the inertia of the moving water column; the pumps will continue to slowly revolve and allow sufficient water into the system to reduce the problems of column separation.

REFERENCES

1. "Steel Pipe Design and Installation." *AWWA*, Manual No. 11 published by the American Waterworks Association, Inc., 1964.

2. PARMAKIAN, JOHN *Water Hammer Analysis*. New York: Dover Publications Inc., 1963.

3. PATTERSON, A. B., F. C. HOPPER, and P. S. CHISHOLM, "Water Hammer Protection, Lake Huron to London Aqueduct." Engineering Journal, page 24, August 1966. Published by the Engineering Institute of Canada, Montreal.

27

RESERVOIRS

City reservoirs serve three principal functions:

(a) equalization between average day and peak hourly demand
(b) fire protection reserve, and
(c) emergency storage.

The total capacity of a reservoir to fulfill these three requirements can be estimated separately for each of the three functions and totalled together. Water stored for fire protection purposes must not be used to supply the peak hour demand. And, water required to maintain a service to the community in the event of an emergency when a supply pump has failed, or a section of the supply main is temporarily out of service, should obviously not be used to meet the normal day to day requirements. Methods of determining the individual requirements of each component for a particular community can be calculated.

Equalizing Storage

Figure 1-4 shows a typical hour by hour demand fluctuation for most domestic communities. The maximum day demand may be anywhere from 2.0 to more than 6 times the average daily flow. It is customary to provide a supply to the reservoir, sufficient to meet the maximum 24 hour demand and to rely on the equalizing storage capacity of the reservoir to bridge the gap between maximum day and peak hour. Several methods of calculating the amount of equalizing storage are described in the textbooks.[1,2,3] The equalizing storage supplies the difference between the normal domestic and industrial peak demand over and above the rate of reservoir filling. The size of this storage can be calculated from reference to graphs such as Figure 1-5, plus experience factors depending on the size of the community, whether there is an industrial demand such as packing houses, food factories, whether or not domestic services are metered, and the summer climatic conditions where there may be an appreciable lawn sprinkling demand.

Fire Protection Reserve

The fire protection reserve is always in addition to any other storage requirement, almost as if a reservoir should have two or more outlet connections at various reservoir levels with the domestic demand outlet say 15 feet above the bottom of the tank. When this outlet runs dry there would still be 15 feet of water left in the bottom for fire protection. Municipal water reservoirs are not designed in this way, but some are operated close to this concept by reducing the pressure in the distribution system when the level in the reservoir drops to the minimum required for fire protection. But this is a dangerous practice, because some parts of the distribution system may be under reduced pressure and subjected to backsiphonage.

The volume of water reserved for fire protection is based on the local fire underwriters' recommendations. This is related to the size of the buildings, the number and area of the floors, rather than the older system of population. It is a more logical approach since it is the buildings that burn, hopefully, not the people. Therefore the fire risk should be based on inflammable areas. It is also emphasized that providing sufficient water can be got to the scene of the fire quickly enough, less reservoir storage will be required.

Emergency Reserve

The magnitude of this component depends:

(a) upon the possible frequency and period of interruption of the service water supply into the reservoir, and

(b) the time necessary to repair any part of the supply system most likely to fail.

These factors are impossible to completely evaluate and it is usual to provide between 33 and 50% standby storage capacity for use in the event of supply failure. Some utilities have dual power supplies to their pump stations from alternative power sources, but few systems have duplicate water supply mains. The National Board of Fire Underwriters bases their rating schemes on 5 days storage at maximum flow as being ideal, but it is unlikely that many growing communities can afford storage facilities of this magnitude.

TYPE OF RESERVOIR

All service distribution system reservoirs should be covered. Every precaution should be taken to prevent the ingress of birds, small animals, insects, human beings, and the like. It is not uncommon, particularly in colder climates, to cover the concrete top and sides with 2–3 feet of soil and plant grass to provide some heat insulation and to blend the structure into the surrounding landscape and camouflage its appearance. In other cases architecturally designed tennis courts, ornamental gardens, and even parking lots are placed over the reservoirs to utilize the area. Since they are usually located on higher ground, they can make good "lookout locations" for the surrounding countryside. The water engineer must be careful to insure that the roof drains are encased in concrete or designed in such a way that a leaking roof leader will not discharge into the reservoir. The outlet ends of the overflow pipes must not become plugged with debris or overgrown, and must be designed to prevent them from being used as access tunnels for various types of rodents.

Concrete is the preferred material of construction for large ground level reservoirs either in rectangular or cylindrical designs. Steel tanks should also be evaluated, but they require periodic painting both internally and externally, more maintenance and possibly cathodic protection, whereas once a concrete reservoir is completed, it requires little maintenance apart from cleaning.

Villages and small towns have installed elevated wood-stave tanks, but they often leak, and produce a dangerous array of icicles in the winter. The wooden staves sometimes rot if the outsides are painted with waterproof enamels. They are also reported to show high bacterial counts, presumably from rotting wood, and polyethylene liners have been installed in several instances to overcome these difficulties.

Steel tanks suffer from corrosion problems. The conditions inside an elevated storage tank are severe and vary widely from season to season. The temperatures under the steel roof during hot summer days can reach 40°C (104°F) or more. Moist air is in contact with the roof, and the walls are alternately wet and dry as the tank fills and draws, so that corrosion at the "wind and water" line is inevitable. Layers of ice form around the sides which act as a heat insulator, but it is necessary to keep the level fluctuating during the winter months otherwise a very thick layer of ice will form on the surface and may arch across, fall, and damage the tank if the water is drained. An ice plug can form across the surface making a hermetic seal. If the tank is 34 feet above ground level, the tank and stand pipe can be subjected to a full vacuum if the water in the tank is frozen onto the surface and then drawn down during an emergency. In cold climates where temperatures may be substantially below freezing for long periods in the winter, a water heating system should be installed to prevent this occurring.

Elevated reservoirs are an expensive form of storage, and it is unusual to find one that is big enough to hold sufficient water to handle a fire without continuously pumping into it from ground storage. They are also targets for vandalism.

With good pumps, quick starting diesels, it is preferable to install large ground level reservoirs rather than small, inadequate, and costly elevated storage tanks. The pumps can be variable speed to provide a constant pressure in the distribution system for any variation in demand.

VALVING The filling and withdrawal for small reservoirs are usually incorporated in a one pipe system but since they serve two separate functions they should be designed accordingly.

One system is shown in Figure 27-1. In this case the check valve is 18 inches in diameter for outflow only; it will pass 6,000 U.S. gpm with a velocity of 7.5 feet per second, and a head loss as low as 0.15 feet. The altitude valve is 8-inch diameter and capable of filling only at 1500 U.S. gpm with 9 feet head loss. If one combined altitude valve were installed to serve both filling and emptying, it would be totally inefficient at a 1500 U.S. gpm filling rate and have a considerably higher head loss when emptying at 6000 U.S. gpm than the simple check valve, and the capital cost would be higher.

CONTROLS Wherever possible, two independent level controls should be provided. If not, then at least there should be an adequate level indicator or recorder with a readout at a central point in addition to an altitude valve. In other cases a pressure gauge calibrated in feet connected into the side of the tank will serve the purpose. Signals from a pressure transmitter attached to the gauge can be carried in underground cables or by rented telephone wires to the pump station.

DESIGN OF RESERVOIR Textbooks on water engineering have many good chapters on reservoir design.[1,2,3,4,5] There are a number of AWWA Standards and published papers on the design, construction, and operation of reservoirs.[6,7,8,9]

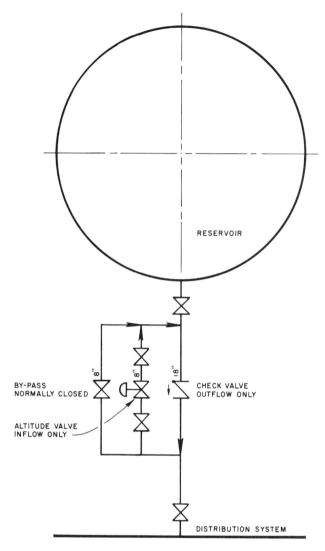

FIGURE 27-1 Typical Valving Arrangement.

REFERENCES

1. FAIR, G. M., J. C. GEYER, and D. A. OKUN, *Water and Wastewater Engineering, Volume 1, Water Supply and Wastewater Removal.* New York: John Wiley & Sons Inc., 1966.
2. CLARK, J. W., W. VIESSMAN, and M. J. HAMMER, *Water Supply and Pollution Control*, 2nd Ed. New York: International Textbook Company, 1971.

3. Babbitt, H. E., J. J. Dolland, and J. L. Cleasby, *Water Supply Engineering*, 6th Ed., New York: McGraw-Hill Book Company, 1962.

4. Skeat, W. O. *Manual of British Water Engineering Practice.* Volume II, (1969). Published for the Institution of Water Engineers, London, by W. Heffer & Sons Ltd., Cambridge, England.

5. Manning, G. P. *Concrete Reservoir and Tanks.* Concrete Publications Limited, 60 Buckingham Gate, London, SW1, 1967.

6. "AWWA Standard for Welded Steel Elevated Tanks, Stand Pipes, and Reservoirs for Water Storage." *AWWA* D100–73. Published by the American Water Works Association, 1973.

7. "*AWWA* Standard for Inspection and Repairing Steel Tanks, Stand Pipes Reservoirs and Elevated Tanks for Water Storage." *AWWA* D101–53. Published by the American Water Works Association, 1953.

8. "*AWWA* Standard for Painting and Repainting Steel Tanks, Stand Pipes Reservoirs and Elevated Tanks for Water Storage." *AWWA* D102–64. Published by the American Water Works Association, 1964.

9. Crowley, F. X. "Prestressed Concrete Tanks—Maintenance Problems and Their Solution," Annual AWWA Conference, Las Vegas 1973.

28

HYDROPNEUMATIC TANKS

For small distribution systems including industrial facilities, a hydropneumatic tank offers many advantages. It is capable of supplying water to the system in a wide range of flow rates at almost constant pressure for short duration peak flows many times greater than the pump capacity. There is, however, a limitation to the size of distribution systems that can be economically serviced by a hydropneumatic tank system and beyond that limit variable-speed driven pumps are more economical with or without hydropneumatic tank storage capacity.

The hydropneumatic tank system consists of one or more pumps, a pressure vessel to contain both water and air, an air compressor, and an air-water level control. The pumps must be designed with sufficient capacity to supply the peak hourly demand and the tank volume provides the short duration demands. Relatively small distribution systems with an average consumption of approximately 100 U.S. gpm would have an anticipated peak hourly demand of at least 350 U.S. gpm with short duration domestic demands exceeding 500 U.S. gpm. If the distribution pressure is 40 psi, then the hydropneumatic tank would operate between 45 and 65 psi with an average tank pressure of 55 psi. It will be appreciated that the overall efficiency of a hydropneumatic tank system becomes less attractive the larger the difference between the average tank pressure and the distribution system pressure, since the pumps will operate up to 65 psi in order to maintain a constant 40 psi in the distribution system.

HYDROPNEU-MATIC TANK SYSTEMS

To avoid pressure fluctuations in the system, it is customary to install pressure reducing valves between the hydropneumatic tank and the distribution system. For example, if the hydropneumatic tank is designed to operate between 45 and 65 psi, it may be convenient to install a pressure reducing valve set to 40 psi immediately downstream of the tank to maintain a constant distribution system pressure. It will be appreciated that the supply pumps will have to discharge against an average $\frac{65 + 45}{2} = 55$ psi in order to supply the system at 40 psi. This difference in pressures may represent a substantial increase in power cost over a variable speed pump, particularly in the larger horsepower installations and both schemes should be evaluated. If fire protection is required, a diesel engine-driven pump should be provided.

HYDROPNEUMATIC TANK CAPACITY[1]

The quality of water available for servicing the distribution system when the water supply pumps are stopped is governed by the volume of air inside the tank which expands from the high pressure (i.e., pump cut-out pressure P_1) to the low pressure (i.e., pump cut-in pressure P_2). Where P is in terms of absolute pressure (psi a) = (psi + 14.7 psi). It is assumed that both the air and water temperatures are constant and that Boyles' Law applies. However, the calculation for determining the quality of water that can be withdrawn from a tank when the pressure drops from (P_1) to (P_2) can be simplified if (V_1) is used to express the volume of

TABLE 28-1 Shows the Relationship between Pressures and Available Volume

Case	(a)	(b)	(c)	(d)	(e)	(f)	(g)	(h)
Pump cut-out P_1 (psi)	40	50	60	70	80	80	90	100
Pump cut-in P_2 (psi)	20	30	40	50	50	60	60	70
Average pressure (psi)	30	40	50	60	65	70	75	85
Ratio $= \frac{P_1 + 14.7}{P_2 + 14.7}$	1.58	1.45	1.37	1.31	1.46	1.27	1.40	1.35
Available water volume in percent of tank capacity when $V_1 = 50\%$ of tank capacity at cut-out P_1	29%	22.5%	18.5%	15.5%	23%	13.5%	20%	17.5%
Available water volume in percent of tank capacity when $V_1 = 40\%$ of tank capacity at cut-out P_1	35%	27%	22.2%	18.6%	27.6%	16.2%	24%	21%

water in the tank at the high pressure (P_1) and likewise if (V_2) is used to express the volume of water in the tank at the lower pressure (P_2). This is a deviation from the usual form of the Boyles' Law equation.

The graph of Figure 28-1 and Table 28-1 show the relationship between tank pressures and available water capacity. It will be noticed that the largest water volume occurs when the tank is filled with water up to 40% of its total volume, and when the ratio of pump cut-out and cut-in pressures are the highest. There are many other computations, but for small municipal systems 40–50% of the tank volume is frequently used.

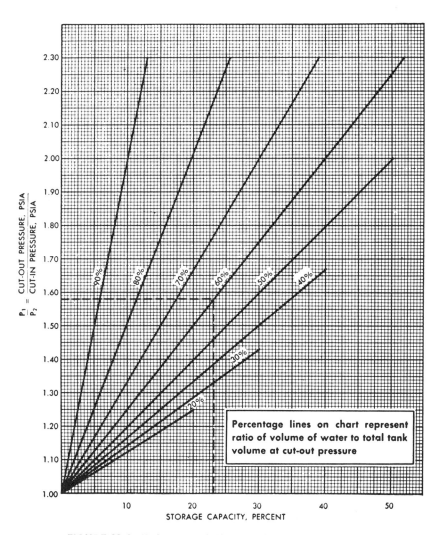

FIGURE 28-1 Hydropneumatic Tanks—Relationship Between Pressure Ranges and Storage Capacity.

To determine the amount of water that can be withdrawn from a tank when the pressure drops from P_1 to P_2 psia use the following equation.

$$V_1 - V_2 = \text{water withdrawn or storage capacity of tank,}$$

$$= \left(\frac{P_1}{P_2} - 1\right)(100 - V_1)$$

In this equation P_1 and P_2 must be expressed in psia—pounds per square inch absolute. V_1 and V_2 are expressed in percent, (see Figure 28-2).

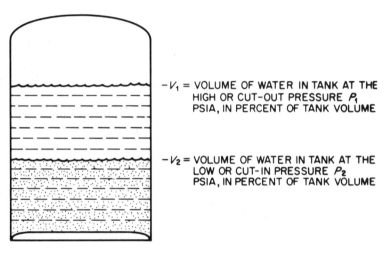

FIGURE 28-2 Hydropneumatic Tank[1].

Example: In a 1000 gal tank the gauge pressure at the cut-out point is 40 psi and the tank is 60% full of water. The cut-in pressure is 20 psi. What is the storage capacity of the tank?

$$\frac{P_1}{P_2} = \frac{40 + 14.7}{20 + 14.7} = \frac{54.7}{34.7} = 1.58$$

$$\text{storage capacity} = (1.58 - 1)(100 - 60) = 23.2\%$$

Therefore in the 1000 gal tank the storage capacity = 1000 × 0.232 = 232 gal.

The storage capacity for tanks in percent can be read directly from the chart Figure 28-1.

PUMP CHARACTERISTICS There are no limits to the number of pumps that can be connected together in parallel to supply a hydropneumatic tank system, but usually two pumps are provided. In the previous example, where the cut-in pressure was 45 psi and the cut-out pressure was 65 psi, it would be preferable to select a pump with the maximum efficiency at 55 to 60 psi. It should be capable of supplying the peak hourly demand with at least a 20% reserve to insure some margin of excess capacity for quick recovery. With hydropneumatic tank systems the pumps start and stop fairly frequently, and the smaller the tank in relationship to the average demand, the more frequently will the pumps operate. The maximum number of "starts and stops" occurs when the system demand is equivalent to half the pump capacity. If the tank is relatively small the pumps may only operate for 5 minutes, stop, and then restart 5 minutes later. This would be equivalent to 6 starts per hour. If the water demand exceeds 50% of the pump capacity, the pump will operate longer and will have shorter rest periods. There is a limit to the number of times an electric motor can be started in any 1 hour and this restriction must be considered when designing a hydropneumatic tank system. If the problem is presented to a motor manufacturer he will recommend the most suitable motor for the job.

SIZE OF HYDROPNEUMATIC TANK

A simple formula has been compiled to encompass the variables discussed as follows:

$$\text{Tank capacity in U.S. gallons} = \frac{\frac{\text{Pump capacity in U.S. gallons}}{2} \times \frac{60}{2 \times \text{number of starts per hour}}}{\text{Percentage available tank capacity}}$$

This can be simplified to:

$$\text{Tank capacity in U.S. gallons} = \frac{(15)(Q)}{(N)(\% \text{ available capacity})}$$

Where:

Q = pump capacity in U.S. gpm
N = number of starts per hour
% available capacity = see Table 28-1

The simplest way of demonstrating the significance of the various factors involved is by example.

EXAMPLE:

Design a hydropneumatic tank system for a small community capable of providing a peak hourly flow of 200 U.S. gpm at not less than 35 psi at the hydropneumatic tank.

The pump should be capable of at least 20% more than the peak hourly flow 200 × 1.20 = 240 U.S. gpm. To insure that the pressure in the distribution system will not drop below 35 psi at the tank, select the 40–60 psi range of pressure. Check available pump curves to insure that a suitable pump can be obtained capable of 240 U.S. gpm at 50 to 55 psi (115–127 ft) with reasonable efficiency. Insure that it has at least 80% of its normal operating capacity at the cut-out pressure of 60 psi (139 ft). Assuming that the efficiency of the pump at 240 U.S. gpm and 115 ft would be not less than 75%, the horsepower would be:

$$HP = \frac{Q \times H}{33,000 \times E}$$

$$HP = \frac{240 \times 8.33 \times 115}{33,000 \times 0.75} = 9.3 \text{ (say 10 hp)}$$

Assuming that a drip proof motor enclosure is satisfactory for the conditions of the environment and that it will be able to operate satisfactorily at six starts per hour:

$$\text{Pressure ratio } \frac{P_1}{P_2} = \frac{60 + 14.7}{40 + 14.7} = \frac{74.7}{54.7} = 1.37$$

See Graph Figure 28-1 and Table 28-1 which indicate that under case (c), a tank operating within the required pressure ranges of 40 to 60 psi will have a pressure ratio of 1.37 and would provide 18.5% available tank capacity if filled with water to 50% of the tank volume at the cut-out pressure (P_1) of 60 psi; or, 22.2% available tank capacity if filled to 40% of the tank volume at cut-out pressure.

(a) Tank Capacity (Filled to 50% of Tank Volume)

$$\text{Tank capacity (U.S. gallons)} = \frac{(15)(Q)}{(N)(\% \text{ available capacity})}$$

Where:

$Q = 240$ U.S. gpm

$N = 6$

% = available capacity for 50% tank volume = 18.5%

$$\text{Tank capacity (U.S. gallons)} = \frac{(15)(240)}{(6)(0.185)} = 3250 \text{ U.S. gallons}$$

(b) Tank Capacity (Filled to 40% of Tank Volume)

Where:

$Q = 240$ U.S. gpm

$N = 6$

$\% =$ available capacity for 40% tank volume $= 22.2\%$

$$\text{Tank capacity (U.S. gallons)} = \frac{(15)(240)}{(6)(0.222)} = 2700 \text{ U.S. gallons}$$

If, on further investigation, it is discovered that only 4 starts per hour can be tolerated for continuous duty then the 2700 gallon tank (in solution (b)) would have to be increased to 4100 U.S. gallons.

In practice two service pumps would be installed, arranged to alternate, thus reducing by half the number of starts per pump per hour. There would also be a second contact on the pressure switch which would start the second pump in parallel with the first if the system demand caused the pressure in the hydropneumatic tank to fall below 40 psi with one pump running. A third identical pump would be installed to work in conjunction with the other two and provide one spare pump so as not to disrupt the sequence of operations when one pump is out of service for repairs and maintenance.

However, if the hydropneumatic tank is supplied from a single well pump, then the capacity of the tank must be determined with respect to the permissible number of motor starts per hour.

HYDROPNEUMATIC TANKS AS PRESSURE VESSELS

Hydropneumatic tanks are unfired pressure vessels containing an expandable fluid (air) under pressure and must be designed and constructed in accordance with the requirements of the American Society of Mechanical Engineers (ASME), Boiler and Pressure Vessel Code Section VIII entitled *Pressure Vessels Division 1*, (latest edition), or equivalent. In most countries it is mandatory for all pressure vessels to be registered with the appropriate authority. The vessel is then stamped with its registration number and a certificate is issued describing the duties it is authorized to perform. To insure that these conditions are complied with, the following procedure is adopted.

(a) All drawings and design information concerning the pressure vessels should be submitted to the local authority (in Canada it would be the Boilers Branch of the Provincial Government) for approval before fabrication is commenced. This would include plate thickness, heads, fittings, openings, material specifications, type of weld, and position of the weld.

(b) When the appropriate authority is satisfied that the proposed tank will meet with their requirements they will issue a registration number. In Canada it is known as a Canadian Registration Number (CRN). The fabricator may then proceed with the construction of the tank.

(c) The local boiler inspector may wish to inspect the tank during construction, but will certainly wish to inspect it on completion and will usually witness the pressure test which must be 50% above the working pressure or 50% above the design pressure, whichever is greater. If the tank is satisfactory, the inspector will issue a serial number which is stamped on the vessel.

(d) In due course a Pressure Vessel Certificate is issued by the Boilers Branch, or equivalent authority, and sent to the fabricator, who in turn sends it to the owner. This document is important and should be retained as evidence that the pressure vessel meets the pressure design codes.

WATER LEVEL CONTROL SYSTEMS

For a hydropneumatic tank to operate satisfactorily the air-water ratio must be maintained between reasonably close tolerances. The tank should be equipped with a sight glass, which covers almost its capacity. If it is a horizontal cylindrical tank the gauge glass should be mounted vertically on one end to include approximately 80% of its vertical diameter. If it is a vertical tank the gauge glass should provide visual inspection facilities for 80% of the vertical side wall depth.

Most hydropneumatic tanks either lose air to the distribution system if the water supplied to the tank is below air saturation, or they will gain air if they are supplied from a deep-well vertical turbine pump that emits a pump column full of air each time it starts. Some well waters have excess free carbon dioxide which may be liberated in the tank with changes in pressure and temperature.

Several equipment manufacturers have developed hydropneumatic tank control systems which will perform the following functions.

(a) Start and stop one or more pumps in accordance with the preset tank pressures, i.e., cut-in pressure (P_2) and cut-out pressure (P_1).

(b) They will sense the water level in the tank at the cut-out pressure (P_1) and compare it with the preset water volume at 40% or 50% of the tank capacity.

(c) If the water level is high, which indicates that the air volume is too low, then the controller will open the air inlet valve and allow compressed air to enter the tank to compensate for the loss.

(d) If the water level is low, this indicates that there is too much air in the tank, and the controller will open the vent valve to exhaust excess air until the correct water level is restored.

With deep-well vertical turbine submersible pumps with drainable columns, there is almost always some air in the pump column or discharge pipe after the pump has stopped. Unless there is an air release valve on the pump discharge which vents the air to atmosphere, as the pump starts and before it is discharging to the hydropneumatic tank, then all the entrapped air in the pump column and riser pipe will enter the tank. If the quantity of air from the column is not excessive, advantage may be taken of this condition to avoid the necessity to install an air compressor, assuming that the hydropneumatic tank level controller and venting system can maintain the optimum air-water ratio under these conditions.

REFERENCE

1. *Hydraulic Handbook*, 4th Ed. Published by Colt Industries-Fairbanks Morse Pump Division, Kansas City, Kansas.

29

PUMPS

COST OF PUMPING Pumping water from one place to another is like mechanical handling, it adds cost to the product and nothing to its value. Pumping costs consist of four components:

1. Installed cost—amortization
2. Power costs
3. Supervision and maintenance
4. Down-time and stand-by equipment

Installed Costs

The capital cost is the least of all, and particularly when it is amortized annually with inflated dollars. Unfortunately, the contracts are normally awarded to the lowest bidder, without thought for the continuing operating costs for the next 10 to 15 years.

Power Costs

This cost can be kept to a minimum by careful pump selection and good efficiency. It is important to insure that the initial efficiency can be maintained during its useful life. Continuing good efficiency is important and is a function of the leakage between the impellers and the casing, shaft speed, and specific speed.

It is important not to oversize the pumps, either the head or the capacity for the required service. For variable conditions consider using multiple pumps in parallel or series, and/or variable speed drives. The horsepower of a pump can be calculated as follows:

$$HP = \frac{Q \times H}{33,000 \times E}$$

Where:

HP = horsepower at the pump coupling.

Q = flow in terms of pounds of fluid per minute, for water:

= U.S. gpm × 8.327

= U.K. gpm × 10.00

H = head in feet.

E = Efficiency.

EXAMPLE

Calculate the horsepower of a pump operating at 1000 U.S. gpm and 230 feet head with an efficiency of 80%

$$HP = \frac{1,000 \times 8.33 \times 230}{33,000 \times 0.80} = 72.56 \text{ (use 75 hp motor)}$$

This horsepower is at the pump coupling. There are additional losses in the electric motor, transformers and switchgear, and a "rule of thumb" method of estimating the approximate power requirements of an electric motor is to assume that:

1 pump horsepower = (approx) 1 kilowatt

The pump would use approximately 73 kilowatts. If a kilowatt-hour cost 3¢ it will cost approximately $2.19 per hour to run. If it is to operate continuously for 8760 hours per year the annual power cost will be $19,184. If the efficiency drops from 80% to 70% the additional cost for the same amount of water pumped at 230 feet head will increase the horsepower to 82.9 and the annual power cost to $21,793, an increase of

$2,609 per year. Having selected a pump on the basis of good efficiency, it is important to insure that the good efficiency can be maintained.

Supervision and Maintenance

These costs are inflationary and are paid for in current dollars. A major overhaul can cost more than the initial cost of the pump. If spare parts are also required it may be cheaper to buy a new pump.

Down-Time and Stand-By Equipment

Municipal services must be reliable, capable of meeting emergency conditions at any time. There must be equipment capable of maintaining the service when one or more pumps are down for maintenance. Obviously the more reliable the equipment is, the less this cost will be, particularly if rented equipment has to be installed under emergency conditions.

SHAFT SPEED

Pumps driven by standard alternating current induction motors rotate close to the synchronous speed determined by the number of poles in the motor and the alternating current cycles (Hertz) as shown in Table 29-1.

TABLE 29-1 Electric Motor Speeds Revolutions per Minute (R.P.M.)

Number of Poles	60 Hertz		50 Hertz	
	Sync.	F.L.	Sync.	F.L.
2	3600	3500	3000	2900
4	1800	1770	1500	1450
6	1200	1170	1000	960
8	900	870	750	720
10	720	690	600	575
12	600	575	500	480
14	514	490	428	410
16	450	430	375	360
18	400	380	333	319
20	360	340	300	285

Sync. = synchronous speed
F.L. = full load speed

It is in the owner and operator's interest to select the low speed pumps wherever possible since wear is proportioned to speed. See Chapter 9 *Ground Water and Wells*.

PUMP TYPES

In water works practice there are:

— Low lift or intake pumps

— High lift or service pumps

— Booster pumps

— Fire pumps

— Transfer and plant services

— Chemical pumps

Low Lift or Intake Pumps

Usually high capacity and low head unless the water treatment facilities are elevated above the source water. Vertical turbines with good suction submergence are preferred for this service. Some form of screening before the pump may be essential to protect it from debris. Good suction conditions are essential with an adequate available Net Positive Suction Head (NPSH) in excess of the pump required (NPSH).[1]

High Lift or Service Pump

These are the pumps which supply the distribution system with treated finished water. They are usually high capacity and high head and consequently high horsepower. Efficiency becomes important and since there is a fluctuating demand it is customary to install several pumps capable of paralleling together to meet the requirements. Where they are operating directly on the distribution system, as opposed to an elevated storage tank or reservoir, one or more pumps should be equipped with variable speed drives. It is preferred to use identical capacity pumps rather than small and large, so that they will all have an equal amount of use and be interchangeable for all conditions of service.

Vertical turbine pumps are preferred since they have submerged suctions and descending H-Q curves which are convenient for parallel operation. Avoid pumps with humped back curves for parallel operation since they are inclined to "hunt" against each other for the load, since there are two points on the curve for most heads (H) with considerable differences in capacities (Q). See Figure 29-1 and 29-2.

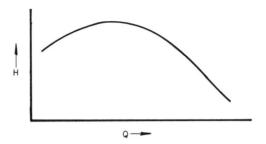

FIGURE 29-1 Humped H.Q. Curve.

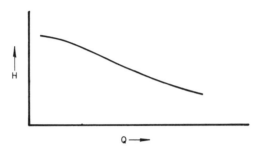

FIGURE 29-2 Descending H.Q. Curve.

Booster Pumps

These are usually installed to increase the head, particularly in water transmission pipe lines where friction losses or increases in elevation necessitate an increase in pressure or in distribution systems at higher elevations. A typical booster station is shown in Fig. 29-3. Normally they have pressure fed suctions and the choice is between end suction, split case centrifugals, or canned rotor vertical turbines. The characteristic curves are important particularly with regard to parallel operation and shut-off head (SOH) when the pumps are operating with no flow.

Where exceptionally high heads are required, multistage split case horizontal centrifugal boiler feed pumps have an application. Since they operate in the range of 10,000 rpm the water must be free from suspended solids and the pump must be under positive suction pressure. Gas or steam turbines or electric motors with speed increasers are used.

Fire Pumps

Fire pumps can be either vertical turbine or horizontal shaft pumps but they should comply with the requirements of the fire insurance underwriters.

In the U.S. *"Grading Schedule for Municipal Fire Protection"* Published by Insurance Services Office, 160 Water Street, New York. N.Y. 10038 (1973).

FIGURE 29-3 Rosslyn Booster Pump Station, Edmonton, Alberta,—with Two 150 H.P. 6 mgd, 125-foot Head, Horizontal Split Case Centrifugal Pumps.

In Canada The Insurers Advisory Organization of Canada (IAO) 42, James St. South., Hamilton, Ontario L8P-2Y4.

Fire pumps are normally manufactured in the sizes shown in the following Table 29-2. Larger units are accepted by some insurance underwriters.

The pumps can be driven by electric motors, gasoline engines, and/or diesel engines, but to be acceptable they must be approved as a complete unit pump and driver, by the insurance underwriters.

Diesel engines are preferable to gasoline or natural gas engines, since they are easier to start, more reliable, and will take full load as soon as they are up to operating speed.

The piping systems are also of interest to the insurance underwriters and facilities must be provided to test the entire unit at full capacity. Special spring loaded—flow calibrated return flow valves are available to enable the fire pumps to be run under operating conditions without the need to rig firehoses and nozzles.

TABLE 29-2 Standard Capacities of Fire Pumps

Capacity (U.S. gpm)	Pressure (psi)
500 U.S. gpm	40–250 psi
750 U.S. gpm	40–340 psi
1000 U.S. gpm	40–210 psi
1500 U.S. gpm	40–175 psi
2000 U.S. gpm	65–150 psi
2500 U.S. gpm	100–150 psi

The head capacity curves of fire pumps must be within specified limits required by the insurance underwriters but the shut-off heads (SOH) may vary appreciably and it is desirable, when evaluating a number of different manufacturers pumps, to plot all the HQ curves on the same graph with the same scales.[1] Preferred characteristics are low shut-off-head (SOH) and good capacity at low head, and a relatively flat curve, so as to minimize shock waves and abnormally high pipe line pressures.

Transfer and Plant Services

Frequently operating on a hydropneumatic tank to provide high pressure services to the plant, these pumps are normally end suction capable of rugged use with little maintenance. Reliability is more important than efficiency since they are relatively small horsepower but must be capable of frequent starting and stopping.

Chemical Pumps

Corrosion and abrasive resistance, freedom from clogging, accurate proportioning, reliability and ease of maintainance are the major factors influencing the selection of these pumps. Power consumption and efficiency are not significantly important, compared to reliability and easy maintenance.

They cover a wide range of diaphragm, plunger, gear, progressing cavity, and rubber-lined end suction centrifugal pumps.

Variable capacity is a required criteria and can be achieved with the diaphragm and plunger pumps by varying the length of stroke and the number of strokes per minute by varying the shaft speed.

Molded rubber impellers in centrifugal pumps used for pumping slurries should not normally exceed the following impeller periphery speeds. (Table 29-3.)

These limits can be exceeded if the temperature of the fluids is less than

30°C (86°F) and if the concentration of solids is less than 10% by weight. Most of these pumps are belt driven and the slower they operate the longer they last.

TABLE 29-3 Molded Rubber Impellers

Impeller Diameter (Ins)	Tip Speed Max (FPM)	Rpm Max (Approx)
7"	4000	2000
10"–12"	4250	1400
14"–17"	4500	1000
18"–31"	5000	600
32"–39"	5500	500

NET POSITIVE SUCTION HEAD (NPSH) (See Chapter 9)

Centrifugal pumps are incapable of "sucking" water in the same context as a plunger pump, and they must be supplied with water at a greater pressure than is indicated by the required NPSH. There are two values for NPSH:

(a) Required NPSH which is specified by the pump manufacturer, and the

(b) Available NPSH which the pump applications engineer must provide.

It is essential that the available NPSH be greater than the required NPSH. If the reverse is true the pump will not perform satisfactorily, but will cavitate and wear rapidly with noise and vibration.

For these reasons vertical turbine pumps are preferred since the first stage impellers are submerged and have a positive suction supply. With end suction and split case centrifugal pumps, care must be taken to insure that they have a positive NPSH. If it is not possible to achieve this condition, then alternative pumps should be considered.

VORTEXING AND SUBMERGENCE

One problem with underwater suction pipes and vertical turbine pumps is the development of vortices. A vortex is a miniature whirlpool which originates at the pump suction or pipe inlet and spirals upward to the surface, with the result that air is drawn into the pump suction. To avoid this condition the sump must be designed in accordance with the recommendations of the Hydraulic Institute.[2] In general the suction bells should be as close to the floor of the sump as possible with sufficient free area to allow unrestricted entry. This is usually equivalent to 0.5 × diameter of the suction bell. They should also be as close to one wall or divider as possible and with a minimum submergence of not less than 2.0 × diameter of the suction bell.

SPECIFIC SPEED (Ns)

Specific speed (*Ns*) is calculated from the equation:

$$Ns = \frac{n\sqrt{Q}}{H^{0.75}}$$

Where:

Ns = specific speed
n = rotative speed
Q = capacity
H = head per stage

Specific speed (*Ns*) is a number which can be used to compare the performance of specific impellers under different conditions of speed, flow, and head per stage. Used for dynamic similarity it expresses the fact that two pumps geometrically similar to each other will have similar performance characteristics.

The units used to calculate specific speed (*Ns*) must be consistent within the numerical system for (*Ns*) to become meaningful.

In U.S. English units:

N = revolutions per minute (rpm)
Q = capacity in (U.S. gpm)
H = head (feet)

In metric units:

N = revolutions per minute (rpm)
Q = capacity in meter^3/second
H = head (meters)

Charts are provided in many published texts[3,4,5] to relate these three parameters.

Specific speed (*Ns*) is of significant importance in the selection of a pump for a particular service. The different types of impellers have various ranges of (*Ns*), with low numbers for high head closed impeller to high (*Ns*) numbers for axial flow pumps of low head and high capacity. The normal range in English units for single suction impeller designs is 500 to 15,000 and in metric units from 25 to 300.

If two or more pumps of different manufacture are being evaluated for a specific job then the specific speed (*Ns*) comparison may be significant since there is a relationship to pump efficiency and internal leakage within the pump[6] as shown in Table 29-4 for double suction split case centrifugals.

TABLE 29-4 Leakage Losses vs Specific[6] Speed for Double Suction Pumps

Specific Speed (U.S. Units) (Ns)	Leakage Losses in Percentage of Power Input
500	9.0%
1000	4.0%
1500	2.2%
2000	1.5%
2500	1.2%
3000	1.0%

The table shows the inherent losses within a double suction pump for example.

PUMP A

A pump with a 14 inch diameter impeller has its best efficiency at 3200 U.S. gpm and 170 feet head when operating at 1800 rpm.

$$Ns = \frac{1800\sqrt{3200}}{170^{0.75}} = 2162$$

This pump would have an inherent loss of approximately 1.4% whereas a higher speed, higher head lower capacity pump would have large percentage internal loss.

PUMP B

A pump designed for 180 U.S. gpm at 250 feet when operating at 3550 rpm

$$Ns = \frac{3550\sqrt{180}}{250^{0.75}} = 757$$

The inherent loss would be approximately 5.5% and the efficiency would be less than the pump A.

Invitation to Tender

PUMP SELECTION

The written specifications should clearly state the performance requirements under all conditions of service, including the location, elevation above sea level, ambient temperatures, fluid temperatures, physical and chemical characteristics of the fluid if other than water, available power supplies, and other pertinent data. The specifications should also specify the type of pumps required and any other special conditions such as limitations in space, NPSH, etc., that may apply.

Pumps can be supplied in various shaft speeds, and the preferred speed should be stated. If two different synchronous speeds are applicable to the situation, then the supplier should be asked to quote on both units; since one will be considerably less expensive it is necessary to carefully evaluate the two quotations.

The standards which apply should be stated.[2,8,9] In order to completely evaluate one set of tender documents with another the suppliers should be asked to complete data sheets supplied with the invitation to tender.[1] In this way a more concise and accurate evaluation can be made. The award should never be given to the lowest bidder until a complete and detailed comparison has been made.[1]

In this way the owner will get the best the market-place can offer to suit the service requirements. Modern up-to-date equipment can be evaluated against more conservative traditional designs on the basis of the overall cost of owning and operating a pump station. The capital cost is usually a small component of the total cost projection.

Testing Procedures

The invitation to tender must state what testing procedures are required:

1. Normal manufacture test
2. Certified performance curves
3. Witnessed factory tests

The published characteristics performance curves are only good approximations, and a manufacturer will only guarantee one or two points on the H-Q curve generally in the range of maximum efficiency. It is impossible to test a pump when it is installed on site in accordance with the *Hydraulic Institute Standards*.[2] The only way in which the owner can be certain that he is getting the performance he specified is to witness the factory test and not to accept the pumps until they are capable of producing the performance characteristics quoted in the manufacturer's quotation when tested in accordance with the accepted standard.[2] If the witnessed factory tests do not come up to the standards anticipated, it is usually possible to polish the impellers and diffuser vanes, until the required results are obtained. If the published characteristics cannot be obtained, the manufacturer should reimburse the owner accordingly. There are firms who specialize in witness testing, the charges are reasonable, but the pump manufacturers will increase their price and extend their delivery. The alternative is to accept the manufacturer's test data in the form of a certified performance curve without witness. Whether or not to insist on a witness test or certified performance curves depends on the size and value of the project and the annual power cost equated to each point percentage of efficiency. Once the pump has been delivered the owner has little or no recourse if it is believed that the performance is less than anticipated.

MAINTENANCE

Since maintenance and downtime are always expensive items any steps taken to reduce them is profitable.

Mechanical Seals

Wherever possible mechanical seals should be used. If they are installed correctly, properly lubricated and operated under suitable conditions, they will give many years of trouble free operation.

Stuffing Boxes

Soft packing stuffing boxes are used where ideal conditions for the proper use of mechanical seals do not exist such as pumps handling gritty liquids, on intermittent duty, or where shaft malalignment or vibration is likely to occur.

Pipe Stresses

Many problems occur due to stresses imposed on pump casings and discharge heads due to pipe stresses. When a pump and its associated piping have been installed it should be possible to turn the shafts as easily as it was before the pipes were installed. It is advisable, when in doubt, to have the pipe flanges uncoupled from the pump casing and determine if there is any spring or thrust imposed on the pumps.

Bedding

Horizontal shaft pumps are normally fitted with flexible couplings but these should also be checked for malalignment both when cold and after they have been in operation.

REFERENCES

1. WALKER, R. *Pump Selection*. Ann Arbor, Michigan: Ann Arbor Science Publishers, Inc., 1972.
2. *Hydraulic Institute Standards for Centrifugal, Rotary and Reciprocating Pumps*, 13th Ed. Cleveland, Ohio: Hydraulic Institute, 1975.

3. KARASSIK, I. J. and R. CARTER, *Centrifugal Pumps.* New York: McGraw-Hill Book Company, 1960.

4. GARTMANN, H. *DeLaval Engineering Handbook*, 3rd Ed. New York: McGraw-Hill Book Company, 1970.

5. *Pump Handbook—KSB.* Published by Klien, Schanzlin and Becker Aktiengesellschaft 1968.

6. *Pumpworld.* Worthington Pump Inc., New Jersey 1 (1), Summer 1975.

7. *Centrifugal Fire Pumps.* Peerless Pump, FMC Corporation. Bulletin B-1500.

8. "American National Standard for Deep Well Vertical Turbine Pumps—Line Shaft and Submersible Types *AWWA* E101–71. Published by the American Water Works Association at Denver, Colo. 1971.

9. *Centrifugal Fire Pumps.* Standard of the National Fire Protection Association, U.S.A. adopted by Canadian Underwriters' Association CUA20 1972.

30

DRIVES—
MECHANICAL AND ELECTRICAL

CHOICE OF DRIVES

The public power supply companies are by far the cheapest sources of continuous energy to the water works industry. There are exceptions where large pumps are infrequently required for short periods to meet peak requirements and where the electrical demand charges for the power supply would exceed the cost of owning and operating engines. Engines may be required for essential services and fire protection when public power supplies fail or are not available. Gas engines have not always been economically competitive with electric motors even when the gas has been available at no cost. In some cases, the problems have been the fault of the engine manufacturers who have over-rated their continuous horsepower capacity at the expense of higher than normal maintenance. On the other hand gas turbines have a successful record in the larger sizes where they have been competitive with electric motors.

ELECTRIC MOTORS

Induction motors starting from rest will draw up to six times their normal full load current, in order to accelerate up to synchronous speed. This starting current is only applied during a very short period, but it heats up the motor windings above normal operating temperatures. If the motor cooling system is incapable of cooling the motor down to the normal running temperature before it is stopped, then the motor will be already heated the next time it starts and more heat will be added. If frequent starting and stopping under these conditions continues, the motor will overheat until eventually it fails to restart because the safety overloads have cut out. To overcome this difficulty, it is necessary to make equipment selections and pump sizes to reduce the number of starts per hour. Some motors may be capable of 6 to 8 starts per hour if they were spaced at 7-½ to 10 minute intervals, but would probably overheat if the 6 to 8 starts occurred within the first few minutes. The number of "starts" per hour that an electric motor can accommodate is dependent on the torque characteristics of the pump and motor as a complete coupled unit. In the case of a vertical turbine with several hundred feet of shafting, bearings at 10 foot intervals and terminated by a large bowl assembly, the torque to accelerate to operating speed will be much greater than an end suction pump located on the surface, of equal capacity. More heat will be generated in the windings of the motor for the vertical turbine pump, hence it will accommodate fewer starts per hour. To insure that the motors and starters will meet conditions of frequent start-stop service, it is necessary to determine the torque characteristic of the pump and motor known as the WR^2 value of the rotating components.

The average net accelerating torque of a motor is calculated from the motor full load torque, the locked rotor torque and the breakdown torque. The average of these values is related in the formula

$$t = \frac{WR^2 \times RPM}{308 \times T}$$

Where:

t = accelerating time in seconds

WR^2 = refers to the momentum of the rotating shaft parts of the pump and motor

RPM = motor full load speed

T = average net accelerating torque of the motor in lb-ft

The average net accelerating torque is the average gross torque less the torque necessary to accelerate the rotor and the fan. The cooling fan can absorb a substantial percentage of the available torque.

From this data and the calculated accelerating time in seconds, the motor manufacturer will be able to advise on the number of starts per hour and the intervals between starts that the motors may be subjected to under continuous operating conditions. As a further precaution, the electric motor selected should be capable of providing adequate horsepower for all operating conditions without encroaching on the 15% service factor which should be reserved for abnormal power supply conditions of low voltage and frequency. Since the number of starts that can be accommodated per hour is a function of the heat generated in accelerating the load, the rate of heat dissipation at full speed is a function of the type of motor enclosure. Different types of enclosures have been developed to suit the many different environmental conditions.

Open Motors

An open motor has a free exchange of air from the surrounding atmosphere which is forced through the windings and back into the environment by a fan attached to the motor shaft. There are four types of open motor enclosures.

- (a) *Drip-proof Motors* They are the least expensive and the most commonly used. The lowered ventilating openings are designed to prevent solid or liquid particles from entering the motor providing they are falling at 15° or less from the vertical. Drip-proof motors are designed for use in applications where there is a clean atmosphere with little dust or water and where there are no explosive elements.
- (b) *Splash-proof Motors* The ventilating openings are designed to prevent liquid or solid particles moving at 100° or less from the vertical, i.e., 10° below the horizontal, from entering the motor.
- (c) *Flood-proof Motors* The ventilating openings are constructed similar to the drip-proof motors, only the starter windings are encapsulated in an epoxy resin material which protects them from the surrounding environment. Flood-proof motors are used in severe applications where repeated flooding by immersion of the motor may occur.
- (d) *Pipe-Ventilated* Pipe-ventilated motors are similar to the standard drip-proof motors except that openings are provided at each end of the motor enclosure for the connection of cooling ducts or pipes. The cooling ducts enable cooling air to be drawn from the normal clean environment remote from the immediate operating area of the motor.

Totally Enclosed Motors

Totally enlosed motors have no free exchange of air from the interior of the motor to the surrounding atmosphere. They are cooled by direct radiation of heat from the outer surface of the motor enclosure, and they may also have a shaft driven fan, mounted in an external housing to the enclosure. The fan draws air into this housing and directs it along the outer surface of the enclosure between cooling fins cast on the stator frame. In both the totally enclosed non-ventilated and the totally enclosed externally fan cooled types, the radiation of heat from the rotor and windings is aided by internal fans formed on each end of the rotor assembly. These fans keep the air within the motor enclosure circulating to improve the heat transfer to the outer fins. There are four types of totally enclosed motors:

- (a) *Totally Enclosed Non-Ventilating* Normally of limited horsepower for use in extremely wet, dirty and dusty conditions. They are being superseded by totally enclosed fan cooled motors with the exception of the small horsepower in extremely adverse conditions. This type of motor would be limited to the least number of starts per hour, because of its poor cooling characteristics.

(b) *Totally Enclosed Fan Cooled (TEFC)*   These are the most common types of enclosed motors and are used for wet, dirty and dusty conditions.

(c) *Explosion Proof*   Explosion proof motors are similar to the TEFC motors but are constructed with wide metal-to-metal joints having close clearances, non-sparking fans enabling them to withstand an explosion inside the motor, and to prevent any sparks, flashes, or explosions generated within the enclosure from igniting the surrounding atmosphere. Explosion proof motors must be used in all locations where hazardous gases or vapor surround the machine. This type of motor should be considered for service in sewage lift stations and other wastewater collection systems that could receive gasoline or other inflammable liquids and vapors.

(d) *Dust Explosion Proof*   Dust explosion proof motors are designed to include ignitable amounts of dust and thus prevent a dust explosion inside the motor. Dust explosion proof motors should be used in all locations where there are hazardous dusts in the atmosphere, for example, where activated carbon or sodium chlorate is used in water and waste treatment processes.

Water Cooled Motors

Water cooled motors have a heat exchanger in their air cooling circuit. The fan draws air through the heat exchanger and blows the cool air through the motor windings and is then ducted back through the water cooled heat exchanger. The motors are enclosed and very much quieter in operation; they can be housed in an enclosed building with little outside ventilation. They are ideal for use in residential areas where the noise of a larger motor would be objectionable.

Reduced Voltage Starting

The standard cross-the-line starting of induction motors draws approximately up to 6 times the normal full load current in accelerating up to synchronous speed. It is important to insure that the motor has more than adequate starting torque to accelerate the load up to full speed as quickly as possible so as to keep the period of high in-rush current down to a minimum.

In certain cases a reduced voltage starter can be installed instead of the standard cross-the-line starter. These starters can reduce the in-rush current from the normal 6 times full load current to approximately 3 times full load current. However, the motor accelerating period will be prolonged and it is important to insure that the starting torque is sufficiently in excess of the torque requirements of the pump, otherwise a stalemate condition will occur, the unit will fail to get up to synchronous speed, and the starter will trip out on overheat.

Wound Rotor Motors

If emergency power generation is necessary to provide essential services during public power supply outages, then the use of wound rotor motors with liquid or wire rheostats, are worth considering since they can be started at less than full load current. Otherwise, the generating capacity needs to be capable of supplying in-rush currents up to 6 times the normal full load current for cross-the-line starters and 3 times for reduced voltage starters, which is only needed for a few milliseconds each time a motor is started.

VARIABLE SPEED DRIVES

Wound Rotor Motors

Wound rotor motors with liquid rheostats are frequently used not only for purposes of low in-rush current but for variable speed drives.

Electro-Magnetic Couplings

They consist of two halves of an electro-magnetic coupling. One is attached to the motor shaft and the other to the pump shaft. The motor runs at synchronous speed but the pump speed is controlled by the magnetic flux between the two halves of the coupling. The higher the magnetic flux, the higher the pump speed. The slip can be increased to reduce the pump speed by lowering the flux.

Hydraulic Couplings

A similar result can be achieved with a split hydraulic coupling, sometimes referred to as a "fluid flywheel." The hydraulic fluid or transmission oil conducts the heat generated to a water cooled heat exchanger.

Variable Frequency Drives (VFD)[2]

This consists of a solid state electronic sine wave converter which is capable of receiving a normal 3-phase, 60 Hertz power input and controlling the output frequency between 24 and 60 Hertz. As a result, the synchronous speed of any standard induction motor can be reduced by 60%. This is more than adequate for most pumping installations. The advantages are many with this type of variable speed control.

- Combined motor and VFD efficiency at full load is approximately 85%.
- No external venting or heat exchangers are required.
- Uses a standard starter and induction motor. Can be installed between an existing series of motors and starters. One VFD unit can be used for several motors. As the demand increases a motor will speed up until it eventually reaches full speed, it will then lock that unit back to its starter, and then select the second pump for variable speed drive until it too is up to full speed.
- Power factor is improved at full speed, and there is good conversion efficiency at all speed ranges.

The only problem with VFD is that it is limited in power and together with the switching equipment it is expensive.

Combustion Engines

Variable speed output is readily available from gas, diesel, and turbines.

ENGINE SELECTION High maintenance costs, noise, vibration, and increased supervision have prevented the use of engines for continuous pumping service in most municipal installations. They are used extensively for standby service when power supplies fail and where they become economic for use during short peak hourly demand periods when it would not be economic to pay the "installed horsepower charges" to the power corporation for only a few hours operating time per month.

Gasoline and natural gas engines have been used extensively for these duties; but diesels are preferred if economically feasible, since they are less temperamental than gas engines. Fuel for both gas and diesel engines must be clean and free from water, particularly in the case of diesel engines. A little water in diesel fuel will seize the fuel injection pumps and "blow the tip" of the fuel injectors. Apart from these problems a diesel has many advantages over gas engines.

- Diesels start much more easily than gas engines, particularly if fitted with glow plugs in each cylinder.
- Diesels will take full load as soon as they are up to operating speeds and all cylinders are firing.
- Fuel is less inflammable and does not deteriorate with storage to the same degree as gasoline.
- Diesels are noisier than gas engines, may require heavier foundations, require more space, and usually are higher in capital cost but much cheaper for fuel and maintenance.

— Gas engines are sometimes difficult to start if the gas reducing valves have not been operated for some time or have not been taken apart for cleaning. A gum deposit may collect on the valve stems and diaphragms preventing them from operating freely.

Maintenance costs on various types of engines are difficult to obtain. Unfortunately, since engines are sold competitively some engine horsepowers have been over rated with the result that many large engine users have adopted derating factors as low as 70%, multiplied by the manufacturer's continuous rating, to arrive at the optimum economical continuous horsepower rating. There is a relationship between shaft speed, continuous horsepower rating, and maintenance costs, and unless there is good operating experience with a particular make and model it is advisable to use a derating factor.

The jet aircraft industry has developed a breed of gas turbines of outstanding reliability and it is understood that the industrial applications of gas turbines have been equally successful. It is possible that future installation in the larger horsepower will be toward gas turbines and to diesels for the smaller horsepowers.

Dual Fuel Engines[3]

An interesting development in the large diesel engine horsepower range is the dual fuel engine. This is basically a compression ignition diesel engine capable of aspirating a weak mixture of natural or sewage gas and air, compressing it, and igniting the mixture with an injection of diesel oil. The engine can operate equally effectively as a straight diesel engine using liquid fuels only, or as a dual fuel engine using 95–96% gas with 4–5% fuel oil. The compression ratio is between 14:1 and 16:1.

Natural or sewage digester gases, which are high in methane, are drawn into the cylinders as relatively weak mixtures and are compressed without detonation. At the end of the compression stroke, diesel oil is injected into the cylinders, immediately ignites, and burns the gas-air mixture. The efficiency of the dual fuel engine is said to be equal or slightly better than a straight gas engine since the superior qualities of compression ignition are more efficient than spark ignition.

They are available from 2500 HP to 6000 HP. Their capital cost is higher than for a straight diesel engine, but they have several advantages in areas where it is economical to use gas on a non-continuous or interruptable basis. Often gas rates are considerably cheaper on this basis, and since the engine can immediately revert to 100% diesel fuel if the gas supply fails, this transition can be automatically accomplished with only a momentary fluctuation in speed. It is also understood that the maintenance costs are appreciably less for a dual fuel engine compared to spark ignition and could be less than straight diesel since natural gas is a much cleaner fuel to burn than diesel oil.

THRUST BEARINGS Whether pumps are driven by electric motors or by engines through right-angle gear boxes it is essential to insure that the thrust bearings are adequate. Split-case centrifugal double suction pumps are relatively evenly balanced and have no axial thrust, however, vertical turbine pumps have an up-thrust when starting, immediately followed by a down-thrust plus the weight of the shaft and impellers.

The magnitude of these thrusts are calculable and can be obtained from the pump manufacturer. The ball and thrust bearings in the motors and gear boxes must also have sufficient thrust-absorbing capacity to sustain the axial loads during the 100,000 hour expected life of the bearing.

REFERENCES
1. "Industrial Induction Motor Manual," published by Canadian General Electric, Peterborough, Ontario, Canada.
2. Autocon Industries Inc., Minneapolis, Minnesota.
3. WALKER, R., *Pump Selection*. Ann Arbor, Michigan: Ann Arbor Science Publishers Inc., 1972.

APPENDICES

TABLE A-1 Gallons (UK)—Gallons (US)—Cubic Meters

UK	US	M3	UK	US	M3
0.833	1	0.00379	34.140	41	0.15520
1.665	2	0.00757	34.972	42	0.15898
2.498	3	0.01136	35.805	43	0.16277
3.331	4	0.01514	36.638	44	0.16655
4.163	5	0.01893	37.470	45	0.17034
4.996	6	0.02271	38.303	46	0.17413
5.829	7	0.02650	39.136	47	0.17791
6.661	8	0.03028	39.968	48	0.18170
7.494	9	0.03407	40.801	49	0.18548
8.327	10	0.03785	41.634	50	0.18927
9.159	11	0.04164	42.466	51	0.19305
9.992	12	0.04542	43.299	52	0.19684
10.825	13	0.04921	44.132	53	0.20062
11.657	14	0.05299	44.964	54	0.20441
12.490	15	0.05678	45.797	55	0.20819
13.323	16	0.06057	46.630	56	0.21198
14.155	17	0.06435	47.462	57	0.21576
14.988	18	0.06814	48.295	58	0.21955
15.821	19	0.07192	49.128	59	0.22333
16.654	20	0.07571	49.961	60	0.22712
17.486	21	0.07949	50.793	61	0.23091
18.319	22	0.08328	51.626	62	0.23469
19.152	23	0.08706	52.459	63	0.23848
19.984	24	0.09085	53.291	64	0.24226
20.817	25	0.09463	54.124	65	0.24605
21.650	26	0.09842	54.957	66	0.24983
22.482	27	0.10220	55.789	67	0.25362
23.315	28	0.10599	56.622	68	0.25740
24.148	29	0.10977	57.455	69	0.26119
24.980	30	0.11356	58.287	70	0.26497
25.813	31	0.11735	59.120	71	0.26876
26.646	32	0.12113	59.953	72	0.27254
27.478	33	0.12492	60.785	73	0.27633
28.311	34	0.12870	61.618	74	0.28011
29.144	35	0.13249	62.451	75	0.28390
29.976	36	0.13627	63.283	76	0.28769
30.809	37	0.14006	64.116	77	0.29147
31.642	38	0.14384	64.949	78	0.29526
32.474	39	0.14763	65.781	79	0.29904
33.307	40	0.15141	66.614	80	0.30283

TABLE A-1 (Continued) Gallons (UK)—Gallons (US)—Cubic Meters

UK	US	M3	UK	US	M3
67.447	81	0.30661	100.754	121	0.45802
68.279	82	0.31040	101.586	122	0.46181
69.112	83	0.31418	102.419	123	0.46560
69.945	84	0.31797	103.252	124	0.46938
70.777	85	0.32175	104.084	125	0.47317
71.610	86	0.32554	104.917	126	0.47695
72.443	87	0.32932	105.750	127	0.48074
73.275	88	0.33311	106.582	128	0.48452
74.108	89	0.33689	107.415	129	0.48831
74.941	90	0.34068	108.248	130	0.49209
75.773	91	0.34447	109.080	131	0.49588
76.606	92	0.34825	109.913	132	0.49966
77.439	93	0.35204	110.746	133	0.50345
78.271	94	0.35582	111.578	134	0.50723
79.104	95	0.35961	112.411	135	0.51102
79.937	96	0.36339	113.244	136	0.51480
80.769	97	0.36718	114.076	137	0.51859
81.602	98	0.37096	114.909	138	0.52238
82.435	99	0.37475	115.742	139	0.52616
83.267	100	0.37853	116.575	140	0.52995
84.100	101	0.38232	117.407	141	0.53373
84.933	102	0.38610	118.240	142	0.53752
85.766	103	0.38989	119.073	143	0.54130
86.598	104	0.39367	119.905	144	0.54509
87.431	105	0.39746	120.738	145	0.54887
88.264	106	0.40124	121.571	146	0.55266
89.096	107	0.40503	122.403	147	0.55644
89.929	108	0.40882	123.236	148	0.56023
90.762	109	0.41260	124.069	149	0.56401
91.594	110	0.41639	124.901	150	0.56780
92.427	111	0.42017	125.734	151	0.57158
93.260	112	0.42396	126.567	152	0.57537
94.092	113	0.42774	127.399	153	0.57916
94.925	114	0.43153	128.232	154	0.58294
95.758	115	0.43531	129.065	155	0.58673
96.590	116	0.43910	129.897	156	0.59051
97.423	117	0.44288	130.730	157	0.59430
98.256	118	0.44667	131.563	158	0.59808
99.088	119	0.45045	132.395	159	0.60187
99.921	120	0.45424	133.228	160	0.60565

TABLE A-1 (Continued)   Gallons (UK)—Gallons (US)—Cubic Meters

UK	US	M3	UK	US	M3
134.061	161	0.60944	167.368	201	0.76085
134.893	162	0.61322	168.200	202	0.76464
135.726	163	0.61701	169.033	203	0.76842
136.559	164	0.62079	169.866	204	0.77221
137.391	165	0.62458	170.698	205	0.77599
138.224	166	0.62836	171.531	206	0.77978
139.057	167	0.63215	172.364	207	0.78356
139.889	168	0.63594	173.196	208	0.78735
140.722	169	0.63972	174.029	209	0.79113
141.555	170	0.64351	174.862	210	0.79492
142.387	171	0.64729	175.694	211	0.79870
143.220	172	0.65108	176.527	212	0.80249
144.053	173	0.65486	177.360	213	0.80628
144.885	174	0.65865	178.192	214	0.81006
145.718	175	0.66243	179.025	215	0.81385
146.551	176	0.66622	179.858	216	0.81763
147.383	177	0.67000	180.690	217	0.82142
148.216	178	0.67379	181.523	218	0.82520
149.049	179	0.67757	182.356	219	0.82899
149.882	180	0.68136	183.189	220	0.83277
150.714	181	0.68514	184.021	221	0.83656
151.547	182	0.68893	184.854	222	0.84034
152.380	183	0.69272	185.687	223	0.84413
153.212	184	0.69650	186.519	224	0.84791
154.045	185	0.70029	187.352	225	0.85170
154.878	186	0.70407	188.185	226	0.85548
155.710	187	0.70786	189.017	227	0.85927
156.543	188	0.71164	189.850	228	0.86306
157.376	189	0.71543	190.683	229	0.86684
158.208	190	0.71921	191.515	230	0.87063
159.041	191	0.72300	192.348	231	0.87441
159.874	192	0.72678	193.181	232	0.87820
160.706	193	0.73057	194.013	233	0.88198
161.539	194	0.73435	194.846	234	0.88577
162.372	195	0.73814	195.679	235	0.88955
163.204	196	0.74192	196.511	236	0.89334
164.037	197	0.74571	197.344	237	0.89712
164.870	198	0.74950	198.177	238	0.90091
165.702	199	0.75328	199.009	239	0.90469
166.535	200	0.75707	199.842	240	0.90848

TABLE A-1 (Continued) Gallons (UK)—Gallons (US)—Cubic Meters

UK	US	M3	UK	US	M3
200.675	241	0.91226	233.982	281	1.06368
201.507	242	0.91605	234.814	282	1.06746
202.340	243	0.91984	235.647	283	1.07125
203.173	244	0.92362	236.480	284	1.07503
204.005	245	0.92741	237.312	285	1.07882
204.838	246	0.93119	238.145	286	1.08260
205.671	247	0.93498	238.978	287	1.08639
206.503	248	0.93876	239.810	288	1.09018
207.336	249	0.94255	240.643	289	1.09396
208.169	250	0.94633	241.476	290	1.09775
209.001	251	0.95012	242.308	291	1.10153
209.834	252	0.95390	243.141	292	1.10532
210.667	253	0.95769	243.974	293	1.10910
211.499	254	0.96147	244.806	294	1.11289
212.332	255	0.96526	245.639	295	1.11667
213.165	256	0.96904	246.472	296	1.12046
213.997	257	0.97283	247.304	297	1.12424
214.830	258	0.97662	248.137	298	1.12803
215.663	259	0.98040	248.970	299	1.13181
216.496	260	0.98419	249.803	300	1.13560
217.328	261	0.98797	250.635	301	1.13938
218.161	262	0.99176	251.468	302	1.14317
218.994	263	0.99554	252.301	303	1.14695
219.826	264	0.99933	253.133	304	1.15074
220.659	265	1.00311	253.966	305	1.15453
221.492	266	1.00690	254.799	306	1.15831
222.324	267	1.01068	255.631	307	1.16210
223.157	268	1.01447	256.464	308	1.16588
223.990	269	1.01825	257.297	309	1.16967
224.822	270	1.02204	258.129	310	1.17345
225.655	271	1.02582	258.962	311	1.17724
226.488	272	1.02961	259.795	312	1.18102
227.320	273	1.03340	260.627	313	1.18481
228.153	274	1.03718	261.460	314	1.18859
228.986	275	1.04097	262.293	315	1.19238
229.818	276	1.04475	263.125	316	1.19616
230.651	277	1.04854	263.958	317	1.19995
231.484	278	1.05232	264.791	318	1.20373
232.316	279	1.05611	265.623	319	1.20752
233.149	280	1.05989	266.456	320	1.21131

TABLE A-1 (Continued) Gallons (UK)—Gallons (US)—Cubic Meters

UK	US	M3	UK	US	M3
267.289	321	1.21509	300.596	361	1.36650
268.121	322	1.21888	301.428	362	1.37029
268.954	323	1.22266	302.261	363	1.37407
269.787	324	1.22645	303.094	364	1.37786
270.619	325	1.23023	303.926	365	1.38165
271.452	326	1.23402	304.759	366	1.38543
272.285	327	1.23780	305.592	367	1.38922
273.117	328	1.24159	306.424	368	1.39300
273.950	329	1.24537	307.257	369	1.39679
274.783	330	1.24916	308.090	370	1.40057
275.615	331	1.25294	308.922	371	1.40436
276.448	332	1.25673	309.755	372	1.40814
277.281	333	1.26051	310.588	373	1.41193
278.113	334	1.26430	311.420	374	1.41571
278.946	335	1.26809	312.253	375	1.41950
279.779	336	1.27187	313.086	376	1.42328
280.611	337	1.27566	313.918	377	1.42707
281.444	338	1.27944	314.751	378	1.43085
282.277	339	1.28323	315.584	379	1.43464
283.110	340	1.28701	316.417	380	1.43843
283.942	341	1.29080	317.249	381	1.44221
284.775	342	1.29458	318.082	382	1.44600
285.608	343	1.29837	318.915	383	1.44978
286.440	344	1.30215	319.747	384	1.45357
287.273	345	1.30594	320.580	385	1.45735
288.106	346	1.30972	321.413	386	1.46114
288.938	347	1.31351	322.245	387	1.46492
289.771	348	1.31729	323.078	388	1.46871
290.604	349	1.32108	323.911	389	1.47249
291.436	350	1.32487	324.743	390	1.47628
292.269	351	1.32865	325.576	391	1.48006
293.102	352	1.33244	326.409	392	1.48385
293.934	353	1.33622	327.241	393	1.48763
294.767	354	1.34001	328.074	394	1.49142
295.600	355	1.34379	328.907	395	1.49521
296.432	356	1.34758	329.739	396	1.49899
297.265	357	1.35136	330.572	397	1.50278
298.098	358	1.35515	331.405	398	1.50656
298.930	359	1.35893	332.237	399	1.51035
299.763	360	1.36272	333.070	400	1.51413

TABLE A-1 (Continued)   Gallons (UK)—Gallons (US)—Cubic Meters

UK	US	M3	UK	US	M3
333.903	401	1.51792	367.210	441	1.66933
334.735	402	1.52170	368.042	442	1.67312
335.568	403	1.52549	368.875	443	1.67690
336.401	404	1.52927	369.708	444	1.68069
337.233	405	1.53306	370.540	445	1.68447
338.066	406	1.53684	371.373	446	1.68826
338.899	407	1.54063	372.206	447	1.69204
339.731	408	1.54441	373.038	448	1.69583
340.564	409	1.54820	373.871	449	1.69961
341.397	410	1.55199	374.704	450	1.70340
342.229	411	1.55577	375.536	451	1.70718
343.062	412	1.55956	376.369	452	1.71097
343.895	413	1.56334	377.202	453	1.71475
344.727	414	1.56713	378.034	454	1.71854
345.560	415	1.57091	378.867	455	1.72233
346.393	416	1.57470	379.700	456	1.72611
347.225	417	1.57848	380.532	457	1.72990
348.058	418	1.58227	381.365	458	1.73368
348.891	419	1.58605	382.198	459	1.73747
349.723	420	1.58984	383.031	460	1.74125
350.556	421	1.59362	383.863	461	1.74504
351.389	422	1.59741	384.696	462	1.74882
352.222	423	1.60119	385.529	463	1.75261
353.054	424	1.60498	386.361	464	1.75639
353.887	425	1.60877	387.194	465	1.76018
354.720	426	1.61255	388.027	466	1.76396
355.552	427	1.61634	388.859	467	1.76775
356.385	428	1.62012	389.692	468	1.77153
357.218	429	1.62391	390.525	469	1.77532
358.050	430	1.62769	391.357	470	1.77911
358.883	431	1.63148	392.190	471	1.78289
359.716	432	1.63526	393.023	472	1.78668
360.548	433	1.63905	393.855	473	1.79046
361.381	434	1.64283	394.688	474	1.79425
362.214	435	1.64662	395.521	475	1.79803
363.046	436	1.65040	396.353	476	1.80182
363.879	437	1.65419	397.186	477	1.80560
364.712	438	1.65797	398.019	478	1.80939
365.544	439	1.66176	398.851	479	1.81317
366.377	440	1.66555	399.684	480	1.81696

TABLE A-1 (Continued) Gallons (UK)—Gallons (US)—Cubic Meters

UK	US	M3	UK	US	M3
400.517	481	1.82074	433.824	521	1.97216
401.349	482	1.82453	434.656	522	1.97594
402.182	483	1.82831	435.489	523	1.97973
403.015	484	1.83210	436.322	524	1.98351
403.847	485	1.83589	437.154	525	1.98730
404.680	486	1.83967	437.987	526	1.99108
405.513	487	1.84346	438.820	527	1.99487
406.345	488	1.84724	439.652	528	1.99865
407.178	489	1.85103	440.485	529	2.00244
408.011	490	1.85481	441.318	530	2.00622
408.843	491	1.85860	442.150	531	2.01001
409.676	492	1.86238	442.983	532	2.01380
410.509	493	1.86617	443.816	533	2.01758
411.341	494	1.86995	444.648	534	2.02137
412.174	495	1.87374	445.481	535	2.02515
413.007	496	1.87752	446.314	536	2.02894
413.839	497	1.88131	447.146	537	2.03272
414.672	498	1.88509	447.979	538	2.03651
415.505	499	1.88888	448.812	539	2.04029
416.338	500	1.89266	449.645	540	2.04408
417.170	501	1.89645	450.477	541	2.04786
418.003	502	1.90024	451.310	542	2.05165
418.836	503	1.90402	452.143	543	2.05543
419.668	504	1.90781	452.975	544	2.05922
420.501	505	1.91159	453.808	545	2.06300
421.334	506	1.91538	454.641	546	2.06679
422.166	507	1.91916	455.473	547	2.07058
422.999	508	1.92295	456.306	548	2.07436
423.832	509	1.92673	457.139	549	2.07815
424.664	510	1.93052	457.971	550	2.08193
425.497	511	1.93430	458.804	551	2.08572
426.330	512	1.93809	459.637	552	2.08950
427.162	513	1.94187	460.469	553	2.09329
427.995	514	1.94566	461.302	554	2.09707
428.828	515	1.94944	462.135	555	2.10086
429.660	516	1.95323	462.967	556	2.10464
430.493	517	1.95702	463.800	557	2.10843
431.326	518	1.96080	464.633	558	2.11221
432.158	519	1.96459	465.465	559	2.11600
432.991	520	1.96837	466.298	560	2.11978

TABLE A-1 (Continued) Gallons (UK)—Gallons (US)—Cubic Meters

UK	US	M3	UK	US	M3
467.131	561	2.12357	500.438	601	2.27498
467.963	562	2.12736	501.270	602	2.27877
468.796	563	2.13114	502.103	603	2.28255
469.629	564	2.13493	502.936	604	2.28634
470.461	565	2.13871	503.768	605	2.29012
471.294	566	2.14250	504.601	606	2.29391
472.127	567	2.14628	505.434	607	2.29770
472.959	568	2.15007	506.266	608	2.30148
473.792	569	2.15385	507.099	609	2.30527
474.625	570	2.15764	507.932	610	2.30905
475.457	571	2.16142	508.764	611	2.31284
476.290	572	2.16521	509.597	612	2.31662
477.123	573	2.16899	510.430	613	2.32041
477.955	574	2.17278	511.262	614	2.32419
478.788	575	2.17656	512.095	615	2.32798
479.621	576	2.18035	512.928	616	2.33176
480.453	577	2.18414	513.760	617	2.33555
481.286	578	2.18792	514.593	618	2.33933
482.119	579	2.19171	515.426	619	2.34312
482.951	580	2.19549	516.258	620	2.34690
483.784	581	2.19928	517.091	621	2.35069
484.617	582	2.20306	517.924	622	2.35448
485.450	583	2.20685	518.757	623	2.35826
486.282	584	2.21063	519.589	624	2.36205
487.115	585	2.21442	520.422	625	2.36583
487.948	586	2.21820	521.255	626	2.36962
488.780	587	2.22199	522.087	627	2.37340
489.613	588	2.22577	522.920	628	2.37719
490.446	589	2.22956	523.753	629	2.38097
491.278	590	2.23334	524.585	630	2.38476
492.111	591	2.23713	525.418	631	2.38854
492.944	592	2.24092	526.251	632	2.39233
493.776	593	2.24470	527.083	633	2.39611
494.609	594	2.24849	527.916	634	2.39990
495.442	595	2.25227	528.749	635	2.40368
496.274	596	2.25606	529.581	636	2.40747
497.107	597	2.25984	530.414	637	2.41126
497.940	598	2.26363	531.247	638	2.41504
498.772	599	2.26741	532.079	639	2.41883
499.605	600	2.27120	532.912	640	2.42261

TABLE A-1 (Continued) Gallons (UK)—Gallons (US)—Cubic Meters

UK	US	M3	UK	US	M3
533.745	641	2.42640	567.052	681	2.57781
534.577	642	2.43018	567.884	682	2.58160
535.410	643	2.43397	568.717	683	2.58538
536.243	644	2.43775	569.550	684	2.58917
537.075	645	2.44154	570.382	685	2.59295
537.908	646	2.44532	571.215	686	2.59674
538.741	647	2.44911	572.048	687	2.60052
539.573	648	2.45289	572.880	688	2.60431
540.406	649	2.45668	573.713	689	2.60809
541.239	650	2.46046	574.546	690	2.61188
542.071	651	2.46425	575.378	691	2.61566
542.904	652	2.46804	576.211	692	2.61945
543.737	653	2.47182	577.044	693	2.62323
544.569	654	2.47561	577.876	694	2.62702
545.402	655	2.47939	578.709	695	2.63080
546.235	656	2.48318	579.542	696	2.63459
547.067	657	2.48696	580.374	697	2.63838
547.900	658	2.49075	581.207	698	2.64216
548.733	659	2.49453	582.040	699	2.64595
549.565	660	2.49832	582.873	700	2.64973
550.398	661	2.50210	583.705	701	2.65352
551.231	662	2.50589	584.538	702	2.65730
552.064	663	2.50967	585.371	703	2.66109
552.896	664	2.51346	586.203	704	2.66487
553.729	665	2.51724	587.036	705	2.66866
554.562	666	2.52103	587.869	706	2.67244
555.394	667	2.52482	588.701	707	2.67623
556.227	668	2.52860	589.534	708	2.68001
557.060	669	2.53239	590.367	709	2.68380
557.892	670	2.53617	591.199	710	2.68758
558.725	671	2.53996	592.032	711	2.69137
559.558	672	2.54374	592.865	712	2.69515
560.390	673	2.54753	593.697	713	2.69894
561.223	674	2.55131	594.530	714	2.70273
562.056	675	2.55510	595.363	715	2.70651
562.888	676	2.55888	596.195	716	2.71030
563.721	677	2.56267	597.028	717	2.71408
564.554	678	2.56645	597.861	718	2.71787
565.386	679	2.57024	598.693	719	2.72165
566.219	680	2.57402	599.526	720	2.72544

TABLE A-1 (Continued) Gallons (UK)—Gallons (US)—Cubic Meters

UK	US	M3	UK	US	M3
600.359	721	2.72922	633.660	761	2.88064
601.191	722	2.73301	634.498	762	2.88442
602.024	723	2.73679	635.331	763	2.88821
602.857	724	2.74058	636.164	764	2.89199
603.689	725	2.74436	636.996	765	2.89578
604.522	726	2.74815	637.829	766	2.89956
605.355	727	2.75193	638.662	767	2.90335
606.187	728	2.75572	639.494	768	2.90713
607.020	729	2.75951	640.327	769	2.91092
607.853	730	2.76329	641.160	770	2.91470
608.685	731	2.76708	641.992	771	2.91849
609.518	732	2.77086	642.825	772	2.92227
610.351	733	2.77465	643.658	773	2.92606
611.183	734	2.77843	644.490	774	2.92985
612.016	735	2.78222	645.323	775	2.93363
612.849	736	2.78600	646.156	776	2.93742
613.681	737	2.78979	646.988	777	2.94120
614.514	738	2.79357	647.821	778	2.94499
615.347	739	2.79736	648.654	779	2.94877
616.180	740	2.80114	649.487	780	2.95256
617.012	741	2.80493	650.319	781	2.95634
617.845	742	2.80871	651.152	782	2.96013
618.678	743	2.81250	651.985	783	2.96391
619.510	744	2.81629	652.817	784	2.96770
620.343	745	2.82007	653.650	785	2.97148
621.176	746	2.82386	654.483	786	2.97527
622.008	747	2.82764	655.315	787	2.97905
622.841	748	2.83143	656.148	788	2.98284
623.674	749	2.83521	656.981	789	2.98663
624.506	750	2.83900	657.813	790	2.99041
625.339	751	2.84278	658.646	791	2.99420
626.172	752	2.84657	659.479	792	2.99798
627.004	753	2.85035	660.311	793	3.00177
627.837	754	2.85414	661.144	794	3.00555
628.670	755	2.85792	661.977	795	3.00934
629.502	756	2.86171	662.809	796	3.01312
630.335	757	2.86549	663.642	797	3.01691
631.168	758	2.86928	664.475	798	3.02069
632.000	759	2.87307	665.307	799	3.02448
632.833	760	2.87685	666.140	800	3.02826

Appendices

TABLE A-1 (Continued)   Gallons (UK)—Gallons (US)—Cubic Meters

UK	US	M3	UK	US	M3
666.973	801	3.03205	700.280	841	3.18346
667.805	802	3.03583	701.112	842	3.18725
668.638	803	3.03962	701.945	843	3.19103
669.471	804	3.04341	702.778	844	3.19482
670.303	805	3.04719	703.610	845	3.19860
671.136	806	3.05098	704.443	846	3.20239
671.969	807	3.05476	705.276	847	3.20617
672.801	808	3.05855	706.108	848	3.20996
673.634	809	3.06233	706.941	849	3.21375
674.467	810	3.06612	707.774	850	3.21753
675.299	811	3.06990	708.606	851	3.22132
676.132	812	3.07369	709.439	852	3.22510
676.965	813	3.07747	710.272	853	3.22889
677.797	814	3.08126	711.104	854	3.23267
678.630	815	3.08504	711.937	855	3.23646
679.463	816	3.08883	712.770	856	3.24024
680.295	817	3.09261	713.602	857	3.24403
681.128	818	3.09640	714.435	858	3.24781
681.961	819	3.10019	715.268	859	3.25160
682.794	820	3.10397	716.101	860	3.25538
683.626	821	3.10776	716.933	861	3.25917
684.459	822	3.11154	717.766	862	3.26295
685.292	823	3.11533	718.599	863	3.26674
686.124	824	3.11911	719.431	864	3.27053
686.957	825	3.12290	720.264	865	3.27431
687.790	826	3.12668	721.097	866	3.27810
688.622	827	3.13047	721.929	867	3.28188
689.455	828	3.13425	722.762	868	3.28567
690.288	829	3.13804	723.595	869	3.28945
691.120	830	3.14182	724.427	870	3.29324
691.953	831	3.14561	725.260	871	3.29702
692.786	832	3.14939	726.093	872	3.30081
693.618	833	3.15318	726.925	873	3.30459
694.451	834	3.15697	727.758	874	3.30838
695.284	835	3.16075	728.591	875	3.31216
696.116	836	3.16454	729.423	876	3.31595
696.949	837	3.16832	730.256	877	3.31973
697.782	838	3.17211	731.089	878	3.32352
698.614	839	3.17589	731.921	879	3.32731
699.447	840	3.17968	732.754	880	3.33109

TABLE A-1 (Continued) Gallons (UK)—Gallons (US)—Cubic Meters

UK	US	M3	UK	US	M3
733.587	881	3.33488	766.894	921	3.48629
734.419	882	3.33866	767.726	922	3.49007
735.252	883	3.34245	768.559	923	3.49386
736.085	884	3.34623	769.392	924	3.49764
736.917	885	3.35002	770.224	925	3.50143
737.750	886	3.35380	771.057	926	3.50522
738.583	887	3.35759	771.890	927	3.50900
739.415	888	3.36137	772.722	928	3.51279
740.248	889	3.36516	773.555	929	3.51657
741.081	890	3.36894	774.388	930	3.52036
741.913	891	3.37273	775.220	931	3.52414
742.746	892	3.37651	776.053	932	3.52793
743.579	893	3.38030	776.886	933	3.53171
744.411	894	3.38409	777.718	934	3.53550
745.244	895	3.38787	778.551	935	3.53928
746.077	896	3.39166	779.384	936	3.54307
746.909	897	3.39544	780.216	937	3.54685
747.742	898	3.39923	781.049	938	3.55064
748.575	899	3.40301	781.882	939	3.55442
749.408	900	3.40680	782.715	940	3.55821
750.240	901	3.41058	783.547	941	3.56200
751.073	902	3.41437	784.380	942	3.56578
751.906	903	3.41815	785.213	943	3.56957
752.738	904	3.42194	786.045	944	3.57335
753.571	905	3.42572	786.878	945	3.57714
754.404	906	3.42951	787.711	946	3.58092
755.236	907	3.43329	788.543	947	3.58471
756.069	908	3.43708	789.376	948	3.58849
756.902	909	3.44086	790.209	949	3.59228
757.734	910	3.44465	791.041	950	3.59606
758.567	911	3.44844	791.874	951	3.59985
759.400	912	3.45222	792.707	952	3.60363
760.232	913	3.45601	793.539	953	3.60742
761.065	914	3.45979	794.372	954	3.61120
761.898	915	3.46358	795.205	955	3.61499
762.730	916	3.46736	796.037	956	3.61878
763.563	917	3.47115	796.870	957	3.62256
764.396	918	3.47493	797.703	958	3.62635
765.228	919	3.47872	798.535	959	3.63013
766.061	920	3.48250	799.368	960	3.63392

Appendices

TABLE A-1 (Continued)   Gallons (UK)—Gallons (US)—Cubic Meters

UK	US	M3	UK	US	M3
800.201	961	3.63770	833.508	1001	3.78912
801.033	962	3.64149	834.340	1002	3.79290
801.866	963	3.64527	835.173	1003	3.79669
802.699	964	3.64906	836.006	1004	3.80047
803.531	965	3.65284	836.838	1005	3.80426
804.364	966	3.65663	837.671	1006	3.80804
805.197	967	3.66041	838.504	1007	3.81183
806.029	968	3.66420	839.336	1008	3.81561
806.862	969	3.66798	840.169	1009	3.81940
807.695	970	3.67177	841.002	1010	3.82318
808.527	971	3.67556	841.834	1011	3.82697
809.360	972	3.67934	842.667	1012	3.83075
810.193	973	3.68313	843.500	1013	3.83454
811.025	974	3.68691	844.332	1014	3.83832
811.858	975	3.69070	845.165	1015	3.84211
812.691	976	3.69448	845.998	1016	3.84590
813.523	977	3.69827	846.830	1017	3.84968
814.356	978	3.70205	847.663	1018	3.85347
815.189	979	3.70584	848.496	1019	3.85725
816.021	980	3.70962	849.328	1020	3.86104
816.854	981	3.71341	850.161	1021	3.86482
817.687	982	3.71719	850.994	1022	3.86861
818.520	983	3.72098	851.827	1023	3.87239
819.352	984	3.72476	852.659	1024	3.87618
820.185	985	3.72855	853.492	1025	3.87996
821.018	986	3.73234	854.325	1026	3.88375
821.850	987	3.73612	855.157	1027	3.88753
822.683	988	3.73991	855.990	1028	3.89132
823.516	989	3.74369	856.823	1029	3.89510
824.348	990	3.74748	857.655	1030	3.89889
825.181	991	3.75126	858.488	1031	3.90268
826.014	992	3.75505	859.321	1032	3.90646
826.846	993	3.75883	860.153	1033	3.91025
827.679	994	3.76262	860.986	1034	3.91403
828.512	995	3.76640	861.819	1035	3.91782
829.344	996	3.77019	862.651	1036	3.92160
830.177	997	3.77397	863.484	1037	3.92539
831.010	998	3.77776	864.317	1038	3.92917
831.842	999	3.78154	865.149	1039	3.93296
832.675	1000	3.78533	865.982	1040	3.93674

TABLE A-1 (Continued) Gallons (UK)—Gallons (US)—Cubic Meters

UK	US	M3	UK	US	M3
866.815	1041	3.94053	900.122	1081	4.09194
867.647	1042	3.94431	900.954	1082	4.09573
868.480	1043	3.94810	901.787	1083	4.09951
869.313	1044	3.95188	902.620	1084	4.10330
870.145	1045	3.95567	903.452	1085	4.10708
870.978	1046	3.95946	904.285	1086	4.11087
871.811	1047	3.96324	905.118	1087	4.11465
872.643	1048	3.96703	905.950	1088	4.11844
873.476	1049	3.97081	906.783	1089	4.12222
874.309	1050	3.97460	907.616	1090	4.12601
875.141	1051	3.97838	908.448	1091	4.12980
875.974	1052	3.98217	909.281	1092	4.13358
876.807	1053	3.98595	910.114	1093	4.13737
877.639	1054	3.98974	910.946	1094	4.14115
878.472	1055	3.99352	911.779	1095	4.14494
879.305	1056	3.99731	912.612	1096	4.14872
880.137	1057	4.00109	913.444	1097	4.15251
880.970	1058	4.00488	914.277	1098	4.15629
881.803	1059	4.00866	915.110	1099	4.16008
882.636	1060	4.01245	915.943	1100	4.16386
883.468	1061	4.01624	916.775	1101	4.16765
884.301	1062	4.02002	917.608	1102	4.17143
885.134	1063	4.02381	918.441	1103	4.17522
885.966	1064	4.02759	919.273	1104	4.17900
886.799	1065	4.03138	920.106	1105	4.18279
887.632	1066	4.03516	920.939	1106	4.18657
888.464	1067	4.03895	921.771	1107	4.19036
889.297	1068	4.04273	922.604	1108	4.19415
890.130	1069	4.04652	923.437	1109	4.19793
890.962	1070	4.05030	924.269	1110	4.20172
891.795	1071	4.05409	925.102	1111	4.20550
892.628	1072	4.05787	925.935	1112	4.20929
893.460	1073	4.06166	926.767	1113	4.21307
894.293	1074	4.06544	927.600	1114	4.21686
895.126	1075	4.06923	928.433	1115	4.22064
895.958	1076	4.07302	929.265	1116	4.22443
896.791	1077	4.07680	930.098	1117	4.22821
897.624	1078	4.08059	930.931	1118	4.23200
898.456	1079	4.08437	931.763	1119	4.23578
899.289	1080	4.08816	932.596	1120	4.23957

TABLE A-2 Cubic Meters—Gallons (UK)—Gallons (US)

M3	UK	US	M3	UK	US
0.00455	1	1.201	0.18638	41	49.239
0.00909	2	2.402	0.19093	42	50.440
0.01364	3	3.603	0.19548	43	51.641
0.01818	4	4.804	0.20002	44	52.842
0.02273	5	6.005	0.20457	45	54.043
0.02728	6	7.206	0.20911	46	55.244
0.03182	7	8.407	0.21366	47	56.445
0.03637	8	9.608	0.21821	48	57.646
0.04091	9	10.809	0.22275	49	58.847
0.04546	10	12.009	0.22730	50	60.047
0.05001	11	13.210	0.23184	51	61.248
0.05455	12	14.411	0.23639	52	62.449
0.05910	13	15.612	0.24094	53	63.650
0.06364	14	16.813	0.24548	54	64.851
0.06819	15	18.014	0.25003	55	66.052
0.07274	16	19.215	0.25457	56	67.253
0.07728	17	20.416	0.25912	57	68.454
0.08183	18	21.617	0.26367	58	69.655
0.08637	19	22.818	0.26821	59	70.856
0.09092	20	24.019	0.27276	60	72.057
0.09547	21	25.220	0.27730	61	73.258
0.10001	22	26.421	0.28185	62	74.459
0.10456	23	27.622	0.28640	63	75.660
0.10910	24	28.823	0.29094	64	76.861
0.11365	25	30.024	0.29549	65	78.062
0.11819	26	31.225	0.30003	66	79.263
0.12274	27	32.426	0.30458	67	80.464
0.12729	28	33.627	0.30913	68	81.665
0.13183	29	34.828	0.31367	69	82.866
0.13638	30	36.028	0.31822	70	84.066
0.14092	31	37.229	0.32276	71	85.267
0.14547	32	38.430	0.32731	72	86.468
0.15002	33	39.631	0.33186	73	87.669
0.15456	34	40.832	0.33640	74	88.870
0.15911	35	42.033	0.34095	75	90.071
0.16365	36	43.234	0.34549	76	91.272
0.16820	37	44.435	0.35004	77	92.473
0.17275	38	45.636	0.35458	78	93.674
0.17729	39	46.837	0.35913	79	94.875
0.18184	40	48.038	0.36368	80	96.076

TABLE A-2 (Continued) Cubic Meters—Gallons (UK)—Gallons (US)

M3	UK	US	M3	UK	US
0.36822	81	97.277	0.55006	121	145.315
0.37277	82	98.478	0.55461	122	146.516
0.37731	83	99.679	0.55915	123	147.717
0.38186	84	100.880	0.56370	124	148.918
0.38641	85	102.081	0.56825	125	150.119
0.39095	86	103.282	0.57279	126	151.320
0.39550	87	104.483	0.57734	127	152.521
0.40004	88	105.684	0.58188	128	153.722
0.40459	89	106.885	0.58643	129	154.923
0.40914	90	108.085	0.59097	130	156.123
0.41368	91	109.286	0.59552	131	157.324
0.41823	92	110.487	0.60007	132	158.525
0.42277	93	111.688	0.60461	133	159.726
0.42732	94	112.889	0.60916	134	160.927
0.43187	95	114.090	0.61370	135	162.128
0.43641	96	115.291	0.61825	136	163.329
0.44096	97	116.492	0.62280	137	164.530
0.44550	98	117.693	0.62734	138	165.731
0.45005	99	118.894	0.63189	139	166.932
0.45460	100	120.095	0.63643	140	168.133
0.45914	101	121.296	0.64098	141	169.334
0.46369	102	122.497	0.64553	142	170.535
0.46823	103	123.698	0.65007	143	171.736
0.47278	104	124.899	0.65462	144	172.937
0.47733	105	126.100	0.65916	145	174.138
0.48187	106	127.301	0.66371	146	175.339
0.48642	107	128.502	0.66826	147	176.540
0.49096	108	129.703	0.67280	148	177.741
0.49551	109	130.904	0.67735	149	178.941
0.50006	110	132.104	0.68189	150	180.142
0.50460	111	133.305	0.68644	151	181.343
0.50915	112	134.506	0.69099	152	182.544
0.51369	113	135.707	0.69553	153	183.745
0.51824	114	136.908	0.70008	154	184.946
0.52279	115	138.109	0.70462	155	186.147
0.52733	116	139.310	0.70917	156	187.348
0.53188	117	140.511	0.71372	157	188.549
0.53642	118	141.712	0.71826	158	189.750
0.54097	119	142.913	0.72281	159	190.951
0.54552	120	144.114	0.72735	160	192.152

Appendices

TABLE A-2 (Continued) Cubic Meters—Gallons (UK)—Gallons (US)

Please God, Let me have Peggy Breen!

I would appreciate it very much.

M3	UK	US	M3	UK	US
0.73190	161	193.353	0.91374	201	241.391
0.73645	162	194.554	0.91828	202	242.592
0.74099	163	195.755	0.92283	203	243.793
0.74554	164	196.956	0.92738	204	244.994
0.75008	165	198.157	0.93192	205	246.195
0.75463	166	199.358	0.93647	206	247.396
0.75918	167	200.559	0.94101	207	248.597
0.76372	168	201.760	0.94556	208	249.798
0.76827	169	202.960	0.95011	209	250.998
0.77281	170	204.161	0.95465	210	252.199
0.77736	171	205.362	0.95920	211	253.400
0.78191	172	206.563	0.96374	212	254.601
0.78645	173	207.764	0.96829	213	255.802
0.79100	174	208.965	0.97284	214	257.003
0.79554	175	210.166	0.97738	215	258.204
0.80009	176	211.367	0.98193	216	259.405
0.80463	177	212.568	0.98647	217	260.606
0.80918	178	213.769	0.99102	218	261.807
0.81373	179	214.970	0.99557	219	263.008
0.81827	180	216.171	1.00011	220	264.209
0.82282	181	217.372	1.00466	221	265.410
0.82736	182	218.573	1.00920	222	266.611
0.83191	183	219.774	1.01375	223	267.812
0.83646	184	220.975	1.01829	224	269.013
0.84100	185	222.176	1.02284	225	270.214
0.84555	186	223.377	1.02739	226	271.415
0.85009	187	224.578	1.03193	227	272.615
0.85464	188	225.779	1.03648	228	273.816
0.85919	189	226.979	1.04102	229	275.017
0.86373	190	228.180	1.04557	230	276.218
0.86828	191	229.381	1.05012	231	277.419
0.87282	192	230.582	1.05466	232	278.620
0.87737	193	231.783	1.05921	233	279.821
0.88192	194	232.984	1.06375	234	281.022
0.88646	195	234.185	1.06830	235	282.223
0.89101	196	235.386	1.07285	236	283.424
0.89555	197	236.587	1.07739	237	284.625
0.90010	198	237.788	1.08194	238	285.826
0.90465	199	238.989	1.08648	239	287.027
0.90919	200	240.190	1.09103	240	288.228

TABLE A-2 (Continued) Cubic Meters—Gallons (UK)—Gallons (US)

M3	UK	US	M3	UK	US
1.09558	241	289.429	1.27741	281	337.467
1.10012	242	290.630	1.28196	282	338.668
1.10467	243	291.831	1.28651	283	339.869
1.10921	244	293.031	1.29105	284	341.070
1.11376	245	294.233	1.29560	285	342.271
1.11831	246	295.434	1.30014	286	343.471
1.12285	247	296.635	1.30469	287	344.672
1.12740	248	297.835	1.30924	288	345.873
1.13194	249	299.036	1.31378	289	347.074
1.13649	250	300.237	1.31833	290	348.275
1.14104	251	301.438	1.32287	291	349.476
1.14558	252	302.639	1.32742	292	350.677
1.15013	253	303.840	1.33197	293	351.878
1.15467	254	305.041	1.33651	294	353.079
1.15922	255	306.242	1.34106	295	354.280
1.16377	256	307.443	1.34560	296	355.481
1.16831	257	308.644	1.35015	297	356.682
1.17286	258	309.845	1.35470	298	357.883
1.17740	259	311.046	1.35924	299	359.084
1.18195	260	312.247	1.36379	300	360.285
1.18649	261	313.448	1.36833	301	361.486
1.19104	262	314.649	1.37288	302	362.687
1.19559	263	315.850	1.37743	303	363.888
1.20013	264	317.051	1.38197	304	365.089
1.20468	265	318.251	1.38652	305	366.290
1.20922	266	319.452	1.39106	306	367.490
1.21377	267	320.653	1.39561	307	368.691
1.21832	268	321.854	1.40016	308	369.892
1.22286	269	323.055	1.40470	309	371.093
1.22741	270	324.256	1.40925	310	372.294
1.23195	271	325.457	1.41379	311	373.495
1.23650	272	326.658	1.41834	312	374.696
1.24105	273	327.859	1.42288	313	375.897
1.24559	274	329.060	1.42743	314	377.098
1.25014	275	330.261	1.43198	315	378.299
1.25468	276	331.462	1.43652	316	379.500
1.25923	277	332.663	1.44107	317	380.701
1.26378	278	333.864	1.44561	318	381.902
1.26832	279	335.065	1.45016	319	383.103
1.27287	280	336.266	1.45471	320	384.304

TABLE A-2 (Continued)   Cubic Meters—Gallons (UK)—Gallons (US)

M3	UK	US	M3	UK	US
1.45925	321	385.505	1.64109	361	433.543
1.46380	322	386.706	1.64564	362	434.744
1.46834	323	387.907	1.65018	363	435.945
1.47289	324	389.108	1.65473	364	437.146
1.47744	325	390.309	1.65928	365	438.346
1.48198	326	391.510	1.66382	366	439.547
1.48653	327	392.710	1.66837	367	440.748
1.49107	328	393.911	1.67291	368	441.949
1.49562	329	395.112	1.67746	369	443.150
1.50017	330	396.313	1.68200	370	444.351
1.50471	331	397.514	1.68655	371	445.552
1.50926	332	398.715	1.69110	372	446.753
1.51380	333	399.916	1.69564	373	447.954
1.51835	334	401.117	1.70019	374	449.155
1.52290	335	402.318	1.70473	375	450.356
1.52744	336	403.519	1.70928	376	451.557
1.53199	337	404.720	1.71383	377	452.758
1.53653	338	405.921	1.71837	378	453.959
1.54108	339	407.122	1.72292	379	455.160
1.54563	340	408.323	1.72746	380	456.361
1.55017	341	409.524	1.73201	381	457.562
1.55472	342	410.725	1.73656	382	458.763
1.55926	343	411.926	1.74110	383	459.964
1.56381	344	413.126	1.74565	384	461.165
1.56836	345	414.327	1.75019	385	462.365
1.57290	346	415.529	1.75474	386	463.566
1.57745	347	416.729	1.75929	387	464.767
1.58199	348	417.930	1.76383	388	465.968
1.58654	349	419.131	1.76838	389	467.169
1.59109	350	420.332	1.77292	390	468.370
1.59563	351	421.533	1.77747	391	469.571
1.60018	352	422.734	1.78202	392	470.772
1.60472	353	423.935	1.78656	393	471.973
1.60927	354	425.136	1.79111	394	473.174
1.61382	355	426.337	1.79565	395	474.375
1.61836	356	427.538	1.80020	396	475.576
1.62291	357	428.739	1.80475	397	476.777
1.62745	358	429.940	1.80929	398	477.978
1.63200	359	431.141	1.81384	399	479.179
1.63655	360	432.342	1.81838	400	480.380

TABLE A-2 (Continued) Cubic Meters—Gallons (UK)—Gallons (US)

M3	UK	US	M3	UK	US
1.82293	401	481.581	2.00477	441	529.619
1.82748	402	482.782	2.00931	442	530.820
1.83202	403	483.983	2.01386	443	532.021
1.83657	404	485.184	2.01841	444	533.221
1.84111	405	486.385	2.02295	445	534.422
1.84566	406	487.585	2.02750	446	535.624
1.85021	407	488.786	2.03204	447	536.824
1.85475	408	489.987	2.03659	448	538.025
1.85930	409	491.188	2.04114	449	539.226
1.86384	410	492.389	2.04568	450	540.427
1.86839	411	493.590	2.05023	451	541.628
1.87294	412	494.791	2.05477	452	542.829
1.87748	413	495.992	2.05932	453	544.030
1.88203	414	497.193	2.06387	454	545.231
1.88657	415	498.394	2.06841	455	546.432
1.89112	416	499.595	2.07296	456	547.633
1.89567	417	500.796	2.07750	457	548.834
1.90021	418	501.997	2.08205	458	550.035
1.90476	419	503.198	2.08660	459	551.236
1.90930	420	504.399	2.09114	460	552.437
1.91385	421	505.600	2.09569	461	553.638
1.91840	422	506.801	2.10023	462	554.839
1.92294	423	508.002	2.10478	463	556.040
1.92749	424	509.203	2.10933	464	557.240
1.93203	425	510.404	2.11387	465	558.441
1.93658	426	511.604	2.11842	466	559.642
1.94112	427	512.805	2.12296	467	560.843
1.94567	428	514.006	2.12751	468	562.044
1.95022	429	515.207	2.13206	469	563.245
1.95476	430	516.408	2.13660	470	564.446
1.95931	431	517.609	2.14115	471	565.647
1.96385	432	518.810	2.14569	472	566.848
1.96840	433	520.011	2.15024	473	568.049
1.97295	434	521.212	2.15479	474	569.250
1.97749	435	522.413	2.15933	475	570.451
1.98204	436	523.614	2.16388	476	571.652
1.98658	437	524.815	2.16842	477	572.853
1.99113	438	526.016	2.17297	478	574.054
1.99568	439	527.217	2.17752	479	575.255
2.00022	440	528.418	2.18206	480	576.456

TABLE A-2 (Continued) Cubic Meters—Gallons (UK)—Gallons (US)

M3	UK	US	M3	UK	US
2.18661	481	577.657	2.36845	521	625.695
2.19115	482	578.858	2.37299	522	626.896
2.19570	483	580.059	2.37754	523	628.096
2.20024	484	581.260	2.38208	524	629.298
2.20479	485	582.460	2.38663	525	630.499
2.20934	486	583.661	2.39117	526	631.699
2.21388	487	584.862	2.39572	527	632.900
2.21843	488	586.063	2.40027	528	634.101
2.22297	489	587.264	2.40481	529	635.302
2.22752	490	588.465	2.40936	530	636.503
2.23207	491	589.666	2.41390	531	637.704
2.23661	492	590.867	2.41845	532	638.905
2.24116	493	592.068	2.42300	533	640.106
2.24570	494	593.269	2.42754	534	641.307
2.25025	495	594.470	2.43209	535	642.508
2.25480	496	595.671	2.43663	536	643.709
2.25934	497	596.872	2.44118	537	644.910
2.26389	498	598.073	2.44573	538	646.111
2.26843	499	599.274	2.45027	539	647.312
2.27298	500	600.475	2.45482	540	648.513
2.27753	501	601.676	2.45936	541	649.714
2.28207	502	602.877	2.46391	542	650.915
2.28662	503	604.078	2.46846	543	652.115
2.29116	504	605.279	2.47300	544	653.316
2.29571	505	606.479	2.47755	545	654.517
2.30026	506	607.680	2.48209	546	655.719
2.30480	507	608.881	2.48664	547	656.919
2.30935	508	610.082	2.49119	548	658.120
2.31389	509	611.283	2.49573	549	659.321
2.31844	510	612.484	2.50028	550	660.522
2.32299	511	613.685	2.50482	551	661.723
2.32753	512	614.886	2.50937	552	662.924
2.33208	513	616.087	2.51392	553	664.125
2.33662	514	617.288	2.51846	554	665.326
2.34117	515	618.489	2.52301	555	666.527
2.34572	516	619.690	2.52755	556	667.728
2.35026	517	620.891	2.53210	557	668.929
2.35481	518	622.092	2.53665	558	670.130
2.35935	519	623.293	2.54119	559	671.331
2.36390	520	624.494	2.54574	560	672.532

TABLE A-2 (Continued) Cubic Meters—Gallons (UK)—Gallons (US)

M3	UK	US	M3	UK	US
2.55028	561	673.733	2.73212	601	721.771
2.55483	562	674.934	2.73667	602	722.972
2.55938	563	676.135	2.74121	603	724.173
2.56392	564	677.335	2.74576	604	725.374
2.56847	565	678.536	2.75031	605	726.574
2.57301	566	679.737	2.75485	606	727.775
2.57756	567	680.938	2.75940	607	728.976
2.58210	568	682.139	2.76394	608	730.177
2.58665	569	683.340	2.76849	609	731.378
2.59120	570	684.541	2.77304	610	732.579
2.59574	571	685.742	2.77758	611	733.780
2.60029	572	686.943	2.78213	612	734.981
2.60483	573	688.144	2.78667	613	736.182
2.60938	574	689.345	2.79122	614	737.383
2.61393	575	690.546	2.79576	615	738.584
2.61847	576	691.747	2.80031	616	739.785
2.62302	577	692.948	2.80486	617	740.986
2.62756	578	694.149	2.80940	618	742.187
2.63211	579	695.350	2.81395	619	743.388
2.63666	580	696.551	2.81849	620	744.589
2.64120	581	697.752	2.82304	621	745.790
2.64575	582	698.953	2.82759	622	746.990
2.65029	583	700.154	2.83213	623	748.191
2.65484	584	701.354	2.83668	624	749.393
2.65939	585	702.555	2.84122	625	750.594
2.66393	586	703.756	2.84577	626	751.794
2.66848	587	704.957	2.85032	627	752.995
2.67302	588	706.158	2.85486	628	754.196
2.67757	589	707.359	2.85941	629	755.397
2.68212	590	708.560	2.86395	630	756.598
2.68666	591	709.761	2.86850	631	757.799
2.69121	592	710.962	2.87305	632	759.000
2.69575	593	712.163	2.87759	633	760.201
2.70030	594	713.364	2.88214	634	761.402
2.70485	595	714.565	2.88668	635	762.603
2.70939	596	715.766	2.89123	636	763.804
2.71394	597	716.967	2.89578	637	765.005
2.71848	598	718.168	2.90032	638	766.206
2.72303	599	719.369	2.90487	639	767.407
2.72758	600	720.570	2.90941	640	768.608

TABLE A-2 (Continued) Cubic Meters—Gallons (UK)—Gallons (US)

M3	UK	US	M3	UK	US
2.91396	641	769.809	3.09580	681	817.847
2.91851	642	771.010	3.10034	682	819.048
2.92305	643	772.210	3.10489	683	820.249
2.92760	644	773.411	3.10944	684	821.449
2.93214	645	774.612	3.11398	685	822.650
2.93669	646	775.813	3.11853	686	823.851
2.94124	647	777.014	3.12307	687	825.052
2.94578	648	778.215	3.12762	688	826.253
2.95033	649	779.416	3.13217	689	827.454
2.95487	650	780.617	3.13671	690	828.655
2.95942	651	781.818	3.14126	691	829.856
2.96397	652	783.019	3.14580	692	831.057
2.96851	653	784.220	3.15035	693	832.258
2.97306	654	785.421	3.15490	694	833.459
2.97760	655	786.622	3.15944	695	834.660
2.98215	656	787.823	3.16399	696	835.861
2.98670	657	789.024	3.16853	697	837.062
2.99124	658	790.225	3.17308	698	838.263
2.99579	659	791.426	3.17763	699	839.464
3.00033	660	792.627	3.18217	700	840.665
3.00488	661	793.828	3.18672	701	841.865
3.00943	662	795.029	3.19126	702	843.067
3.01397	663	796.229	3.19581	703	844.268
3.01852	664	797.430	3.20036	704	845.469
3.02306	665	798.631	3.20490	705	846.669
3.02761	666	799.832	3.20945	706	847.870
3.03216	667	801.033	3.21399	707	849.071
3.03670	668	802.234	3.21854	708	850.272
3.04125	669	803.435	3.22309	709	851.473
3.04579	670	804.636	3.22763	710	852.674
3.05034	671	805.837	3.23218	711	853.875
3.05488	672	807.038	3.23672	712	855.076
3.05943	673	808.239	3.24127	713	856.277
3.06398	674	809.440	3.24582	714	857.478
3.06852	675	810.641	3.25036	715	858.679
3.07307	676	811.842	3.25491	716	859.880
3.07761	677	813.043	3.25945	717	861.081
3.08216	678	814.244	3.26400	718	862.282
3.08671	679	815.445	3.26855	719	863.483
3.09125	680	816.646	3.27309	720	864.684

TABLE A-2 (Continued) Cubic Meters—Gallons (UK)—Gallons (US)

M3	UK	US	M3	UK	US
3.27764	721	865.885	3.45948	761	913.923
3.28218	722	867.085	3.46402	762	915.124
3.28673	723	868.286	3.46857	763	916.324
3.29128	724	869.488	3.47311	764	917.525
3.29582	725	870.688	3.47766	765	918.726
3.30037	726	871.889	3.48221	766	919.927
3.30491	727	873.090	3.48675	767	921.128
3.30946	728	874.291	3.49130	768	922.329
3.31400	729	875.492	3.49584	769	923.530
3.31855	730	876.693	3.50039	770	924.731
3.32310	731	877.894	3.50494	771	925.932
3.32764	732	879.095	3.50948	772	927.133
3.33219	733	880.296	3.51403	773	928.334
3.33673	734	881.497	3.51857	774	929.535
3.34128	735	882.698	3.52312	775	930.736
3.34583	736	883.899	3.52767	776	931.937
3.35037	737	885.100	3.53221	777	933.138
3.35492	738	886.301	3.53676	778	934.339
3.35946	739	887.502	3.54130	779	935.540
3.36401	740	888.703	3.54585	780	936.741
3.36856	741	889.904	3.55040	781	937.942
3.37310	742	891.104	3.55494	782	939.143
3.37765	743	892.305	3.55949	783	940.344
3.38219	744	893.506	3.56403	784	941.544
3.38674	745	894.707	3.56858	785	942.745
3.39129	746	895.908	3.57312	786	943.946
3.39583	747	897.109	3.57767	787	945.147
3.40038	748	898.310	3.58222	788	946.348
3.40492	749	899.511	3.58676	789	947.549
3.40947	750	900.712	3.59131	790	948.750
3.41402	751	901.913	3.59585	791	949.951
3.41856	752	903.114	3.60040	792	951.152
3.42311	753	904.315	3.60495	793	952.353
3.42765	754	905.516	3.60949	794	953.554
3.43220	755	906.717	3.61404	795	954.755
3.43675	756	907.918	3.61858	796	955.956
3.44129	757	909.119	3.62313	797	957.157
3.44584	758	910.320	3.62768	798	958.358
3.45038	759	911.521	3.63222	799	959.559
3.45493	760	912.722	3.63677	800	960.760

TABLE A-2 (Continued) Cubic Meters—Gallons (UK)—Gallons (US)

M3	UK	US	M3	UK	US
3.64131	801	961.96	3.82315	841	1010.00
3.64586	802	963.16	3.82770	842	1011.20
3.65041	803	964.36	3.83224	843	1012.40
3.65495	804	965.56	3.83679	844	1013.60
3.65950	805	966.76	3.84134	845	1014.80
3.66404	806	967.97	3.84588	846	1016.00
3.66859	807	969.17	3.85043	847	1017.20
3.67314	808	970.37	3.85497	848	1018.41
3.67768	809	971.57	3.85952	849	1019.61
3.68223	810	972.77	3.86407	850	1020.81
3.68677	811	973.97	3.86861	851	1022.01
3.69132	812	975.17	3.87316	852	1023.21
3.69587	813	976.37	3.87770	853	1024.41
3.70041	814	977.57	3.88225	854	1025.61
3.70496	815	978.77	3.88680	855	1026.81
3.70950	816	979.97	3.89134	856	1028.01
3.71405	817	981.18	3.89589	857	1029.21
3.71860	818	982.38	3.90043	858	1030.41
3.72314	819	983.58	3.90498	859	1031.62
3.72769	820	984.78	3.90953	860	1032.82
3.73223	821	985.98	3.91407	861	1034.02
3.73678	822	987.18	3.91862	862	1035.22
3.74133	823	988.38	3.92316	863	1036.42
3.74587	824	989.58	3.92771	864	1037.62
3.75042	825	990.78	3.93226	865	1038.82
3.75496	826	991.98	3.93680	866	1040.02
3.75951	827	993.19	3.94135	867	1041.22
3.76406	828	994.39	3.94589	868	1042.42
3.76860	829	995.59	3.95044	869	1043.63
3.77315	830	996.79	3.95499	870	1044.83
3.77769	831	997.99	3.95953	871	1046.03
3.78224	832	999.19	3.96408	872	1047.23
3.78679	833	1000.39	3.96862	873	1048.43
3.79133	834	1001.59	3.97317	874	1049.63
3.79588	835	1002.79	3.97772	875	1050.83
3.80042	836	1003.99	3.98226	876	1052.03
3.80497	837	1005.19	3.98681	877	1053.23
3.80951	838	1006.40	3.99135	878	1054.43
3.81406	839	1007.60	3.99590	879	1055.63
3.81861	840	1008.80	4.00044	880	1056.84

TABLE A-2 (Continued) Cubic Meters—Gallons (UK)—Gallons (US)

M3	UK	US	M3	UK	US
4.00499	881	1058.04	4.18683	921	1106.07
4.00954	882	1059.24	4.19137	922	1107.28
4.01408	883	1060.44	4.19592	923	1108.48
4.01863	884	1061.64	4.20047	924	1109.68
4.02317	885	1062.84	4.20501	925	1110.88
4.02772	886	1064.04	4.20956	926	1112.08
4.03227	887	1065.24	4.21410	927	1113.28
4.03681	888	1066.44	4.21865	928	1114.48
4.04136	889	1067.64	4.22320	929	1115.68
4.04590	890	1068.84	4.22774	930	1116.88
4.05045	891	1070.05	4.23229	931	1118.08
4.05500	892	1071.25	4.23683	932	1119.28
4.05954	893	1072.45	4.24138	933	1120.49
4.06409	894	1073.65	4.24593	934	1121.69
4.06863	895	1074.85	4.25047	935	1122.89
4.07318	896	1076.05	4.25502	936	1124.09
4.07773	897	1077.25	4.25956	937	1125.29
4.08227	898	1078.45	4.26411	938	1126.49
4.08682	899	1079.65	4.26866	939	1127.69
4.09136	900	1080.85	4.27320	940	1128.89
4.09591	901	1082.06	4.27775	941	1130.09
4.10046	902	1083.26	4.28229	942	1131.29
4.10500	903	1084.46	4.28684	943	1132.50
4.10955	904	1085.66	4.29139	944	1133.70
4.11409	905	1086.86	4.29593	945	1134.90
4.11864	906	1088.06	4.30048	946	1136.10
4.12319	907	1089.26	4.30502	947	1137.30
4.12773	908	1090.46	4.30957	948	1138.50
4.13228	909	1091.66	4.31412	949	1139.70
4.13682	910	1092.86	4.31866	950	1140.90
4.14137	911	1094.06	4.32321	951	1142.10
4.14592	912	1095.27	4.32775	952	1143.30
4.15046	913	1096.47	4.33230	953	1144.50
4.15501	914	1097.67	4.33685	954	1145.71
4.15955	915	1098.87	4.34139	955	1146.91
4.16410	916	1100.07	4.34594	956	1148.11
4.16864	917	1101.27	4.35048	957	1149.31
4.17319	918	1102.47	4.35503	958	1150.51
4.17774	919	1103.67	4.35958	959	1151.71
4.18228	920	1104.87	4.36412	960	1152.91

Appendices

TABLE A-2 (Continued) Cubic Meters—Gallons (UK)—Gallons (US)

M3	UK	US	M3	UK	US
4.36867	961	1154.11	4.55051	1001	1202.15
4.37321	962	1155.31	4.55505	1002	1203.35
4.37776	963	1156.51	4.55960	1003	1204.55
4.38231	964	1157.72	4.56414	1004	1205.75
4.38685	965	1158.92	4.56869	1005	1206.95
4.39140	966	1160.12	4.57324	1006	1208.16
4.39594	967	1161.32	4.57778	1007	1209.36
4.40049	968	1162.52	4.58233	1008	1210.56
4.40504	969	1163.72	4.58687	1009	1211.76
4.40958	970	1164.92	4.59142	1010	1212.96
4.41413	971	1166.12	4.59597	1011	1214.16
4.41867	972	1167.32	4.60051	1012	1215.36
4.42322	973	1168.52	4.60506	1013	1216.56
4.42776	974	1169.72	4.60960	1014	1217.76
4.43231	975	1170.93	4.61415	1015	1218.96
4.43686	976	1172.13	4.61870	1016	1220.16
4.44140	977	1173.33	4.62324	1017	1221.37
4.44595	978	1174.53	4.62779	1018	1222.57
4.45049	979	1175.73	4.63233	1019	1223.77
4.45504	980	1176.93	4.63688	1020	1224.97
4.45959	981	1178.13	4.64143	1021	1226.17
4.46413	982	1179.33	4.64597	1022	1227.37
4.46868	983	1180.53	4.65052	1023	1228.57
4.47322	984	1181.73	4.65506	1024	1229.77
4.47777	985	1182.94	4.65961	1025	1230.97
4.48232	986	1184.14	4.66416	1026	1232.17
4.48686	987	1185.34	4.66870	1027	1233.38
4.49141	988	1186.54	4.67325	1028	1234.58
4.49595	989	1187.74	4.67779	1029	1235.78
4.50050	990	1188.94	4.68234	1030	1236.98
4.50505	991	1190.14	4.68688	1031	1238.18
4.50959	992	1191.34	4.69143	1032	1239.38
4.51414	993	1192.54	4.69598	1033	1240.58
4.51868	994	1193.74	4.70052	1034	1241.78
4.52323	995	1194.94	4.70507	1035	1242.98
4.52778	996	1196.15	4.70961	1036	1244.18
4.53232	997	1197.35	4.71416	1037	1245.38
4.53687	998	1198.55	4.71871	1038	1246.59
4.54141	999	1199.75	4.72325	1039	1247.79
4.54596	1000	1200.95	4.72780	1040	1248.99

TABLE A-2 (Continued) Cubic Meters—Gallons (UK)—Gallons (US)

M3	UK	US	M3	UK	US
4.73234	1041	1250.19	4.91418	1081	1298.23
4.73689	1042	1251.39	4.91873	1082	1299.43
4.74144	1043	1252.59	4.92327	1083	1300.63
4.74598	1044	1253.79	4.92782	1084	1301.83
4.75053	1045	1254.99	4.93237	1085	1303.03
4.75507	1046	1256.19	4.93691	1086	1304.23
4.75962	1047	1257.39	4.94146	1087	1305.43
4.76417	1048	1258.60	4.94600	1088	1306.63
4.76871	1049	1259.80	4.95055	1089	1307.83
4.77326	1050	1261.00	4.95510	1090	1309.03
4.77780	1051	1262.20	4.95964	1091	1310.24
4.78235	1052	1263.40	4.96419	1092	1311.44
4.78690	1053	1264.60	4.96873	1093	1312.64
4.79144	1054	1265.80	4.97328	1094	1313.84
4.79599	1055	1267.00	4.97783	1095	1315.04
4.80053	1056	1268.20	4.98237	1096	1316.24
4.80508	1057	1269.40	4.98692	1097	1317.44
4.80963	1058	1270.60	4.99146	1098	1318.64
4.81417	1059	1271.81	4.99601	1099	1319.84
4.81872	1060	1273.01	5.00056	1100	1321.04
4.82326	1061	1274.21	5.00510	1101	1322.25
4.82781	1062	1275.41	5.00965	1102	1323.45
4.83236	1063	1276.61	5.01419	1103	1324.65
4.83690	1064	1277.81	5.01874	1104	1325.85
4.84145	1065	1279.01	5.02329	1105	1327.05
4.84599	1066	1280.21	5.02783	1106	1328.25
4.85054	1067	1281.41	5.03238	1107	1329.45
4.85509	1068	1282.61	5.03692	1108	1330.65
4.85963	1069	1283.82	5.04147	1109	1331.85
4.86418	1070	1285.02	5.04602	1110	1333.05
4.86872	1071	1286.22	5.05056	1111	1334.25
4.87327	1072	1287.42	5.05511	1112	1335.46
4.87782	1073	1288.62	5.05965	1113	1336.66
4.88236	1074	1289.82	5.06420	1114	1337.86
4.88691	1075	1291.02	5.06875	1115	1339.06
4.89145	1076	1292.22	5.07329	1116	1340.26
4.89600	1077	1293.42	5.07784	1117	1341.46
4.90055	1078	1294.62	5.08238	1118	1342.66
4.90509	1079	1295.82	5.08693	1119	1343.86
4.90964	1080	1297.03	5.09148	1120	1345.06

TABLE A-3 Gallons (US)—Cubic Meters—Gallons (UK)

US	M3	UK	US	M3	UK
264.2	1	220.0	10831.3	41	9019.0
528.4	2	440.0	11095.5	42	9239.0
792.5	3	659.9	11359.6	43	9458.9
1056.7	4	879.9	11623.8	44	9678.9
1320.9	5	1099.9	11888.0	45	9898.9
1585.1	6	1319.9	12152.2	46	10118.9
1849.2	7	1539.8	12416.4	47	10338.8
2113.4	8	1759.8	12680.5	48	10558.8
2377.6	9	1979.8	12944.7	49	10778.8
2641.8	10	2199.8	13208.9	50	10998.8
2906.0	11	2419.7	13473.1	51	11218.8
3170.1	12	2639.7	13737.2	52	11438.7
3434.3	13	2859.7	14001.4	53	11658.7
3698.5	14	3079.7	14265.6	54	11878.7
3962.7	15	3299.6	14529.8	55	12098.7
4226.8	16	3519.6	14794.0	56	12318.6
4491.0	17	3739.6	15058.1	57	12538.6
4755.2	18	3959.6	15322.3	58	12758.6
5019.4	19	4179.5	15586.5	59	12978.6
5283.6	20	4399.5	15850.7	60	13198.5
5547.7	21	4619.5	16114.8	61	13418.5
5811.9	22	4839.5	16379.0	62	13638.5
6076.1	23	5059.4	16643.2	63	13858.5
6340.3	24	5279.4	16907.4	64	14078.4
6604.4	25	5499.4	17171.6	65	14298.4
6868.6	26	5719.4	17435.7	66	14518.4
7132.8	27	5939.3	17699.9	67	14738.4
7397.0	28	6159.3	17964.1	68	14958.3
7661.2	29	6379.3	18228.3	69	15178.3
7925.3	30	6599.3	18492.4	70	15398.3
8189.5	31	6819.2	18756.6	71	15618.3
8453.7	32	7039.2	19020.8	72	15838.2
8717.9	33	7259.2	19285.0	73	16058.2
8982.0	34	7479.2	19549.2	74	16278.2
9246.2	35	7699.1	19813.3	75	16498.2
9510.4	36	7919.1	20077.5	76	16718.1
9774.6	37	8139.1	20341.7	77	16938.1
10038.8	38	8359.1	20605.9	78	17158.1
10302.9	39	8579.0	20870.0	79	17378.1
10567.1	40	8799.0	21134.2	80	17598.0

TABLE A-3 (Continued)   Gallons (US)—Cubic Meters—Gallons (UK)

US	M3	UK	US	M3	UK
21398.4	81	17818.0	31965.5	121	26617.0
21662.6	82	18038.0	32229.7	122	26837.0
21926.8	83	18258.0	32493.9	123	27057.0
22190.9	84	18477.9	32758.0	124	27277.0
22455.1	85	18697.9	33022.2	125	27496.9
22719.3	86	18917.9	33286.4	126	27716.9
22983.5	87	19137.9	33550.6	127	27936.9
23247.6	88	19357.8	33814.8	128	28156.9
23511.8	89	19577.8	34078.9	129	28376.8
23776.0	90	19797.8	34343.1	130	28596.8
24040.2	91	20017.8	34607.3	131	28816.8
24304.4	92	20237.7	34871.5	132	29036.8
24568.5	93	20457.7	35135.6	133	29256.7
24832.7	94	20677.7	35399.8	134	29476.7
25096.9	95	20897.7	35664.0	135	29696.7
25361.1	96	21117.6	35928.2	136	29916.7
25625.2	97	21337.6	36192.4	137	30136.6
25889.4	98	21557.6	36456.5	138	30356.6
26153.6	99	21777.6	36720.7	139	30576.6
26417.8	100	21997.6	36984.9	140	30796.6
26682.0	101	22217.5	37249.1	141	31016.5
26946.1	102	22437.5	37513.2	142	31236.5
27210.3	103	22657.5	37777.4	143	31456.5
27474.5	104	22877.5	38041.6	144	31676.5
27738.7	105	23097.4	38305.8	145	31896.4
28002.8	106	23317.4	38570.0	146	32116.4
28267.0	107	23537.4	38834.1	147	32336.4
28531.2	108	23757.4	39098.3	148	32556.4
28795.4	109	23977.3	39362.5	149	32776.3
29059.6	110	24197.3	39626.7	150	32996.3
29323.7	111	24417.3	39890.8	151	33216.3
29587.9	112	24637.3	40155.0	152	33436.3
29852.1	113	24857.2	40419.2	153	33656.3
30116.3	114	25077.2	40683.4	154	33876.2
30380.4	115	25297.2	40947.6	155	34096.2
30644.6	116	25517.2	41211.7	156	34316.2
30908.8	117	25737.1	41475.9	157	34536.2
31173.0	118	25957.1	41740.1	158	34756.1
31437.2	119	26177.1	42004.3	159	34976.1
31701.3	120	26397.1	42268.4	160	35196.1

TABLE A-3 (Continued) Gallons (US)—Cubic Meters—Gallons (UK)

US	M3	UK	US	M3	UK
42532.6	161	35416.1	53099.7	201	44215.1
42796.8	162	35636.0	53363.9	202	44435.1
43061.0	163	35856.0	53628.1	203	44655.0
43325.2	164	36076.0	53892.3	204	44875.0
43589.3	165	36296.0	54156.4	205	45095.0
43853.5	166	36515.9	54420.6	206	45315.0
44117.7	167	36735.9	54684.8	207	45534.9
44381.9	168	36955.9	54949.0	208	45754.9
44646.0	169	37175.9	55213.2	209	45974.9
44910.2	170	37395.8	55477.3	210	46194.9
45174.4	171	37615.8	55741.5	211	46414.8
45438.6	172	37835.8	56005.7	212	46634.8
45702.8	173	38055.8	56269.9	213	46854.8
45966.9	174	38275.7	56534.0	214	47074.8
46231.1	175	38495.7	56798.2	215	47294.7
46495.3	176	38715.7	57062.4	216	47514.7
46759.5	177	38935.7	57320.6	217	47734.7
47023.6	178	39155.6	57590.8	218	47954.7
47287.8	179	39375.6	57854.9	219	48174.6
47552.0	180	39595.6	58119.1	220	48394.6
47816.2	181	39815.6	58383.3	221	48614.6
48080.4	182	40035.5	58647.5	222	48834.6
48344.5	183	40255.5	58911.6	223	49054.5
48608.7	184	40475.5	59175.8	224	49274.5
48872.9	185	40695.5	59440.0	225	49494.5
49137.1	186	40915.4	59704.2	226	49714.5
49401.2	187	41135.4	59968.4	227	49934.4
49665.4	188	41355.4	60232.5	228	50154.4
49929.6	189	41575.4	60496.7	229	50374.4
50193.8	190	41795.3	60760.9	230	50594.4
50458.0	191	42015.3	61025.1	231	50814.3
50722.1	192	42235.3	61289.2	232	51034.3
50986.3	193	42455.3	61553.4	233	51254.3
51250.5	194	42675.2	61817.6	234	51474.3
51514.7	195	42895.2	62081.8	235	51694.2
51778.8	196	43115.2	62346.0	236	51914.2
52043.0	197	43335.2	62610.1	237	52134.2
52307.2	198	43555.1	62874.3	238	52354.2
52571.4	199	43775.1	63138.5	239	52574.1
52835.6	200	43995.1	63402.7	240	52794.1

TABLE A-3 (Continued) Gallons (US)—Cubic Meters—Gallons (UK)

US	M3	UK	US	M3	UK
63666.8	241	53014.1	74234.0	281	61813.1
63931.0	242	53234.1	74498.1	282	62033.1
64195.2	243	53454.0	74762.3	283	62253.1
64459.4	244	53674.0	75026.5	284	62473.0
64723.6	245	53894.0	75290.7	285	62693.0
64987.7	246	54114.0	75554.9	286	62913.0
65251.9	247	54333.9	75819.0	287	63133.0
65516.1	248	54553.9	76083.2	288	63352.9
65780.3	249	54773.9	76347.4	289	63572.9
66044.5	250	54993.9	76611.6	290	63792.9
66308.6	251	55213.9	76875.7	291	64012.9
66572.8	252	55433.8	77139.9	292	64232.8
66837.0	253	55653.8	77404.1	293	64452.8
67101.2	254	55873.8	77668.3	294	64672.8
67365.3	255	56093.8	77932.5	295	64892.8
67629.5	256	56313.7	78196.6	296	65112.7
67893.7	257	56533.7	78460.8	297	65332.7
68157.9	258	56753.7	78725.0	298	65552.7
68422.0	259	56973.7	78989.2	299	65772.7
68686.2	260	57193.6	79253.3	300	65992.7
68950.4	261	57413.6	79517.5	301	66212.6
69214.6	262	57633.6	79781.7	302	66432.6
69478.8	263	57853.6	80045.9	303	66652.6
69742.9	264	58073.5	80310.1	304	66872.6
70007.1	265	58293.5	80574.2	305	67092.5
70271.3	266	58513.5	80838.4	306	67312.5
70535.5	267	58733.5	81102.6	307	67532.5
70799.7	268	58953.4	81366.8	308	67752.5
71063.8	269	59173.4	81630.9	309	67972.4
71328.0	270	59393.4	81895.1	310	68192.4
71592.2	271	59613.4	82159.3	311	68412.4
71856.4	272	59833.3	82423.5	312	68632.4
72120.5	273	60053.3	82687.7	313	68852.3
72384.7	274	60273.3	82951.8	314	69072.3
72648.9	275	60493.3	83216.0	315	69292.3
72913.1	276	60713.2	83480.2	316	69512.3
73177.2	277	60933.2	83744.4	317	69732.2
73441.4	278	61153.2	84008.5	318	69952.2
73705.6	279	61373.2	84272.7	319	70172.2
73969.8	280	61593.1	84536.9	320	70392.2

TABLE A-3 (Continued)   Gallons (US)—Cubic Meters—Gallons (UK)

US	M3	UK	US	M3	UK
84801.	321	70612.	95368.	361	79411.
85065.	322	70832.	95632.	362	79631.
85329.	323	71052.	95897.	363	79851.
85594.	324	71272.	96161.	364	80071.
85858.	325	71492.	96425.	365	80291.
86122.	326	71712.	96689.	366	80511.
86386.	327	71932.	96953.	367	80731.
86650.	328	72152.	97217.	368	80951.
86914.	329	72372.	97482.	369	81171.
87179.	330	72592.	97746.	370	81391.
87443.	331	72812.	98010.	371	81611.
87707.	332	73032.	98274.	372	81831.
87971.	333	73252.	98538.	373	82051.
88235.	334	73472.	98802.	374	82271.
88500.	335	73692.	99067.	375	82491.
88764.	336	73912.	99331.	376	82711.
89028.	337	74132.	99595.	377	82931.
89292.	338	74352.	99859.	378	83151.
89556.	339	74572.	100123.	379	83371.
89820.	340	74792.	100388.	380	83591.
90085.	341	75012.	100652.	381	83811.
90349.	342	75232.	100916.	382	84031.
90613.	343	75452.	101180.	383	84251.
90877.	344	75672.	101444.	384	84471.
91141.	345	75892.	101708.	385	84691.
91406.	346	76112.	101973.	386	84911.
91670.	347	76331.	102237.	387	85131.
91934.	348	76551.	102501.	388	85350.
92198.	349	76771.	102765.	389	85570.
92462.	350	76991.	103029.	390	85790.
92726.	351	77211.	103294.	391	86010.
92991.	352	77431.	103558.	392	86230.
93255.	353	77651.	103822.	393	86450.
93519.	354	77871.	104086.	394	86670.
93783.	355	78091.	104350.	395	86890.
94047.	356	78311.	104614.	396	87110.
94311.	357	78531.	104879.	397	87330.
94576.	358	78751.	105143.	398	87550.
94840.	359	78971.	105407.	399	87770.
95104.	360	79191.	105671.	400	87990.

TABLE A-3 (Continued) Gallons (US)—Cubic Meters—Gallons (UK)

US	M3	UK	US	M3	UK
105935.	401	88210.	116502.	441	97009.
106199.	402	88430.	116767.	442	97229.
106464.	403	88650.	117031.	443	97449.
106728.	404	88870.	117295.	444	97669.
106992.	405	89090.	117559.	445	97889.
107256.	406	89310.	117823.	446	98109.
107520.	407	89530.	118087.	447	98329.
107785.	408	89750.	118352.	448	98549.
108049.	409	89970.	118616.	449	98769.
108313.	410	90190.	118880.	450	98989.
108577.	411	90410.	119144.	451	99209.
108841.	412	90630.	119408.	452	99429.
109105.	413	90850.	119673.	453	99649.
109370.	414	91070.	119937.	454	99869.
109634.	415	91290.	120201.	455	100089.
109898.	416	91510.	120465.	456	100309.
110162.	417	91730.	120729.	457	100529.
110426.	418	91950.	120993.	458	100749.
110690.	419	92170.	121258.	459	100969.
110955.	420	92390.	121522.	460	101189.
111219.	421	92610.	121786.	461	101409.
111483.	422	92830.	122050.	462	101629.
111747.	423	93050.	122314.	463	101849.
112011.	424	93270.	122578.	464	102069.
112276.	425	93490.	122843.	465	102289.
112540.	426	93710.	123107.	466	102509.
112804.	427	93930.	123371.	467	102729.
113068.	428	94150.	123635.	468	102949.
113332.	429	94369.	123899.	469	103169.
113596.	430	94589.	124164.	470	103388.
113861.	431	94809.	124428.	471	103608.
114125.	432	95029.	124692.	472	103828.
114389.	433	95249.	124956.	473	104048.
114653.	434	95469.	125220.	474	104268.
114917.	435	95689.	125484.	475	104488.
115182.	436	95909.	125749.	476	104708.
115446.	437	96129.	126013.	477	104928.
115710.	438	96349.	126277.	478	105148.
115974.	439	96569.	126541.	479	105368.
116238.	440	96789.	126805.	480	105588.

TABLE A-3 (Continued) Gallons (US)—Cubic Meters—Gallons (UK)

US	M3	UK	US	M3	UK
127070.	481	105808.	137637.	521	114607.
127334.	482	106028.	137901.	522	114827.
127598.	483	106248.	138165.	523	115047.
127862.	484	106468.	138429.	524	115267.
128126.	485	106688.	138693.	525	115487.
128390.	486	106908.	138958.	526	115707.
128655.	487	107128.	139222.	527	115927.
128919.	488	107348.	139486.	528	116147.
129183.	489	107568.	139750.	529	116367.
129447.	490	107788.	140014.	530	116587.
129711.	491	108008.	140278.	531	116807.
129975.	492	108228.	140543.	532	117027.
130240.	493	108448.	140807.	533	117247.
130504.	494	108668.	141071.	534	117467.
130768.	495	108888.	141335.	535	117687.
131032.	496	109108.	141599.	536	117907.
131296.	497	109328.	141863.	537	118127.
131561.	498	109548.	142128.	538	118347.
131825.	499	109768.	142392.	539	118567.
132089.	500	109988.	142656.	540	118787.
132353.	501	110208.	142920.	541	119007.
132617.	502	110428.	143184.	542	119227.
132881.	503	110648.	143449.	543	119447.
133146.	504	110868.	143713.	544	119667.
133410.	505	111088.	143977.	545	119887.
133674.	506	111308.	144241.	546	120107.
133938.	507	111528.	144505.	547	120327.
134202.	508	111748.	144769.	548	120547.
134467.	509	111968.	145034.	549	120767.
134731.	510	112188.	145298.	550	120987.
134995.	511	112407.	145562.	551	121207.
135259.	512	112627.	145826.	552	121426.
135523.	513	112847.	146090.	553	121646.
135787.	514	113067.	146355.	554	121866.
136052.	515	113287.	146619.	555	122086.
136316.	516	113507.	146883.	556	122306.
136580.	517	113727.	147147.	557	122526.
136844.	518	113947.	147411.	558	122746.
137108.	519	114167.	147675.	559	122966.
137372.	520	114387.	147940.	560	123186.

TABLE A-3 (Continued) Gallons (US)—Cubic Meters—Gallons (UK)

US	M3	UK	US	M3	UK
148204.	561	123406.	158771.	601	132205.
148468.	562	123626.	159035.	602	132425.
148732.	563	123846.	159299.	603	132645.
148996.	564	124066.	159563.	604	132865.
149260.	565	124286.	159828.	605	133085.
149525.	566	124506.	160092.	606	133305.
149789.	567	124726.	160356.	607	133525.
150053.	568	124946.	160620.	608	133745.
150317.	569	125166.	160884.	609	133965.
150581.	570	125386.	161148.	610	134185.
150846.	571	125606.	161413.	611	134405.
151110.	572	125826.	161677.	612	134625.
151374.	573	126046.	161941.	613	134845.
151638.	574	126266.	162205.	614	135065.
151902.	575	126486.	162469.	615	135285.
152166.	576	126706.	162734.	616	135505.
152431.	577	126926.	162998.	617	135725.
152695.	578	127146.	163262.	618	135945.
152959.	579	127366.	163526.	619	136165.
153223.	580	127586.	163790.	620	136385.
153487.	581	127806.	164054.	621	136605.
153751.	582	128026.	164319.	622	136825.
154016.	583	128246.	164583.	623	137045.
154280.	584	128466.	164847.	624	137265.
154544.	585	128686.	165111.	625	137485.
154808.	586	128906.	165375.	626	137705.
155072.	587	129126.	165639.	627	137925.
155337.	588	129346.	165904.	628	138145.
155601.	589	129566.	166168.	629	138365.
155865.	590	129786.	166432.	630	138585.
156129.	591	130006.	166696.	631	138805.
156393.	592	130225.	166960.	632	139025.
156657.	593	130445.	167225.	633	139244.
156922.	594	130665.	167489.	634	139464.
157186.	595	130885.	167753.	635	139684.
157450.	596	131105.	168017.	636	139904.
157714.	597	131325.	168281.	637	140124.
157978.	598	131545.	168545.	638	140344.
158243.	599	131765.	168810.	639	140564.
158507.	600	131985.	169074.	640	140784.

Appendices

TABLE A-3 (Continued) Gallons (US)—Cubic Meters—Gallons (UK)

US	M3	UK	US	M3	UK
169338.	641	141004.	179905.	681	149803.
169602.	642	141224.	180169.	682	150023.
169866.	643	141444.	180433.	683	150243.
170131.	644	141664.	180698.	684	150463.
170395.	645	141884.	180962.	685	150683.
170659.	646	142104.	181226.	686	150903.
170923.	647	142324.	181490.	687	151123.
171187.	648	142544.	181754.	688	151343.
171451.	649	142764.	182019.	689	151563.
171716.	650	142984.	182283.	690	151783.
171980.	651	143204.	182547.	691	152003.
172244.	652	143424.	182811.	692	152223.
172508.	653	143644.	183075.	693	152443.
172772.	654	143864.	183339.	694	152663.
173036.	655	144084.	183604.	695	152883.
173301.	656	144304.	183868.	696	153103.
173565.	657	144524.	184132.	697	153323.
173829.	658	144744.	184396.	698	153543.
174093.	659	144964.	184660.	699	153763.
174357.	660	145184.	184924.	700	153983.
174622.	661	145404.	185189.	701	154203.
174886.	662	145624.	185453.	702	154423.
175150.	663	145844.	185717.	703	154643.
175414.	664	146064.	185981.	704	154863.
175678.	665	146284.	186245.	705	155083.
175942.	666	146504.	186510.	706	155303.
176207.	667	146724.	186774.	707	155523.
176471.	668	146944.	187038.	708	155743.
176735.	669	147164.	187302.	709	155963.
176999.	670	147384.	187566.	710	156183.
177263.	671	147604.	187830.	711	156403.
177527.	672	147824.	188095.	712	156623.
177792.	673	148044.	188359.	713	156843.
178056.	674	148263.	188623.	714	157063.
178320.	675	148483.	188887.	715	157282.
178584.	676	148703.	189151.	716	157502.
178848.	677	148923.	189415.	717	157722.
179113.	678	149143.	189680.	718	157942.
179377.	679	149363.	189944.	719	158162.
179641.	680	149583.	190208.	720	158382.

TABLE A-3 (Continued)   Gallons (US)—Cubic Meters—Gallons (UK)

US	M3	UK	US	M3	UK
190472.	721	158602.	201039.	761	167401.
190736.	722	158822.	201303.	762	167621.
191001.	723	159042.	201568.	763	167841.
191265.	724	159262.	201832.	764	168061.
191529.	725	159482.	202096.	765	168281.
191793.	726	159702.	202360.	766	168501.
192057.	727	159922.	202624.	767	168721.
192321.	728	160142.	202889.	768	168941.
192586.	729	160362.	203153.	769	169161.
192850.	730	160582.	203417.	770	169381.
193114.	731	160802.	203681.	771	169601.
193378.	732	161022.	203945.	772	169821.
193642.	733	161242.	204209.	773	170041.
193907.	734	161462.	204474.	774	170261.
194171.	735	161682.	204738.	775	170481.
194435.	736	161902.	205002.	776	170701.
194699.	737	162122.	205266.	777	170921.
194963.	738	162342.	205530.	778	171141.
195227.	739	162562.	205795.	779	171361.
195492.	740	162782.	206059.	780	171581.
195756.	741	163002.	206323.	781	171801.
196020.	742	163222.	206587.	782	172021.
196284.	743	163442.	206851.	783	172241.
196548.	744	163662.	207115.	784	172461.
196812.	745	163882.	207380.	785	172681.
197077.	746	164102.	207644.	786	172901.
197341.	747	164322.	207908.	787	173121.
197605.	748	164542.	208172.	788	173341.
197869.	749	164762.	208436.	789	173561.
198133.	750	164982.	208700.	790	173781.
198398.	751	165202.	208965.	791	174001.
198662.	752	165422.	209229.	792	174221.
198926.	753	165642.	209493.	793	174441.
199190.	754	165862.	209757.	794	174661.
199454.	755	166082.	210021.	795	174881.
199718.	756	166301.	210286.	796	175100.
199983.	757	166521.	210550.	797	175320.
200247.	758	166741.	210814.	798	175540.
200511.	759	166961.	211078.	799	175760.
200775.	760	167181.	211342.	800	175980.

TABLE A-3 (Continued) Gallons (US)—Cubic Meters—Gallons (UK)

US	M3	UK	US	M3	UK
211606.	801	176200.	222174.	841	184999.
211871.	802	176420.	222438.	842	185219.
212135.	803	176640.	222702.	843	185439.
212399.	804	176860.	222966.	844	185659.
212663.	805	177080.	223230.	845	185879.
212927.	806	177300.	223494.	846	186099.
213191.	807	177520.	223759.	847	186319.
213456.	808	177740.	224023.	848	186539.
213720.	809	177960.	224287.	849	186759.
213984.	810	178180.	224551.	850	186979.
214248.	811	178400.	224815.	851	187199.
214512.	812	178620.	225079.	852	187419.
214777.	813	178840.	225344.	853	187639.
215041.	814	179060.	225608.	854	187859.
215305.	815	179280.	225872.	855	188079.
215569.	816	179500.	226136.	856	188299.
215833.	817	179720.	226400.	857	188519.
216097.	818	179940.	226665.	858	188739.
216362.	819	180160.	226929.	859	188959.
216626.	820	180380.	227193.	860	189179.
216890.	821	180600.	227457.	861	189399.
217154.	822	180820.	227721.	862	189619.
217418.	823	181040.	227985.	863	189839.
217683.	824	181260.	228250.	864	190059.
217947.	825	181480.	228514.	865	190279.
218211.	826	181700.	228778.	866	190499.
218475.	827	181920.	229042.	867	190719.
218739.	828	182140.	229306.	868	190939.
219003.	829	182360.	229571.	869	191159.
219268.	830	182580.	229835.	870	191379.
219532.	831	182800.	230099.	871	191599.
219796.	832	183020.	230363.	872	191819.
220060.	833	183240.	230627.	873	192039.
220324.	834	183460.	230891.	874	192259.
220588.	835	183680.	231156.	875	192479.
220853.	836	183900.	231420.	876	192699.
221117.	837	184119.	231684.	877	192919.
221381.	838	184339.	231948.	878	193138.
221645.	839	184559.	232212.	879	193358.
221909.	840	184779.	232476.	880	193578.

TABLE A-3 (Continued)   Gallons (US)—Cubic Meters—Gallons (UK)

US	M3	UK	US	M3	UK
232741.	881	193798.	243308.	921	202597.
233005.	882	194018.	243572.	922	202817.
233269.	883	194238.	243836.	923	203037.
233533.	884	194458.	244100.	924	203257.
233797.	885	194678.	244364.	925	203477.
234062.	886	194898.	244629.	926	203697.
234326.	887	195118.	244893.	927	203917.
234590.	888	195338.	245157.	928	204137.
234854.	889	195558.	245421.	929	204357.
235118.	890	195778.	245685.	930	204577.
235382.	891	195998.	245950.	931	204797.
235647.	892	196218.	246214.	932	205017.
235911.	893	196438.	246478.	933	205237.
236175.	894	196658.	246742.	934	205457.
236439.	895	196878.	247006.	935	205677.
236703.	896	197098.	247270.	936	205897.
236967.	897	197318.	247535.	937	206117.
237232.	898	197538.	247799.	938	206337.
237496.	899	197758.	248063.	939	206557.
237760.	900	197978.	248327.	940	206777.
238024.	901	198198.	248591.	941	206997.
238288.	902	198418.	248855.	942	207217.
238553.	903	198638.	249120.	943	207437.
238817.	904	198858.	249384.	944	207657.
239081.	905	199078.	249648.	945	207877.
239345.	906	199298.	249912.	946	208097.
239609.	907	199518.	250176.	947	208317.
239873.	908	199738.	250441.	948	208537.
240138.	909	199958.	250705.	949	208757.
240402.	910	200178.	250969.	950	208977.
240666.	911	200398.	251233.	951	209197.
240930.	912	200618.	251497.	952	209417.
241194.	913	200838.	251761.	953	209637.
241459.	914	201058.	252026.	954	209857.
241723.	915	201278.	252290.	955	210077.
241987.	916	201498.	252554.	956	210297.
242251.	917	201718.	252818.	957	210517.
242515.	918	201938.	253082.	958	210737.
242779.	919	202157.	253347.	959	210957.
243044.	920	202377.	253611.	960	211176.

TABLE A-3 (Continued) Gallons (US)—Cubic Meters—Gallons (UK)

US	M3	UK	US	M3	UK
253875.	961	211396.	264442.	1001	220195.
254139.	962	211616.	264706.	1002	220415.
254403.	963	211836.	264970.	1003	220635.
254667.	964	212056.	265235.	1004	220855.
254932.	965	212276.	265499.	1005	221075.
255196.	966	212496.	265763.	1006	221295.
255460.	967	212716.	266027.	1007	221515.
255724.	968	212936.	266291.	1008	221735.
255988.	969	213156.	266555.	1009	221955.
256252.	970	213376.	266820.	1010	222175.
256517.	971	213596.	267084.	1011	222395.
256781.	972	213816.	267348.	1012	222615.
257045.	973	214036.	267612.	1013	222835.
257309.	974	214256.	267876.	1014	223055.
257573.	975	214476.	268140.	1015	223275.
257838.	976	214696.	268405.	1016	223495.
258102.	977	214916.	268669.	1017	223715.
258366.	978	215136.	268933.	1018	223935.
258630.	979	215356.	269197.	1019	224155.
258894.	980	215576.	269461.	1020	224375.
259158.	981	215796.	269726.	1021	224595.
259423.	982	216016.	269990.	1022	224815.
259687.	983	216236.	270254.	1023	225035.
259951.	984	216456.	270518.	1024	225255.
260215.	985	216676.	270782.	1025	225475.
260479.	986	216896.	271046.	1026	225695.
260743.	987	217116.	271311.	1027	225915.
261008.	988	217336.	271575.	1028	226135.
261272.	989	217556.	271839.	1029	226355.
261536.	990	217776.	272103.	1030	226575.
261800.	991	217996.	272367.	1031	226795.
262064.	992	218216.	272631.	1032	227015.
262329.	993	218436.	272896.	1033	227235.
262593.	994	218656.	273160.	1034	227455.
262857.	995	218876.	273424.	1035	227675.
263121.	996	219096.	273688.	1036	227895.
263385.	997	219316.	273952.	1037	228115.
263649.	998	219536.	274217.	1038	228335.
263914.	999	219756.	274481.	1039	228555.
264178.	1000	219976.	274745.	1040	228775.

TABLE A-3 (Continued) Gallons (US)—Cubic Meters—Gallons (UK)

US	M3	UK	US	M3	UK
275009.	1041	228994.	285576.	1081	237794.
275273.	1042	229214.	285840.	1082	238013.
275537.	1043	229434.	286105.	1083	238233.
275802.	1044	229654.	286369.	1084	238453.
276066.	1045	229874.	286633.	1085	238673.
276330.	1046	230094.	286897.	1086	238893.
276594.	1047	230314.	287161.	1087	239113.
276858.	1048	230534.	287425.	1088	239333.
277123.	1049	230754.	287690.	1089	239553.
277387.	1050	230974.	287954.	1090	239773.
277651.	1051	231194.	288218.	1091	239993.
277915.	1052	231414.	288482.	1092	240213.
278179.	1053	231634.	288746.	1093	240433.
278443.	1054	231854.	289011.	1094	240653.
278708.	1055	232074.	289275.	1095	240873.
278972.	1056	232294.	289539.	1096	241093.
279236.	1057	232514.	289803.	1097	241313.
279500.	1058	232734.	290067.	1098	241533.
279764.	1059	232954.	290331.	1099	241753.
280028.	1060	233174.	290596.	1100	241973.
280293.	1061	233394.	290860.	1101	242193.
280557.	1062	233614.	291124.	1102	242413.
280821.	1063	233834.	291388.	1103	242633.
281085.	1064	234054.	291652.	1104	242853.
281349.	1065	234274.	291916.	1105	243073.
281614.	1066	234494.	292181.	1106	243293.
281878.	1067	234714.	292445.	1107	243513.
282142.	1068	234934.	292709.	1108	243733.
282406.	1069	235154.	292973.	1109	243953.
282670.	1070	235374.	293237.	1110	244173.
282934.	1071	235594.	293502.	1111	244393.
283199.	1072	235814.	293766.	1112	244613.
283463.	1073	236034.	294030.	1113	244833.
283727.	1074	236254.	294294.	1114	245053.
283991.	1075	236474.	294558.	1115	245273.
284255.	1076	236694.	294822.	1116	245493.
284519.	1077	236914.	295087.	1117	245713.
284784.	1078	237134.	295351.	1118	245933.
285048.	1079	237354.	295615.	1119	246153.
285312.	1080	237574.	295879.	1120	246373.

Appendices

TABLE A-4 Melting and boiling points, and atomic weights of the elements, based on the assigned relative atomic mass of $C = 12$ (adapted from R. C. Weast, *Handbook of Chemistry & Physics*, published by CRC Press, 8901 Cranwood Parkway, Cleveland, Ohio. Used by permission of CRC Press).

THE FOLLOWING VALUES APPLY TO ELEMENTS AS THEY EXIST IN MATERIALS OF TERRESTRIAL ORIGIN AND TO CERTAIN ARTIFICIAL ELEMENTS. WHEN USED WITH THE FOOTNOTES THEY ARE RELIABLE TO ± 1 IN THE LAST DIGIT, OR ± 3 IF THAT DIGIT IS IN SMALL TYPE.

Name	Symbol	Atomic number	International at. wt.†	Specific gravity (or density)	Melting point, C	Boiling point, C	Specific heat at 25 C	Thermal conductivity, watt/cm C
Actinium	Ac	89	(227)		1050	3200.		
Aluminum	Al	13	26.9815	(10.02)	660	2441.	0.215	2.37
Americium	Am	95	(243)	2.70	900.			
Antimony (Stibium)	Sb	51	121.75	11.7	630.	1440	0.050	0.185
Argon	Ar	18	39.948	6.69	−189.	−186.	0.125	1.75×10^{-4}
Arsenic	As	33	74.9216	1.78 g/l	815.†	613. (subl.)	0.079	
Astatine	At	85	(210)	5.73 (gray)	729.	2125.		
Barium	Ba	56	137.34	3.5	725.	1630	0.046	
Berkelium	Bk	97	(247)					
Beryllium	Be	4	9.0122	1.85	1285.	2475.	0.436	2.18
Bismuth	Bi	83	208.980	9.75	271.4	1660.	0.030	0.084
Boron	B	5	10.811	2.35	2100.	3800.	0.245	
Bromine	Br	35	79.904	3.12 (liq.)	−7.2	58.8	0.11	0.45×10^{-4}
Cadmium	Cd	48	112.40	8.65	321.	767.	0.055	0.97
Calcium	Ca	20	40.08	1.55	840.	1485.		1.3
Californium	Cf	98	(251)					
Carbon	C	6	12.01115					
Diamond				3.5	>3800.	4827.	0.124	$1.5 (0^\circ)$
Graphite				2.1	>3500.	4200.	0.170	0.24
Cerium	Ce	58	140.12	6.77	798.	3257.	0.047	0.11
Cesium	Cs	55	132.905	1.87	28.6	690.	0.057	
Chlorine	Cl	17	35.453	3.21 g/l	−101.	−34.6	0.114	0.86×10^{-4}
Chromium	Cr	24	51.996	7.2	1860.	2670.	0.110	0.91
Cobalt	Co	27	58.9332	8.9	1495.	2925.	0.10	0.69
Copper	Cu	29	63.546	8.96	1084.	2575.	0.092	3.98
Curium	Cm	96	(247)					
Dysprosium	Dy	66	162.50	8.54	1409.	2335.	0.0414	
Einsteinium	Es	99	(254)					
Erbium	Er	68	167.26	9.05	1522.	2510.	0.04	0.10
Europium	Eu	63	151.96	5.25	822.	1597.	0.042	
Fermium	Fm	100	(257)					
Fluorine	F	9	18.9984	1.11 (liq.)	−219.6	−188.	0.197	
Francium	Fr	87	(223)					
Gadolinium	Gd	64	157.25	7.90	1311.	3233.	0.055	0.096
Gallium	Ga	31	69.72	5.91	29.8	2300.	0.089	0.088
Germanium	Ge	32	72.59	5.32	937.	2830.	0.077	$0.29 - 0.38$
Gold (Aurum)	Au	79	196.967	19.32	1063.	2800.	0.031	3.15
Hafnium	Hf	72	178.49	13.29	2220.	4700.	0.035	0.220
Helium	He	2	4.0026	0.177 g/l		−269.	1.24	14.8×10^{-4}
Holmium	Ho	67	164.930	8.78	1470.	2720.	0.039	
Hydrogen	H	1	1.00797	0.0899 g/l	−259.	−253.	3.41	18.4×10^{-4}
Indium	In	49	114.82	7.31	156.	2050.	0.056	0.24
Iodine	I	53	126.9044	4.93	113.5	184.4	0.102	43.5×10^{-4}
Iridium	Ir	77	192.2	22.42	2450.	4390.	0.031	1.47
Iron (Ferrum)	Fe	26	55.847	7.87	1536.	2870.	0.108	0.803
Krypton	Kr	36	83.80	3.73 g/l	−157.	−152.	0.059	0.94×10^{-4}
Lanthanum	La	57	138.91	6.17	920.	3454.	0.047	0.14
Lawrencium	Lr	103	(257)					
Lead (Plumbum)	Pb	82	207.19	11.35	327.5	1750.	0.031	0.352
Lithium	Li	3	6.939	0.53	180.	1317.	0.84	0.71
Lutetium	Lu	71	174.97	9.84	1656.	3315.	0.037	
Magnesium	Mg	12	24.312	1.74	650.	1090.	0.243	1.56

†At 28 atm.

Name	Symbol	Atomic number	International at. wt.†	Specific gravity (or density)	Melting point, C	Boiling point, C	Specific heat at 25 C	Thermal conductivity, watt/cm C
Manganese	Mn	25	54.9380	7.21–7.44	1244.	2060.	0.114	
Mendelevium	Md	101	(256)					
Mercury (Hydrargyrum)	Hg	80	200.59	13.546	−38.86	356.55	0.033	0.0839
Molybdenum	Mo	42	95.94	10.22	2620.	4651.	0.060	1.38
Neodymium	Nd	60	144.24	7.00	1010.	3127.	0.049	0.13
Neon	Ne	10	20.183	0.90 g/l	−249.	−246.	0.246	4.77×10^{-4}
Neptunium	Np	93	(237)	18.0–20.45			0.296	
Nickel	Ni	28	58.71	8.90	1453.	2800.	0.106	0.905
Niobium (Columbium)	Nb	41	92.906	8.57	2467.	4740.	0.064	0.53
Nitrogen	N	7	14.0067	1.251 g/l	−210.	−201.	0.249	2.55×10^{-4}
Nobelium	No	102	(254)					
Osmium	Os	76	190.2	22.57	3025.	4225.	0.031	0.61
Oxygen	O	8	15.9994	1.43 g/l	−218.4	−183.	0.220	2.61×10^{-4}
Palladium	Pd	46	106.4	12.02	1550.	2927.	0.058	0.71
Phosphorus, white	P	15	30.9738	1.82	44.1	280.	0.18	
Platinum	Pt	78	195.09	21.45	1770.	3825.	0.032	0.73
Plutonium	Pu	94	(244)	19.84	640.	3230.	0.032	0.08
Polonium	Po	84	(209)	9.32	254.	962.	0.030	
Potassium (Kalium)	K	19	39.102	0.86	63.3	760.	0.180	0.99
Praseodymium	Pr	59	140.907	6.77	931.	3212.	0.046	0.12
Promethium	Pm	61	(145)		1080.	2460.	0.044	
Protactinium	Pa	91	(231)	(15.37)			0.029	
Radium	Ra	88	(226)		700.	1700.	0.029	
Radon	Rn	86	(222)	9.73 g/l	−71.	−62.	0.0224	0.005
Rhenium	Re	75	186.2	21.0	3180.	5650.	0.033	0.71
Rhodium	Rh	45	102.905	12.41	1965.	3700.	0.058	1.50
Rubidium	Rb	37	85.47	1.532	39.	700.	0.086	
Ruthenium	Ru	44	101.07	12.4	2400.	4100.	0.057	
Samarium	Sm	62	150.35	7.54	1072.	1778.	0.047	
Scandium	Sc	21	44.956	2.99	1539.	2832.	0.135	
Selenium	Se	34	78.96	4.8	217.	700.	0.077	0.835
Silicon	Si	14	28.086	2.33	1411.	3280.	0.17	
Silver (Argentum)	Ag	47	107.868	10.50	961.	2212.	0.057	4.27
Sodium (Natrium)	Na	11	22.9898	0.97	97.83	884.	0.293	1.34
Strontium	Sr	38	87.62	2.55	770.	1375.	0.072	
Sulfur	S	16	32.064	1.96–2.07	113.	445.	0.175	26.4×10^{-4}
Tantalum	Ta	73	180.948	16.6	2980.	5365.	0.034	0.575
Technetium	Tc	43	(97)	(11.50)			0.058	
Tellurium	Te	52	127.60	6.24	450.	990.	0.05	
Terbium	Tb	65	158.924	8.23	1360.	3041.	0.0435	0.059
Thallium	Tl	81	204.37	11.85	304.	1480.	0.031	0.39
Thorium	Th	90	232.038	11.7	1750.	4800.	0.03	0.41
Thulium	Tm	69	168.934	9.31	1545.	1727.	0.0385	
Tin (Stannum)	Sn	50	118.69	7.31	232.	2600.	0.054	0.67
Titanium	Ti	22	47.90	4.54	1670.	3290.	0.125	0.22
Tungsten (Wolfram)	W	74	183.85	19.3	3400.	5550.	0.032	1.78
Uranium	U	92	238.03	18.8	1132.	4140.	0.028	0.25
Vanadium	V	23	50.942	6.1	1900.	3400.	0.116	0.60
Xenon	Xe	54	131.30	5.89 g/l	−112.	−107.	0.038	5.2×10^{-4}
Ytterbium	Yb	70	173.04	6.97	824.	1193.	0.071	
Yttrium	Y	39	88.905	4.46	1523.	3337.	0.0925	0.15
Zinc	Zn	30	65.37	7.	419.5	910.	0.093	1.21
Zirconium	Zr	40	91.22	6.53	1852.	4400.	0.067	0.227

†A value in parentheses is the mass number of the most stable isotope of the element.

TABLE A-5 Periodic Table of the Elements (adapted from R. C. Weast, *Handbook of Chemistry & Physics*, 55th Ed., 1974–75, published by CRC Press 8901 Cranwood Parkway, Cleveland, Ohio. Used by permission of CRC Press).

1a	2a	3b	4b	5b	6b	7b	8			1b	2b	3a	4a	5a	6a	7a	0	Orbit
1 H +1, −1 1.00797 1																	2 He 0 4.0026 2	K
3 Li +1 6.939 2-1	4 Be +2 9.0122 2-2											5 B +3 10.811 2-3	6 C +2,+4,−4 12.01115 2-4	7 N +1,+2,+3,+4,+5,−1,−2,−3 14.0067 2-5	8 O −2 15.9994 2-6	9 F −1 18.9984 2-7	10 Ne 0 20.183 2-8	K-L
11 Na +1 22.9898 2-8-1	12 Mg +2 24.312 2-8-2											13 Al +3 26.9815 2-8-3	14 Si +2,+4,−4 28.086 2-8-4	15 P +3,+5,−3 30.9738 2-8-5	16 S +4,+6,−2 32.064 2-8-6	17 Cl +1,+5,+7,−1 35.453 2-8-7	18 Ar 0 39.948 2-8-8	K-L-M
19 K +1 39.102 -8-8-1	20 Ca +2 40.08 8-8-2	21 Sc +3 44.956 -8-9-2	22 Ti +2,+3,+4 47.90 8-10-2	23 V +2,+3,+4,+5 50.942 8-11-2	24 Cr +2,+3,+6 51.996 -8-13-1	25 Mn +2,+3,+4,+6,+7 54.9380 -8-13-2	26 Fe +2,+3 55.847 -8-14-2	27 Co +2,+3 58.9332 -8-15-2	28 Ni +2,+3 58.71 -8-16-2	29 Cu +1,+2 63.546 -8-18-1	30 Zn +2 65.37 -8-18-2	31 Ga +3 69.72 -8-18-3	32 Ge +2,+4 72.59 -8-18-4	33 As +3,+5,−3 74.9216 -8-18-5	34 Se +4,+6,−2 78.96 -8-18-6	35 Br +1,+5,−1 79.904 -8-18-7	36 Kr 0 83.80 -8-18-8	-L-M-N
37 Rb +1 85.47 -18-8-1	38 Sr +2 87.62 -18-8-2	39 Y +3 88.905 -18-9-2	40 Zr +4 91.22 -18-10-2	41 Nb +3,+5 92.906 -18-12-1	42 Mo +6 95.94 -18-13-1	43 Tc +4,+6,+7 (97) -18-13-2	44 Ru +3 101.07 -18-15-1	45 Rh +3 102.905 -18-16-1	46 Pd +2,+4 106.4 -18-18-0	47 Ag +1 107.868 -18-18-1	48 Cd +2 112.40 -18-18-2	49 In +3 114.82 -18-18-3	50 Sn +2,+4 118.69 -18-18-4	51 Sb +3,+5,−3 121.75 -18-18-5	52 Te +4,+6,−2 127.60 -18-18-6	53 I +1,+5,+7,−1 126.9044 -18-18-7	54 Xe 0 131.30 -18-18-8	-M-N-O
55 Cs +1 132.905 -18-8-1	56 Ba +2 137.34 -18-8-2	57* La +3 138.91 -18-9-2	72 Hf +4 178.49 -32-10-2	73 Ta +5 180.948 -32-11-2	74 W +6 183.85 -32-12-2	75 Re +4,+6,+7 186.2 -32-13-2	76 Os +3,+4 190.2 -32-14-2	77 Ir +3,+4 192.2 -32-15-2	78 Pt +2,+4 195.09 -32-16-2	79 Au +1,+3 196.967 -32-18-1	80 Hg +1,+2 200.59 -32-18-2	81 Tl +1,+3 204.37 -32-18-3	82 Pb +2,+4 207.19 -32-18-4	83 Bi +3,+5 208.980 -32-18-5	84 Po +2,+4 (209) -32-18-6	85 At (210) -32-18-7	86 Rn 0 (222) -32-18-8	-N-O-P
87 Fr +1 (223) -18-8-1	88 Ra +2 (226) -18-8-2	89** Ac +3 (227) -18-9-2																-O-P-Q

KEY TO CHART: 50 Sn +2,+4 118.69 -18-18-4 — Atomic Number, Symbol, Atomic Weight ← Oxidation States ← Electron Configuration

Transition Elements — Group 8 — Transition Elements

*Lanthanides

58 Ce +3,+4 140.12 -20-8-2	59 Pr +3 140.907 -21-8-2	60 Nd +3 144.24 -22-8-2	61 Pm +3 (145) -23-8-2	62 Sm +2,+3 150.35 -24-8-2	63 Eu +2,+3 151.96 -25-8-2	64 Gd +3 157.25 -25-9-2	65 Tb +3 158.924 -27-8-2	66 Dy +3 162.50 -28-8-2	67 Ho +3 164.930 -29-8-2	68 Er +3 167.26 -30-8-2	69 Tm +3 168.934 -31-8-2	70 Yb +2,+3 173.04 -32-8-2	71 Lu +3 174.97 -32-9-2

N O P

**Actinides

90 Th +4 (232) -18-10-2	91 Pa +5,+4 (231) -20-9-2	92 U +3,+4,+5,+6 (238) -21-9-2	93 Np +3,+4,+5,+6 (237) -22-9-2	94 Pu +3,+4,+5,+6 (244) -24-8-2	95 Am +3,+4,+5,+6 (243) -25-8-2	96 Cm +3 (247) -25 9-2	97 Bk +3,+4 (247) -27-8-2	98 Cf +3 (251) -28-8-2	99 Es (254) -29-8-2	100 Fm (257) -30-8-2	101 Md (256) -31-8-2	102 No (254) -32-8-2	103 Lw +3 (257) -32-9-2

104 —

O P Q

Numbers in parentheses are mass numbers of the most stable isotope of that element.
*From: "CRC Handbook of Chemistry and Physics", 50th ed., R. C. Weast, Ed. The Chemical Rubber Co., 1969.

TABLE A-6 Temperature Conversions

FIND GIVEN VALUE IN MIDDLE COLUMN; IF IN DEGREES CENTIGRADE, READ FAHRENHEIT EQUIVALENT IN RIGHT HAND COLUMN; IF IN DEGREES FAHRENHEIT READ CENTIGRADE EQUIVALENT IN LEFT HAND COLUMN.

C		F	C		F	C		F	C		F	C		F	C		F	C		F
−273	−459.4		−3.9	25	77.0	38	100	212	427	800	1472	843	1550	2822	1260	2300	4172
−268	−450		−3.3	26	78.8	43	110	230	432	810	1490	849	1560	2840	1266	2310	4190
−262	−440		−2.8	27	80.6	49	120	248	438	820	1508	854	1570	2858	1271	2320	4208
−257	−430		−2.2	28	82.4	54	130	266	443	830	1526	860	1580	2876	1277	2330	4226
−251	−420		−1.7	29	84.2	60	140	284	449	840	1544	866	1590	2894	1282	2340	4244
−246	−410		−1.1	30	86.0	66	150	302	454	850	1562	871	1600	2912	1288	2350	4262
−240	−400		−0.6	31	87.8	71	160	320	460	860	1580	877	1610	2930	1293	2360	4280
−234	−390		0.0	32	89.6	77	170	338	466	870	1598	882	1620	2948	1299	2370	4298
−229	−380		0.6	33	91.4	82	180	356	471	880	1616	888	1630	2966	1304	2380	4316
−223	−370		1.1	34	93.2	88	190	374	477	890	1634	893	1640	2984	1310	2390	4334
−218	−360		1.7	35	95.0	93	200	392	482	900	1652	899	1650	3002	1316	2400	4352
−212	−350		2.2	36	96.8	99	210	410	488	910	1670	904	1660	3020	1321	2410	4370
−207	−340		2.8	37	98.6				493	920	1688	910	1670	3038	1327	2420	4388
−201	−330		3.3	38	100.4				499	930	1706	916	1680	3056	1332	2430	4406
−196	−320		3.9	39	102.2	100	212	413.6	504	940	1724	921	1690	3074	1338	2440	4424
−190	−310		4.4	40	104.0				510	950	1742	927	1700	3092	1343	2450	4442
−184	−300		5.0	41	105.8				516	960	1760	932	1710	3110	1349	2460	4460
−179	−290		5.6	42	107.6	104	220	428	521	970	1778	938	1720	3128	1354	2470	4478
−173	−280		6.1	43	109.4	110	230	446	527	980	1796	943	1730	3146	1360	2480	4496
−169	−273	−459.4	6.7	44	111.2	116	240	464	532	990	1814	949	1740	3164	1366	2490	4514
−168	−270	−454	7.2	45	113.0	121	250	482	538	1000	1832	954	1750	3182	1371	2500	4532
−162	−260	−436	7.8	46	114.8	127	260	500	543	1010	1850	960	1760	3200	1377	2510	4550
−157	−250	−418	8.3	47	116.6	132	270	518	549	1020	1868	966	1770	3218	1382	2520	4568
−151	−240	−400	8.9	48	118.4	138	280	536	554	1030	1886	971	1780	3236	1388	2530	4586
−146	−230	−382	9.4	49	120.2	143	290	554	560	1040	1904	977	1790	3254	1393	2540	4604
−140	−220	−364	10.0	50	122.0	149	300	572	566	1050	1922	982	1800	3272	1399	2550	4622
−134	−210	−346	10.6	51	123.8	154	310	590	571	1060	1940	988	1810	3290	1404	2560	4640
−129	−200	−328	11.1	52	125.6	160	320	608	577	1070	1958	993	1820	3308	1410	2570	4658
−123	−190	−310	11.7	53	127.4	166	330	626	582	1080	1976	999	1830	3326	1416	2580	4676
−118	−180	−292	12.2	54	129.2	171	340	644	588	1090	1994	1004	1840	3344	1421	2590	4694
−112	−170	−274	12.8	55	131.0	177	350	662	593	1100	2012	1010	1850	3362	1427	2600	4712
−107	−160	−256	13.3	56	132.8	182	360	680	599	1110	2030	1016	1860	3380	1432	2610	4730
−101	−150	−238	13.9	57	134.6	188	370	698	604	1120	2048	1021	1870	3398	1438	2620	4748
−96	−140	−220	14.4	58	136.4	193	380	716	610	1130	2066	1027	1880	3416	1443	2630	4766
−90	−130	−202	15.0	59	138.2	199	390	734	616	1140	2084	1032	1890	3434	1449	2640	4784
−84	−120	−184	15.6	60	140.0	204	400	752	621	1150	2102	1038	1900	3452	1454	2650	4802
−79	−110	−166	16.1	61	141.8	210	410	770	627	1160	2120	1043	1910	3470	1460	2660	4820
−73	−100	−148	16.7	62	143.6	216	420	788	632	1170	2138	1049	1920	3488	1466	2670	4838
−68	−90	−130	17.2	63	145.4	221	430	806	638	1180	2156	1054	1930	3506	1471	2680	4856
−62	−80	−112	17.8	64	147.2	227	440	824	643	1190	2174	1060	1940	3524	1477	2690	4874
−57	−70	−94	18.3	65	149.0	232	450	842	649	1200	2192	1066	1950	3542	1482	2700	4892
−51	−60	−76	18.9	66	150.8	238	460	860	654	1210	2210	1071	1960	3560	1488	2710	4910
−46	−50	−58	19.4	67	152.6	243	470	878	660	1220	2228	1077	1970	3578	1493	2720	4928
−40	−40	−40	20.0	68	154.4	249	480	896	666	1230	2246	1082	1980	3596	1499	2730	4946
−34	−30	−22	20.6	69	156.2	254	490	914	671	1240	2264	1088	1990	3614	1504	2740	4964
−29	−20	−4	21.1	70	158.0	260	500	932	677	1250	2282	1093	2000	3632	1510	2750	4982
−23	−10	14	21.7	71	159.8	266	510	950	682	1260	2300	1099	2010	3650	1516	2760	5000
			22.2	72	161.6	271	520	968	688	1270	2318	1104	2020	3668	1521	2770	5018
			22.8	73	163.4	277	530	986	693	1280	2336	1110	2030	3686	1527	2780	5036
−17.8	0	32	23.3	74	165.2	282	540	1004	699	1290	2354	1116	2040	3704	1532	2790	5054
			23.9	75	167.0	288	550	1022	704	1300	2372	1121	2050	3722	1538	2800	5072
−17.2	1	33.8	24.4	76	168.8	293	560	1040	710	1310	2390	1127	2060	3740	1543	2810	5090
−16.7	2	35.6	25.0	77	170.6	299	570	1058	716	1320	2408	1132	2070	3758	1549	2820	5108
−16.1	3	37.4	25.6	78	172.4	304	580	1076	721	1330	2426	1138	2080	3776	1554	2830	5126
−15.6	4	39.2	26.1	79	174.2	310	590	1094	727	1340	2444	1143	2090	3794	1560	2840	5144
−15.0	5	41.0	26.7	80	176.0	316	600	1112	732	1350	2462	1149	2100	3812	1566	2850	5162
−14.4	6	42.8	27.2	81	177.8	321	610	1130	738	1360	2480	1154	2110	3830	1571	2860	5180
−13.9	7	44.6	27.8	82	179.6	327	620	1148	743	1370	2498	1160	2120	3848	1577	2870	5198
−13.3	8	46.4	28.3	83	181.4	332	630	1166	749	1380	2516	1166	2130	3866	1582	2880	5216
−12.8	9	48.2	28.9	84	183.2	338	640	1184	754	1390	2534	1171	2140	3884	1588	2890	5234
−12.2	10	50.0	29.4	85	185.0	343	650	1202	760	1400	2552	1177	2150	3902	1593	2900	5252
−11.7	11	51.8	30.0	86	186.8	349	660	1220	766	1410	2570	1182	2160	3920	1599	2910	5270
−11.1	12	53.6	30.6	87	188.6	354	670	1238	771	1420	2588	1188	2170	3938	1604	2920	5288
−10.6	13	55.4	31.1	88	190.4	360	680	1256	777	1430	2606	1193	2180	3956	1610	2930	5306
−10.0	14	57.2	31.7	89	192.2	366	690	1274	782	1440	2624	1199	2190	3974	1616	2940	5324
−9.4	15	59.0	32.2	90	194.0	371	700	1292	788	1450	2642	1204	2200	3992	1621	2950	5342
−8.9	16	60.8	32.8	91	195.8	377	710	1310	793	1460	2660	1210	2210	4010	1627	2960	5360
−8.3	17	62.6	33.3	92	197.6	382	720	1328	799	1470	2678	1216	2220	4028	1632	2970	5378
−7.8	18	64.4	33.9	93	199.4	388	730	1346	804	1480	2696	1221	2230	4046	1638	2980	5396
−7.2	19	66.2	34.4	94	201.2	393	740	1364	810	1490	2714	1227	2240	4064	1643	2990	5414
−6.7	20	68.0	35.0	95	203.0	399	750	1382	816	1500	2732	1232	2250	4082	1649	3000	5432
−6.1	21	69.8	35.6	96	204.8	404	760	1400	821	1510	2750	1238	2260	4100			
−5.6	22	71.6	36.1	97	206.6	410	770	1418	827	1520	2768	1243	2270	4118			
−5.0	23	73.4	36.7	98	208.4	416	780	1436	832	1530	2786	1249	2280	4136			
−4.4	24	75.2	37.2	99	210.2	421	790	1454	838	1540	2804	1254	2290	4154			

TABLE A-7 Chemical Feed Requirements for Flows Measured in US gpm

U.S. g.p.m.	1	2	3	4	5	6	7	8	9	10
1	0.000,499,62	0.000,999,24	0.001,498,860	0.001,998,48	0.002,498,10	0.002,997,720	0.003,497,340	0.003,996,960	0.004,496,580	0.004,996,200
2	0.000,999,24	0.001,998,48	0.002,997,720	0.003,996,96	0.004,996,20	0.005,995,440	0.006,994,680	0.007,993,920	0.008,993,160	0.009,992,400
3	0.001,498,86	0.002,997,72	0.004,496,580	0.005,995,44	0.007,494,30	0.008,993,160	0.010,492,020	0.011,990,880	0.013,489,740	0.014,988,600
4	0.001,998,48	0.003,996,96	0.005,995,440	0.007,993,92	0.009,992,40	0.011,990,880	0.013,989,360	0.015,987,840	0.017,986,320	0.019,984,800
5	0.002,498,10	0.004,996,20	0.007,494,300	0.009,992,40	0.012,490,50	0.014,988,600	0.017,486,700	0.019,984,800	0.022,482,900	0.024,981,000
6	0.002,997,72	0.005,995,44	0.008,993,160	0.011,990,88	0.014,988,60	0.017,986,320	0.020,984,040	0.023,981,760	0.026,979,480	0.029,977,200
7	0.003,497,34	0.006,994,68	0.010,492,020	0.013,989,36	0.017,486,70	0.020,984,040	0.024,481,380	0.027,978,720	0.031,476,060	0.034,973,400
8	0.003,996,96	0.007,993,92	0.011,990,880	0.015,987,84	0.019,984,80	0.023,981,760	0.027,978,720	0.031,975,680	0.035,972,640	0.039,969,600
9	0.004,496,58	0.008,993,16	0.013,489,740	0.017,986,32	0.022,482,90	0.026,979,480	0.031,476,060	0.035,972,640	0.040,469,220	0.044,965,800
10	0.004,996,20	0.009,992,40	0.014,988,600	0.019,984,80	0.024,981,00	0.029,977,200	0.034,973,400	0.039,969,600	0.044,965,800	0.049,962,000

TABLE A-8 Chemical Feed Requirements for Flows Measured in UK gpm

POUNDS OF CHEMICAL PER HOUR FOR GIVEN DOSAGE IN MILLIGRAMS PER LITERS (mg/ℓ)

U.K. g.p.m.	1	2	3	4	5	6	7	8	9	10
1	0.000,6	0.001,2	0.001,8	0.002,4	0.003,0	0.003,60	0.004,20	0.004,80	0.005,40	0.006,00
2	0.001,2	0.002,4	0.003,6	0.004,8	0.006,0	0.007,20	0.008,40	0.009,60	0.010,80	0.012,00
3	0.001,8	0.003,6	0.005,4	0.007,2	0.009,0	0.010,80	0.012,60	0.014,40	0.016,20	0.018,00
4	0.002,4	0.004,8	0.007,2	0.009,6	0.012,0	0.014,40	0.016,80	0.019,20	0.021,60	0.024,00
5	0.003,0	0.006,0	0.009,0	0.012,0	0.015,0	0.018,00	0.021,00	0.024,00	0.027,00	0.030,00
6	0.003,6	0.007,2	0.010,8	0.014,4	0.018,0	0.021,60	0.025,20	0.028,80	0.032,40	0.036,00
7	0.004,2	0.008,4	0.012,6	0.016,8	0.021,0	0.025,20	0.029,40	0.033,60	0.037,80	0.042,00
8	0.004,8	0.009,6	0.014,4	0.019,2	0.024,0	0.028,80	0.033,60	0.038,40	0.043,20	0.048,00
9	0.005,4	0.010,8	0.016,2	0.021,6	0.027,0	0.032,40	0.037,80	0.043,20	0.048,60	0.054,00
10	0.006,0	0.012,0	0.018,0	0.024,0	0.030,0	0.036,00	0.042,00	0.048,00	0.054,00	0.060,00

TABLE A-9 Chemical Feed Requirements for Flows Measured in Liters Per Second

GRAMS OF CHEMICAL PER HOUR FOR GIVEN DOSAGE IN MILLIGRAMS PER LITER (mg/ℓ)

LITERS PER SECOND	1	2	3	4	5	6	7	8	9	10
1	3.6	7.2	10.8	14.4	18.0	21.6	25.2	28.8	32.4	36.0
2	7.2	14.4	21.6	28.8	36.0	43.2	50.4	57.6	64.8	72.0
3	10.8	21.6	32.4	43.2	54.0	64.8	75.6	86.4	97.2	108.0
4	14.4	28.8	43.2	57.6	72.0	86.4	100.8	115.2	129.6	144.0
5	18.0	36.0	54.0	72.0	90.0	108.0	126.0	144.0	162.0	180.0
6	21.6	43.2	64.8	86.4	108.0	129.6	151.2	172.8	194.6	216.0
7	25.2	50.4	75.6	100.8	126.0	151.2	176.4	201.6	226.8	252.0
8	28.8	57.6	86.4	115.2	144.0	172.8	201.6	230.4	259.2	288.0
9	32.4	64.8	97.2	129.6	162.0	194.4	226.8	259.2	291.6	324.0
10	36.0	72.0	108.0	144.0	180.0	216.0	252.0	288.0	324.0	360.0

TABLE A-10 Chemical Feed Requirements for Flows Measured in Cubic Metres Per Day

GRAMS OF CHEMICAL PER HOUR FOR GIVEN DOSAGE IN MILLIGRAMS PER LITER (mg/ℓ)

CUBIC METRES PER DAY	1	2	3	4	5	6	7	8	9	10
1	0.04166	0.08333	0.12500	0.16666	0.20833	0.25000	0.29167	0.33333	0.37500	0.41666
2	0.08333	0.16666	0.25000	0.33333	0.41666	0.50000	0.58333	0.66666	0.75000	0.83333
3	0.12500	0.25000	0.37500	0.50000	0.62500	0.75000	0.87500	1.00000	1.12500	1.25000
4	0.16666	0.33333	0.50000	0.66666	0.83333	1.00000	1.16666	1.33333	1.50000	1.66666
5	0.20833	0.41666	0.62500	0.83333	1.04166	1.25000	1.45833	1.66666	1.87500	2.08333
6	0.25000	0.50000	0.75000	1.00000	1.25000	1.50000	1.75000	2.00000	2.25000	2.50000
7	0.29167	0.58333	0.87500	1.16666	1.45833	1.75000	2.04166	2.33333	2.62500	2.91667
8	0.33333	0.66666	1.00000	1.33333	1.66666	2.00000	2.33333	2.66666	3.00000	3.33333
9	0.37500	0.75000	1.12500	1.50000	1.87500	2.25000	2.62500	3.00000	3.37500	3.75000
10	0.41666	0.83333	1.25000	1.66666	2.08333	2.50000	2.91667	3.33333	3.75000	4.16666

TABLE A-11 Chemical Feed Requirements for Flows Measured in Cubic Metres Per Day

KILOGRAMS OF CHEMICAL PER DAY FOR GIVEN DOSAGE IN MILLIGRAM PER LITER (mg/ℓ)

CUBIC METRES PER DAY	1	2	3	4	5	6	7	8	9	10
1	0.001	0.002	0.003	0.004	0.005	0.006	0.007	0.008	0.009	0.010
2	0.002	0.004	0.006	0.008	0.010	0.012	0.014	0.016	0.018	0.020
3	0.003	0.006	0.009	0.012	0.015	0.018	0.021	0.024	0.027	0.030
4	0.004	0.008	0.012	0.016	0.020	0.024	0.028	0.032	0.036	0.040
5	0.005	0.010	0.015	0.020	0.025	0.030	0.035	0.040	0.045	0.050
6	0.006	0.012	0.018	0.024	0.030	0.036	0.042	0.048	0.054	0.060
7	0.007	0.014	0.021	0.028	0.035	0.042	0.049	0.056	0.063	0.070
8	0.008	0.016	0.024	0.032	0.040	0.048	0.056	0.064	0.072	0.080
9	0.009	0.018	0.027	0.036	0.045	0.054	0.063	0.072	0.081	0.090
10	0.010	0.020	0.030	0.040	0.050	0.060	0.070	0.080	0.090	0.100

TABLE A-12 Conversion of Feet and Meters

Find given value in the middle column; if the value is in meters and the required equivalent is to be in feet, read to the left; if the value is in feet and the required equivalent is to be in meters, read to the right.

FEET		METERS	FEET		METERS	FEET		METERS
3.281	1	0.3048	85.302	26	7.9248	167.322	51	15.5448
6.562	2	0.6096	88.582	27	8.2296	170.603	52	15.8496
9.842	3	0.9144	91.863	28	8.5344	173.884	53	16.1544
13.123	4	1.2192	95.144	29	8.8392	177.165	54	16.4592
16.404	5	1.5240	98.425	30	9.1440	180.446	55	16.7640
19.685	6	1.8288	101.706	31	9.4488	183.727	56	17.0688
22.966	7	2.1336	104.987	32	9.7536	187.007	57	17.3736
26.247	8	2.4384	108.267	33	10.0584	190.288	58	17.6784
29.527	9	2.7432	111.548	34	10.3632	193.569	59	17.9832
32.808	10	3.0480	114.829	35	10.6680	196.850	60	18.2880
36.089	11	3.3528	118.110	36	10.9728	200.131	61	18.5928
39.370	12	3.6576	121.391	37	11.2776	203.412	62	18.8976
42.651	13	3.9624	124.672	38	11.5824	206.692	63	19.2024
45.932	14	4.2672	127.952	39	11.8872	209.973	64	19.5072
49.212	15	4.5720	131.233	40	12.1920	213.254	65	19.8120
52.493	16	4.8768	134.514	41	12.4968	216.535	66	20.1168
55.774	17	5.1816	137.795	42	12.8016	219.816	67	20.4216
59.055	18	5.4864	141.076	43	13.1064	223.097	68	20.7264
62.336	19	5.7912	144.357	44	13.4112	226.377	69	21.0312
65.617	20	6.0960	147.637	45	13.7160	229.658	70	21.3360
68.897	21	6.4008	150.918	46	14.0208	232.939	71	21.6408
72.178	22	6.7056	154.199	47	14.3256	236.220	72	21.9456
75.459	23	7.0104	157.480	48	14.6304	239.501	73	22.2504
78.740	24	7.3152	160.761	49	14.9352	242.782	74	22.5552
82.021	25	7.6200	164.042	50	15.2400	246.062	75	22.8000

FEET		METERS
249.343	76	23.1648
252.624	77	23.4696
255.905	78	23.7744
259.186	79	24.0792
262.467	80	24.3840
265.747	81	24.6888
269.028	82	24.9936
272.309	83	25.2984
275.590	84	25.6032
278.871	85	25.9080
282.152	86	26.2128
285.432	87	26.5176
288.713	88	26.8224
291.994	89	27.1272
295.275	90	27.4320
298.556	91	27.7368
301.837	92	28.0416
305.117	93	28.3464
308.398	94	28.6512
311.679	95	28.9560
314.960	96	29.2608
318.241	97	29.5656
321.522	98	29.8704
324.802	99	30.1752
328.083	100	30.4800

TABLE A-13 Pressure Conversion: Pounds Per Square Inch (PSI) and Kilo Pascals (kPa)

psi		kPa	psi		kPa
0.01450	0.1	0.68947	3.04580	21	144.790
0.02901	0.2	1.37895	3.19084	22	151.684
0.04351	0.3	2.06842	3.33587	23	158.579
0.05802	0.4	2.75790	3.48091	24	165.474
0.07252	0.5	3.44737	3.62595	25	172.369
0.08702	0.6	4.13685	3.77099	26	179.263
0.10153	0.7	4.82632	3.91603	27	186.158
0.11603	0.8	5.51579	4.06106	28	193.053
0.13053	0.9	6.20527	4.20610	29	199.948
			4.35114	30	206.842
0.14504	1	6.8947	4.49618	31	213.737
0.29008	2	13.7895	4.64122	32	220.632
0.43511	3	20.6842	4.78625	33	227.527
0.58015	4	27.5790	4.93129	34	234.421
0.72519	5	34.4737	5.07633	35	241.316
0.87023	6	41.3685	5.22137	36	248.211
1.01527	7	48.2632	5.36641	37	255.106
1.16030	8	55.1580	5.51144	38	262.000
1.30534	9	62.0527	5.65648	39	268.895
1.45038	10	68.9474	5.80152	40	275.790
1.59542	11	75.8422	5.94656	41	282.684
1.74046	12	82.7369	6.09160	42	289.579
1.88549	13	89.6317	6.23663	43	296.474
2.03053	14	96.5264	6.38167	44	303.369
2.17557	15	103.421	6.52671	45	310.263
2.32061	16	110.316	6.67175	46	317.158
2.46565	17	117.211	6.81679	47	324.053
2.61068	18	124.105	6.96182	48	330.948
2.75572	19	131.000	7.10686	49	337.842
2.90076	20	137.895	7.25190	50	344.737

TABLE A-13 (Continued) Pressure Conversion: Pounds Per Square Inch (PSI) and Kilo Pascals (kPa)

psi		kPa	psi		kPa
7.39694	51	351.632	11.7481	81	558.474
7.54198	52	358.527	11.8931	82	565.369
7.68701	53	365.421	12.0382	83	572.264
7.83205	54	372.316	12.1832	84	579.158
7.97709	55	379.211	12.3282	85	586.053
8.12213	56	386.106	12.4733	86	592.948
8.26717	57	393.000	12.6183	87	599.843
8.41220	58	399.895	12.7633	88	606.738
8.55724	59	406.790	12.9084	89	613.632
8.70228	60	413.685	13.0534	90	620.527
8.84732	61	420.579	13.1985	91	627.422
8.99236	62	427.474	13.3435	92	634.316
9.13739	63	434.369	13.4885	93	641.211
9.28243	64	441.264	13.6336	94	648.106
9.42747	65	448.158	13.7786	95	655.001
9.57251	66	455.053	13.9236	96	661.896
9.71755	67	461.948	14.0687	97	668.790
9.86258	68	468.843	14.2137	98	675.685
10.0076	69	475.737	14.3588	99	682.580
10.1527	70	482.632	14.5038	100	689.474
10.2977	71	489.527	14.6488	101	696.369
10.4427	72	496.422	14.7939	102	703.264
10.5878	73	503.316	14.9389	103	710.159
10.7328	74	510.211	15.0840	104	717.053
10.8779	75	517.106	15.2290	105	723.948
11.0229	76	524.000	15.3740	106	730.843
11.1679	77	530.895	15.5191	107	737.738
11.3130	78	537.790	15.6641	108	744.632
11.4580	79	544.685	15.8091	109	751.527
11.6030	80	551.580	15.9542	110	758.422

TABLE A-13 (Continued) Pressure Conversion: Pounds Per Square Inch (PSI) and Kilo Pascals (kPa)

psi		kPa	psi		kPa
16.0992	111	765.317	20.4503	141	972.16
16.2442	112	772.211	20.5954	142	979.05
16.3893	113	779.106	20.7404	143	985.95
16.5343	114	786.001	20.8855	144	992.84
16.6794	115	792.896	21.0305	145	999.74
16.8244	116	799.790	21.1755	146	1006.63
16.9694	117	806.685	21.3206	147	1013.53
17.1145	118	813.580	21.4656	148	1020.42
17.2595	119	820.475	21.6107	149	1027.32
17.4046	120	827.369	21.7557	150	1034.21
17.5496	121	834.264	21.9007	151	1041.11
17.6946	122	841.159	22.0458	152	1048.00
17.8397	123	848.053	22.1908	153	1054.90
17.9847	124	854.948	22.3358	154	1061.79
18.1297	125	861.843	22.4809	155	1068.69
18.2748	126	868.738	22.6259	156	1075.58
18.4198	127	875.633	22.7710	157	1082.47
18.5649	128	882.527	22.9160	158	1089.37
18.7099	129	889.422	23.0610	159	1096.26
18.8549	130	896.317	23.2061	160	1103.16
19.0000	131	903.211	23.3511	161	1110.05
19.1450	132	910.106	23.4962	162	1116.95
19.2901	133	917.001	23.6412	163	1123.84
19.4351	134	923.896	23.7862	164	1130.74
19.5801	135	930.791	23.9313	165	1137.63
19.7252	136	937.685	24.0763	166	1144.53
19.8702	137	944.580	24.2213	167	1151.42
20.0152	138	951.475	24.3664	168	1158.32
20.1603	139	958.369	24.5114	169	1165.21
20.3053	140	965.264	24.6564	170	1172.11

TABLE A-13 (Continued) Pressure Conversion: Pounds Per Square Inch (PSI) and Kilo Pascals (kPa)

psi		kPa	psi		kPa
24.8015	171	1179.00	29.1526	201	1385.84
24.9465	172	1185.90	29.2977	202	1392.74
25.0916	173	1192.79	29.4427	203	1399.63
25.2366	174	1199.69	29.5878	204	1406.53
25.3816	175	1206.58	29.7328	205	1413.42
25.5267	176	1213.48	29.8778	206	1420.32
25.6717	177	1220.37	30.0229	207	1427.21
25.8168	178	1227.26	30.1679	208	1434.11
25.9618	179	1234.16	30.3129	209	1441.00
26.1068	180	1241.05	30.4580	210	1447.90
26.2519	181	1247.95	30.6030	211	1454.79
26.3969	182	1254.84	30.7480	212	1461.69
26.5419	183	1261.74	30.8931	213	1468.58
26.6870	184	1268.63	31.0381	214	1475.48
26.8320	185	1275.53	31.1832	215	1482.37
26.9771	186	1282.42	31.3282	216	1489.26
27.1221	187	1289.32	31.4732	217	1496.16
27.2671	188	1296.21	31.6183	218	1503.05
27.4122	189	1303.11	31.7633	219	1509.95
27.5572	190	1310.00	31.9084	220	1516.84
27.7023	191	1316.90	32.0534	221	1523.74
27.8473	192	1323.79	32.1984	222	1530.63
27.9923	193	1330.69	32.3435	223	1537.53
28.1374	194	1337.58	32.4885	224	1544.42
28.2824	195	1344.48	32.6335	225	1551.32
28.4274	196	1351.37	32.7786	226	1558.21
28.5725	197	1358.26	32.9236	227	1565.11
28.7175	198	1365.16	33.0687	228	1572.00
28.8625	199	1372.05	33.2137	229	1578.90
29.0076	200	1378.95	33.3587	230	1585.79

TABLE A-13 (Continued) Pressure Conversion: Pounds Per Square Inch (PSI) and Kilo Pascals (kPa)

psi		kPa	psi		kPa
33.5038	231	1592.69	37.8549	261	1799.53
33.6488	232	1599.58	38.0000	262	1806.42
33.7939	233	1606.48	38.1450	263	1813.32
33.9389	234	1613.37	38.2900	264	1820.21
34.0839	235	1620.26	38.4351	265	1827.11
34.2290	236	1627.16	38.5801	266	1834.00
34.3740	237	1634.05	38.7251	267	1840.90
34.5190	238	1640.95	38.8702	268	1847.79
34.6641	239	1647.84	39.0152	269	1854.69
34.8091	240	1654.74	39.1602	270	1861.58
34.9541	241	1661.63	39.3053	271	1868.48
35.0992	242	1668.53	39.4503	272	1875.37
35.2442	243	1675.42	39.5954	273	1882.27
35.3893	244	1682.32	39.7404	274	1889.16
35.5343	245	1689.21	39.8855	275	1896.05
35.6793	246	1696.11	40.0305	276	1902.95
35.8244	247	1703.00	40.1755	277	1909.84
35.9694	248	1709.90	40.3206	278	1916.74
36.1145	249	1716.79	40.4656	279	1923.63
36.2595	250	1723.69	40.6106	280	1930.53
36.4045	251	1730.58	40.7557	281	1937.42
36.5496	252	1737.48	40.9007	282	1944.32
36.6946	253	1744.37	41.0457	283	1951.21
36.8396	254	1751.27	41.1908	284	1958.11
36.9847	255	1758.16	41.3358	285	1965.00
37.1297	256	1765.05	41.4809	286	1971.90
37.2748	257	1771.95	41.6259	287	1978.79
37.4198	258	1778.84	41.7709	288	1985.69
37.5648	259	1785.74	41.9160	289	1992.58
37.7099	260	1792.63	42.0610	290	1999.48

TABLE A-13 (Continued) Pressure Conversion: Pounds Per Square Inch (PSI) and Kilo Pascals (kPa)

psi		kPa	psi		kPa
42.2061	291	2006.37	46.5572	321	2213.21
42.3511	292	2013.27	46.7022	322	2220.11
42.4961	293	2020.16	46.8473	323	2227.00
42.6412	294	2027.05	46.9923	324	2233.90
42.7862	295	2033.95	47.1373	325	2240.79
42.9312	296	2040.84	47.2824	326	2247.69
43.0763	297	2047.74	47.4274	327	2254.58
43.2213	298	2054.63	47.5725	328	2261.48
43.3664	299	2061.53	47.7175	329	2268.37
43.5114	300	2068.42	47.8625	330	2275.27
43.6564	301	2075.32	48.0076	331	2282.16
43.8015	302	2082.21	48.1526	332	2289.06
43.9465	303	2089.11	48.2977	333	2295.95
44.0916	304	2096.00	48.4427	334	2302.84
44.2366	305	2102.90	48.5877	335	2309.74
44.3816	306	2109.79	48.7328	336	2316.63
44.5267	307	2116.69	48.8778	337	2323.53
44.6717	308	2123.58	49.0228	338	2330.42
44.8167	309	2130.48	49.1679	339	2337.32
44.9618	310	2137.37	49.3129	340	2344.21
45.1068	311	2144.27	49.4579	341	2351.11
45.2518	312	2151.16	49.6030	342	2358.00
45.3969	313	2158.05	49.7480	343	2364.90
45.5419	314	2164.95	49.8931	344	2371.79
45.6870	315	2171.84	50.0381	345	2378.69
45.8320	316	2178.74	50.1831	346	2385.58
45.9770	317	2185.63	50.3282	347	2392.48
46.1221	318	2192.53	50.4732	348	2399.37
46.2671	319	2199.42	50.6183	349	2406.27
46.4122	320	2206.32	50.7633	350	2413.16

TABLE A-13 (Continued) Pressure Conversion: Pounds Per Square Inch (PSI) and Kilo Pascals (kPa)

psi		kPa	psi		kPa
50.9083	351	2420.06	55.2595	381	2626.90
51.0534	352	2426.95	55.4045	382	2633.79
51.1984	353	2433.84	55.5495	383	2640.69
51.3434	354	2440.74	55.6946	384	2647.58
51.4885	355	2447.63	55.8396	385	2654.48
51.6335	356	2454.53	55.9847	386	2661.37
51.7786	357	2461.42	56.1297	387	2668.27
51.9236	358	2468.32	56.2747	388	2675.16
52.0686	359	2475.21	56.4198	389	2682.06
52.2137	360	2482.11	56.5648	390	2688.95
52.3587	361	2489.00	56.7099	391	2695.85
52.5038	362	2495.90	56.8549	392	2702.74
52.6488	363	2502.79	56.9999	393	2709.63
52.7938	364	2509.69	57.1450	394	2716.53
52.9389	365	2516.58	57.2900	395	2723.42
53.0839	366	2523.48	57.4350	396	2730.32
53.2289	367	2530.37	57.5801	397	2737.21
53.3740	368	2537.27	57.7251	398	2744.11
53.5190	369	2544.16	57.8702	399	2751.00
53.6641	370	2551.06	58.0152	400	2757.90
53.8091	371	2557.95	58.1602	401	2764.79
53.9541	372	2564.84	58.3053	402	2771.69
54.0992	373	2571.74	58.4503	403	2778.58
54.2442	374	2578.63	58.5954	404	2785.48
54.3893	375	2585.53	58.7404	405	2792.37
54.5343	376	2592.42	58.8854	406	2799.27
54.6793	377	2599.32	59.0305	407	2806.16
54.8244	378	2606.21	59.1755	408	2813.06
54.9694	379	2613.11	59.3205	409	2819.95
55.1144	380	2620.00	59.4656	410	2826.85

TABLE A-13 (Continued) Pressure Conversion: Pounds Per Square Inch (PSI) and Kilo Pascals (kPa)

psi		kPa	psi		kPa
59.6106	411	2833.74	63.9618	441	3040.58
59.7556	412	2840.63	64.1068	442	3047.48
59.9007	413	2847.53	64.2518	443	3054.37
60.0457	414	2854.42	64.3969	444	3061.27
60.1908	415	2861.32	64.5419	445	3068.16
60.3358	416	2868.21	64.6870	446	3075.06
60.4808	417	2875.11	64.8320	447	3081.95
60.6259	418	2882.00	64.9770	448	3088.85
60.7709	419	2888.90	65.1221	449	3095.74
60.9160	420	2895.79	65.2671	450	3102.64
61.0610	421	2902.69	65.4121	451	3109.53
61.2060	422	2909.58	65.5572	452	3116.42
61.3511	423	2916.48	65.7022	453	3123.32
61.4961	424	2923.37	65.8472	454	3130.21
61.6411	425	2930.27	65.9923	455	3137.11
61.7862	426	2937.16	66.1373	456	3144.00
61.9312	427	2944.06	66.2824	457	3150.90
62.0763	428	2950.95	66.4274	458	3157.79
62.2213	429	2957.85	66.5724	459	3164.69
62.3663	430	2964.74	66.7175	460	3171.58
62.5114	431	2971.64	66.8625	461	3178.48
62.6564	432	2978.53	67.0076	462	3185.37
62.8015	433	2985.42	67.1526	463	3192.27
62.9465	434	2992.32	67.2976	464	3199.16
63.0915	435	2999.21	67.4427	465	3206.06
63.2366	436	3006.11	67.5877	466	3212.95
63.3816	437	3013.00	67.7327	467	3219.85
63.5266	438	3019.90	67.8778	468	3226.74
63.6717	439	3026.79	68.0228	469	3233.64
63.8167	440	3033.69	68.1679	470	3240.53

TABLE A-13 (Continued) Pressure Conversion: Pounds Per Square Inch (PSI) and Kilo Pascals (kPa)

psi		kPa	psi		kPa
68.3129	471	3247.42	72.6640	501	3454.27
68.4579	472	3254.32	72.8091	502	3461.16
68.6030	473	3261.21	72.9541	503	3468.06
68.7480	474	3268.11	73.0992	504	3474.95
68.8931	475	3275.00	73.2442	505	3481.85
69.0381	476	3281.90	73.3892	506	3488.74
69.1831	477	3288.79	73.5343	507	3495.64
69.3282	478	3295.69	73.6793	508	3502.53
69.4732	479	3302.58	73.8243	509	3509.43
69.6182	480	3309.48	73.9694	510	3516.32
69.7633	481	3316.37	74.1144	511	3523.21
69.9083	482	3323.27	74.2595	512	3530.11
70.0533	483	3330.16	74.4045	513	3537.00
70.1984	484	3337.06	74.5495	514	3543.90
70.3434	485	3343.95	74.6946	515	3550.79
70.4885	486	3350.85	74.8396	516	3557.69
70.6335	487	3357.74	74.9846	517	3564.58
70.7785	488	3364.64	75.1297	518	3571.48
70.9236	489	3371.53	75.2747	519	3578.37
71.0686	490	3378.42	75.4198	520	3585.27
71.2137	491	3385.32	75.5648	521	3592.16
71.3587	492	3392.21	75.7098	522	3599.06
71.5037	493	3399.11	75.8549	523	3605.95
71.6488	494	3406.00	75.9999	524	3612.85
71.7938	495	3412.90	76.1449	525	3619.74
71.9388	496	3419.79	76.2900	526	3626.64
72.0839	497	3426.69	76.4350	527	3633.53
72.2289	498	3433.58	76.5801	528	3640.43
72.3740	499	3440.48	76.7251	529	3647.32
72.5190	500	3447.37	76.8701	530	3654.21

TABLE A-13 (Continued) Pressure conversion: Pounds Per Square Inch (PSI) and Kilo Pascals (kPa)

psi		kPa	psi		kPa
77.0152	531	3661.11	81.3663	561	3867.95
77.1602	532	3668.00	81.5114	562	3874.85
77.3053	533	3674.90	81.6564	563	3881.74
77.4503	534	3681.79	81.8014	564	3888.64
77.5953	535	3688.69	81.9465	565	3895.53
77.7404	536	3695.58	82.0915	566	3902.43
77.8854	537	3702.48	82.2365	567	3909.32
78.0304	538	3709.37	82.3816	568	3916.22
78.1755	539	3716.27	82.5266	569	3923.11
78.3205	540	3723.16	82.6717	570	3930.00
78.4656	541	3730.06	82.8167	571	3936.90
78.6106	542	3736.95	82.9617	572	3943.79
78.7556	543	3743.85	83.1068	573	3950.69
78.9007	544	3750.74	83.2518	574	3957.58
79.0457	545	3757.64	83.3969	575	3964.48
79.1908	546	3764.53	83.5419	576	3971.37
79.3358	547	3771.43	83.6869	577	3978.27
79.4808	548	3778.32	83.8320	578	3985.16
79.6259	549	3785.21	83.9770	579	3992.06
79.7709	550	3792.11	84.1220	580	3998.95
79.9159	551	3799.00	84.2671	581	4005.85
80.0610	552	3805.90	84.4121	582	4012.74
80.2060	553	3812.79	84.5571	583	4019.64
80.3510	554	3819.69	84.7022	584	4026.53
80.4961	555	3826.58	84.8472	585	4033.43
80.6411	556	3833.48	84.9923	586	4040.32
80.7862	557	3840.37	85.1373	587	4047.22
80.9312	558	3847.27	85.2823	588	4054.11
81.0762	559	3854.16	85.4274	589	4061.00
81.2213	560	3861.06	85.5724	590	4067.90

INDEX

INDEX

A

Acre—feet, 113
Act—safe drinking water, 88
Actinomycetes, 60
Activated carbon, 73
Activated silica, 92
Adits, 31
Advantages of ground water supplies, 109
Aeration: 139
   disadvantages, 143
   forced, 142
   odor removal, 140
   theoretical concept, 140
Air vent valves, 291
Albuminoid ammonia, 89
Algae: 46
   blue green, 47
   chlorine on algae, 66
   clean water (Fig. 5-4), 54-55
   cyanophyta, 60
   diatoms, 47
   filter clogging algae (Fig. 5-2), 50-51, (Table 5-3), 64
   flagellates, 47
   green, 47, 61
   growing on reservoir walls (Fig. 5-6), 58-59
   important in water supplies (Fig. 5-1), 48-49
   photosynthesis, 112
   plankton and other surface water algae (Fig. 5-5), 56-57
   polluted water algae (Fig. 5-3), 52-53
   tastes and odor (Table 5-1), 47
Alkalinity: 78
   bicarbonate $(HCO_3)^-$, 78
   carbonate $(CO_3)^{--}$, 78
   hydroxide $(OH)^-$, 78
   relationships (Table 6-6), 80
Allievi's equation, 305
Aluminum, 89, 91
Amebiasis, 26, 35
Amebic colitis, 35
Amebic dysentery, 35
Amebic enteritis, 35
Ammonia, 89
Anabaena, 60
Anions: 76
   exchanger, 271

removal, 266
Anterior poliomyelitis, 38
APHA Units, 71
Aphanizomenon, 60
Approved air gap, 292
Aquifers types (Fig. 9-1), 121
Arsenic (As), 83
Artesian wells, 121-23
Aseptic meningitis, 37
Asiatic cholera, 27
Atomic fallout, 101
Atomic weights, 392

B

Bacillary dysentery, 27
    shigellosis, 27
Backflow, 289
Backsiphonage, 290
Bacteria, 26
Bacteriology: 18
    1892 cholera epidemic in Hamburg, 18
    Metropolis Water Act of 1852, 19
Barium (Ba), 83
Base exchange softening, 266
Bicarbonates, 78
Bilharzial dysentery, 34
Bilharziasis mansoni, 34
"Bill Harris" disease, 34
Biochemical oxygen demand (B.O.D.), 95
Biocides in waters, 87
    pesticides limitations (Table 6-7), 88
Blood fluke trematodes, 34
Blood pressure, 84
Blue babies—methemoglobinemia, 86
Baron (B), 84
Brain disease, 110
Breakpoint chlorination, 222
Buffalo Pound lake, 60
"Buffered" waters, 79

C

Cadmium, 84
Calcium: 90
    carbonate solubility, 82
Canadian Drinking Water Standards, 69
Cancer—possible causes, 83
Canicola fever, 28
Carbon dioxide, solubility, 184

Cardiovascular diseases, 81
Capacity of hydropneumatic tanks, 320
Carbonate, 78
Catarrhal jaundice, 38
Cathodic protection, 301
Cations: 76
    hydrogen ion exchanger, 270
    removal, 266
Chalybeate waters, 92
Chelating (see Sequestering), 72, 206
Chemical commercial, 209
Chemical feed tables: 210
    grams per hour/cubic meters per day, 398
    grams per hour/liters per second, 397
    kilograms per day/cubic meters per day, 399
    pounds per hour/UK gpm, 396
    pounds per hour/US gpm, 395
Chemicals & chemical feeding, 208
Chemicals in drinking water—limitations (Table 6-3), 70
Chemical oxygen demand (C.O.D.), 95
Cholera, 23, 26-27
Chloramines, 90
Chlordane, 87-88
Chlorides, 91
Chlorinated phenols, 94
Chlorination: 219
    pH, 221
    retention time, 221
Chlorine dioxide, 91, 224
Chlorine free residual, 90
Chromium, 85
City of Moose Jaw (Saskatchewan), 60
Clark or English degree of hardness, 82
Clarification, 155
Clogging algae, 50-51
Clean distribution systems, 25
Clean water algae (Fig. 5-4), 54-55
Coagulation: 152
    control, 154
    theory, 152
Colebrook-White equation, 298
Color: 71
    definition, 167
    organic acids, 167
    removal process, 168
Commercial chemicals, 209
Common water borne diseases, 26
Computer programs, 283
Conductivity, specific, 94
Conversion of feet and meters, 401
Conversions (see Equivalents)
Copper, 91

Copper sulfate, 91
Crenothrix, 42, 92, 110
Cross connection control: 25, 289
    summary of problems, 290
Croydon typhoid outbreak, 31
Cyanide, 85
Cyclops, 33

D

Dams, spillway capacity, 117
Dangerous chemicals, handling, 216
D.D.T., 88
Deaeration: 143
    steam, 143
    vacuum technology, 144
    vacuum tower, 143
Deer fly fever, 29
Demineralized water costs, 275
Demineralization, methods, 262
Desalting, freeze process, 273
Desalination (*see* Demineralization, 262)
De-sulfovibrio, 96
Death rates, 18
Detergents, synthetic (syndets), 81
Diatoms, 47, 62
Diatomaceous earth filters, 197
Dilutions: 213
    formula method, 215
    rectangular method, 213
Disadvantages of ground water supplies, 110
Disinfection: 24, 218
    breakpoint chlorination, 222
    chlorination, 219
    chlorine dioxide, 224
    criteria for ultra-violet light (u.v.), 230
    free and combined chlorine residuals, 221
    ozone, 137, 225
    potassium permanganate ($KMnO_4$), 232
    silver, 232
    turbidity, 221
    wells, 132
Disinfection barriers: 17, 24
    removal of turbidity, 24
    source of water, 24
Disolved oxygen, 112
Distribution systems: 279
    clean, 25
    cross connection control, 25
    monitoring the distribution system, 26
    water leakage (Table 1-2), 8
Divining, water, 108

Dog tapeworm, 33
Domestic water demand: 7
    advantages of metering, 7
    broken mains, 7
    distribution system leakage, 7
    leakage detection, 8
    leakage rates, 8
    maximum day, 9
    meters, 7
    peak hour, 9
    per capita per day demand, 7
    unaccounted distribution system losses, 7
    unmetered services, 7
Double check valves, 293
Dracontiasis, 26, 33
Drinking water quality, 68
Dysentery, 26

E

Echinococcosis, 33
Echo virus diseases, 26, 37
Electric motors: 341
    open, 343
        drip-proof, 343
    totally enclosed, 343
        fan cooled (T.E.F.C.), 344
    reduced voltage starting, 344
    water cooled, 344
    wound rotor, 345
Electrodialysis (E.D.), 263
El Tor (cholera), 27
Electromagnetic couplings, 345
Engine selection: 346
    diesels and gas, 346
    dual fuel, 347
Enteric fever, 28-29
Epidemic jaundice, 38
Epsom salts, 93
Equivalents:
    acre-feet, 113
    feet and meters, 400
    cubic meters—gallons (UK)—gallons (US), 364
    degrees of hardness, 82
    flow conversions (Table 11-4), 146
    gallons (UK)—gallons (US)—cubic meters, 350
    gallons (US)—cubic meters—gallons (UK), 378
    grains per gallon, 82
    pressure conversion (Table 11-2), 145
    volume conversions (Table 11-3), 146
    psi/kilo Pascals (kPa), 401
Escherichia coli (E. Coli), 30

Eutrophication, 46
Evaporation: 272
   distillation, 272
   solar, 274
   vapor recompression, 273

F

Fecal coliforms, 17
Ferruginous waters, 92
Filters: 186
   air scour, 194
   backwashing, 193
   biflow, 188
   contamination, 197
   clogging algae, 50, 64
   diatomaceous earth (D.E.), 197
   dual media, 188
   mixed or multi-media, 188
   presses, 244
   upflow, 188
Filtration: 186
   constant rate, 195
   declining rate, 195
   high rate, 196
   lime softened waters, 178
   slow sand, 196
   theory, 186
   vacuum, 244
Fish: 134
   screens, 134
   sensitivity, 85
Fire protection: 314
   pumps, 332
Flagellates, 47, 62
Flash floods, 117
Flocculation, 151
Floods, 117
Fluoridation: 232
   chemicals used, 234
      fluosilicic acid, 236
      sodium fluoride, 235
      sodium silicofluoride, 236
      optimum fluoride levels, 232
Fluosilicic acid, 236
Forced aeration, 142
Francisella tularensis, 29
Frazil ice, 75
Freeze:
   desalting, 273
   sludges, 243, 246
Free and combined chlorine residual, 90, 221
Frozen pipes, 284

G

Gallionella, 42, 92, 110
Giardia: 36
   enteritis, 36
   lamblia, 36
Giardiasis, 26, 36
Grains per Gallon, 82
Granulosus, 33
Green algae, 47, 61
Ground water: 107
   advantages, 109
   disadvantages, 110
   recharge, 108
   wells, 118
      pumps, 118

H

Handling dangerous chemicals, 216
Hardness: 80
   carbonate, 83
   degrees, 82
   health aspects—heart disease, 81
   noncarbonate, 82
   permanent, 83, 175
   temporary, 175
   water quality effects (Table 6-4), 71
Hardy Cross, 282
Hazard of fluoride chemicals, 237
Hazen-Williams equation, 297
Hazen color standards, 71
Heating water in distribution system, 286
Helminthic diseases (parasitic worms), 26, 33
Hemorrhagic jaundice, 28
Hexameta phosphate, 92
HOCl and OCl$^-$ (Fig. 19.1), 220
Holy Cross College, 26
Horizontal collectors, 131
Humic acids (Table 13.1), 168
Hydatidosis, 33
Hydrants (cross connection control), 292
Hydrogren cyanide (HCN), 85
Hydrogen ion, 77
Hydrological cycle, 16, 104
Hydropneumatic tanks: 319
   capacity (Table 28-1), 320
   pressure vessels, 325
   pump characteristics, 322
   sizes, 323
   storage (Fig. 28.1), 321
   surge suppression, 312
   systems, 319
   water level control, 326

Hypochlorinators, self contained, 224
Hydroxide (OH)$^-$, 78

I

Immigration, 4
Indian cholera, 27
Infantile:
　diarrhea, 37
　mortality, 2
　paralysis, 38
Infectious hepatitis, 23, 26, 36, 38
Intestinal bilharziasis, 34
Invitation to tender, pumps, 110
Ion exchange, 265
Iron (in water), 92
Iron bacteria, 41
Iron and manganese removal: 200
　aeration, 202
　chlorine dioxide, 202
　ion exchange, 205
　lime softening, 203
　organic or chelated iron, 203
　potassium permanganate, 203
Irrigation, 13

J

Jackson candle turbidimeter, 74
Jackson turbidity unit (J.T.U.), 74

K

Katadyn process, 87, 232

L

Lake and river water supplies, 111
Lagoons, 243
Lambliasis, 36
Langelier index, 258
Laundry stains, 94
Law, 12
Lead, 86
Leakage surveys, 286
Leakage, pumps, (Table 29.4), 337
Legal aspects, distribution systems, 279
Lepthothrix, 42, 92, 110
Leptospirosis, 26, 28

Lime soda softening: 172
　filtration, 178
　limitations, 174

M

Magnesium: 93
　hydroxide solubility, 82
　sulfate, 93
Maintenance of pumps, 339
Manganese: 93
　removal with potassium permanganate, 204
Manson's intestinal schistosomiasis, 34
Marble chip test, 257
Maximum contamination levels (MCL), 69
Mechanical seals, 339
Medicinal waters, 17
Mercury contamination, 113
Metering, 7, 285
Methomoglobinemia (blue babies), 86
Methyl orange alkalinity (M.O.Alk.), 78
Methylene blue active substance (MBAS), 94
Microorganisms:
　coliform, 17
　pathogenic, 17
　populations, 17
Microstrainers, 60, 135-37
Mixed media filters, 66, 190
Mixing, 148
Molded rubber impellers (Table 29.3), 335
Monitoring radioactivity, 102
Mud fever, 28
Multimedia filters, 66, 190
Municipal services, 1

N

Natural and forced draft aerators, 141
Natural waters, 17
Nematodes, 44
Nephelometers, 74
Neptune MicroFLOC, 66, 156, 158, 159, 160, 190, 192
Net positive suction head (NPSH), 125, 126, 335
Network studies, 282
Nitrates and nitrites, 86
Noncarbonate hardness, 82, 83
Nonpathogenic organisms, 41
Nontoxic chemicals, 89

O

Odor: 72
   algae, 47, 63
   removal, 140
Organic:
   acids and color, 167
   chemicals, 95
   contamination (Table 6-8), 89
   deposits, 260
"Organic iron", 203
Oscillatoria, 60
Oxygen:
   sag, 112
   solubility (Table 11.1), 141
Ozone: 225
   dosages, 225
   physical and chemical properties, 227

P

Palmer, C.M., 62
Paralysis, 39
Parasitic worms (helminthic diseases), 26
Paratyphoid, 26, 28
Pasteurella tularensis, 29
Periodic Table, 393
Permanent hardness, 82
Pesticides and herbicides, 87
pH: 76
   chorination, 221
   and OH values, 76-77
   lime softening (Fig. 14.1), 173
Phenol alkalnity (P. Alk.), 78
Phenolic substances, 94
Phosphates, 94
Photogrammetry, 105
Pipes—roughness values (Tables 25-1), 299
Plankton and other surface water algae (Fig. 5-5), 56-57
Plant effluents, 107
Plugged well screens, 130
Poliomyelitis, 26, 38
Polluted water algae (Fig. 5-3), 52, 53
Pollution, 91
Population: 2
   estimate of world population (Table 1.1), 3
Potassium permanganate, 204
Presses, filter, 244
Pressure conversions:
   pounds per square inch/kilo Pascals, 401
   vacuum systems (Table 11-2), 145

Pressure strainers, 137
Pretreatment:
   microstrainers, 137
   pressure strainers, 137
Probability:
   method, 113
   of recurrence (Fig. 8.1) 114-15
Properties of water, 16
Protozoa: 26
   diseases, 35
Pumps:
   capacity, 119
   characteristics, 332
   cost of pumping, 328
   hydropneumatic tanks, 322
   selection for wells, 124
   shaft speed, 330
   speeds (Table 29-1), 330
   suction location, 129
   testing procedures, 338
   thrust bearings, 348
   wear in relation to speed, 121

Q

Quality of drinking water, 68
Quantity of sludge, 245

R

Radionuclides: 99
   fallout, 101
   monitoring, 102
   removal by settling, 102
   sources, 99
   units of radioactivity, 100
Recarbonation: 180
   liquid carbon dioxide (CO_2), 181
   production of carbon dioxide, 185
   submerged combustion burners, 180
Rectangular, method of dilution, 213
Reduced pressure backflow preventors, (RPBP), 293
Reduced voltage starting, 344
Regina, City of, 60
Regulations, drinking water, 68
Relationship between pH and ions (Table 6-5), 77
Removal:
   bacterial slimes, 43
   radionuclides, 101-02
   turbidity, 24
Resin capacity, 268

Reverse osmosis (R.O.), 262
Reservoirs: 313
　algae on walls, 58-59
　emergency storage, 314
　equalizing storage, 313
　fire protection, 314
　valving (Fig. 27.1), 317
Reynold's number (Re), 161
Riparian owner, 106
River and lake water supplies, 111
Roughness of pipe walls (Table 25-1), 299
Rubber pump impellers (Table 29-3), 335
Runoff, 113

S

Safe Drinking Water Act, 88
Salmonella paratyphi A, 28
Salmonella paratyphi B, 28
Salmonella paratyphi C, 28
Salmonella typhi, 29
Sampling of water, 111
Sand, well pumps, 130
Scale formation, 90
Schistosomiasis, 26, 34
Screens, 134
Selenium, 86
Sequestering of iron and manganese, 206
Shaftspeed, pumps, 119, 121, 330
Shigella dysenteriae, 27
Silver: 87, 232
　disinfection, 87
Sludge quantity, 245
Slime growths, 43, 92
Slow sand filters, 18
Sodium:
　fluoride, 234-235
　silicofluoride, 234, 236
Softening: 172
　disinfection and virus inactivation, 175
　effect of inhibitors, 177
　equipment used, 175
　health aspects, 174
　should water be soft, 172
Soft waters, 80-81
Solar evaporation, 274
Solubility of magnesium hydroxide, 82
Solution strength, 216
Specific conductivity, 94, 259
Specific speed (N_s), 336
Spillway capacity, 117
Spray aerators, 142

Stable waters, 79
Stabilization: 256
　Langelier index, 258
　marble chip test, 257
　Ryznar index, 259
　saturation indices, 258
　well waters, 256
Stationary screens, 134
Steam deaeration, 143
Straining and screening, 134, 137
Stream flows (Table 8.1), 114
Strong basic-anion exchanger, 271
Submergence, 335
Submersible pumps, 128
Sulfamic acid, 136
Sulfate—reducing bacteria, 41
Sulfates, 96
Sulfides, 96
Surface aeration, 142
Surge tanks, 311
Surveys, leakage, 286
Swineherd's disease, 28
Synergistic effect, 97
Synthetic detergents (syndets), 81
Synthesis of water, 16
Synura, 62

T

Tastes and odors, 47, 62, 63, 73, 92, 95
Temperature: 75
　conversion (Table A-6), 394
Tender, invitation, 337
Temporary hardness, 82
Thawing frozen pipes, 284
Thickner design, 247
Threshold odor number (T.O.N.), 62, 72
Threshold treatment, 185
Thrust bearings, 348
Tolerance limit (T. lm), 88
Tongue sensations (Table 5-2), 63
Torr, 144
Total dissolved solids (T.D.S.), 94
Total organic carbon (T.O.C.), 95
Toxic chemicals, 70, 83
Trash racks, 134
Traveling screens, 135
Tri-sodium phosphate, 92
Tube settlers, 156
Tularemia, 26, 29
Turbidity, 73
Typhoid, 23, 26, 29, 31

U

Units of radioactivity, 100
Universal pipe friction (Fig. 25.1, *see* the fold-out)
Uranyl ion, 97
Ultraviolet irradiation, 37
Ultraviolet light (U.V.): 227
  water borne pathogens (Table 19.2), 229
USA Proposed Drinking Water Standards, 68

V

Vacuum:
  breakers, 293
  technology, 144
  tower deaerator, 144
Vancouver, City of, 279
Vapor recompression, 273
Variable speed drives:
  electromagnetic couplings, 345
  hydraulic couplings, 345
  variable frequency drives (VFD), 345
  wound rotor motors, 345
Velocity gradients, 149
Vertical turbine pumps: 127
  NPSH (Fig. 9.3), 125
Viruses diseases, 26, 36
Viscosity of water (Table 12.1), 150
Vortexing, 335

W

Waste disposal: 241
  alum recovery, 246
  centrifuges, 244
  coagulated clarifier sludges, 243
  drying beds, 246
  filter backwash water, 242
  filter presses, 244
  freeze drying, 243
  lagoons, 243
  quantity of sludge, 245
  size of thickener, 247
  softening sludges, 246
  thickening, 247
  vacuum filtration, 244
Water borne diseases: 20
  epidemiologist, 21
  outbreaks, 22
  pathogenic organisms, 21
  source of water, 24
Water consumption, 69
Water cooled motors, 344
Water demand: 1
  backflow prevention devices, 12
  cross connection control, 12
  Edmonton, City of, 9
  domestic, 7
  fire protection, 11
  industrial requirements, 11
  irrigation, 13
  main flushing, 13
  polluted ground water, 9
  reduced pressure conditions, 9
Water divining, 108
Water falls, 142
Water hammer, 303
Water properties, 16
Water quality based on hardness, 71
Water resources: 104
  probability method, 113
  ground water, 107
  water sheds, 104
Water reuse, 14
Water sampling, 111
Water sheds, 104
Water standards for drinking, 68
Water supplies, choice, 106
Weils disease, 28
Weirs, and water falls, aeration, 142
Wells: 118
  alignment, 127
  development, 132
  disinfection, 132
  horizontal collectors, 131
  location, 109
  pump operation, 129
  pump selection, 124
  screens, 120
    plugging, 130
White-Colebrook equation, 298
William and Hazen equation (*see* Hazen-William equation), 297
World populations, 3
Worms (*see* Helminthic diseases): 26
  nematodes, 44
Wound motor electric motors, 345

Z

Zinc, 97